T0338046

Pesticide Risk Assessment for Pollinators

Other Titles from the Society of Environmental Toxicology and Chemistry (SETAC)

ESCORT3: Linking Non-Target Arthropod Testing and Risk Assessment with Protection Goals
Alix, Bakker, Barrett, Brühl, Coulson, Hoy, Jansen, Jepson, Lewis, Neumann, Süßenbach, van Vliet
2012

Ecotoxicology of Amphibians and Reptiles, 2nd ed.
Sparling, Linder, Bishop, Krest
2010

Ecological Assessment of Selenium in the Aquatic Environment
Chapman, Adams, Brooks, Delos, Luoma, Maher, Ohlendorf, Presser, Shaw
2010

Soil Quality Standards for Trace Elements
Merrington, Schoeters
2010

Semi-Field Methods for the Environmental Risk Assessment of Pesticides in Soil
Schäffer, van den Brink, Heimbach, Hoy, de Jong, Römbke, Roß-Nickoll, Sousa
2010

Mixture Toxicity: Linking Approaches from Ecological and Human Toxicology
van Gestel, Jonker, Kammenga, Laskowski, Svendsen
2010

Application of Uncertainty Analysis to Ecological Risks of Pesticides
Warren-Hicks, Hart
2010

Risk Assessment Tools: Software and User's Guide
Mayer, Ellersieck, Asfaw
2009

Derivation and Use of Environmental Quality and Human Health Standards for Chemical Substances in Water and Soil
Crane, Matthiessen, Maycock, Merrington, Whitehouse
2009

Linking Aquatic Exposure and Effects: Risk Assessment of Pesticides
Brock, Alix, Brown, Capri, Gottesbüren, Heimbach, Lythgo, Schulz, Streloke
2009

Aquatic Macrophyte Risk Assessment for Pesticides
Maltby, Arnold, Arts, Davies, Heimbach, Pickl, Poulsen
2009

Ecological Models for Regulatory Risk Assessments of Pesticides: Developing a Strategy for the Future
Thorbek, Forbes, Heimbach, Hommen, Thulke, van den Brink, Wogram, Grimm
2009

Veterinary Medicines in the Environment
Crane, Boxall, Barrett
2008

Relevance of Ambient Water Quality Criteria for Ephemeral and Effluent-dependent Watercourses of the Arid Western United States
Gensemer, Meyerhof, Ramage, Curley
2008

Extrapolation Practice for Ecotoxicological Effect Characterization of Chemicals
Solomon, Brock, De Zwart, Dyer, Posthuma, Richards, Sanderson, Sibley, van den Brink
2008

Environmental Life Cycle Costing
Hunkeler, Lichtenvort, Rebitzer
2008

Pesticide Risk Assessment for Pollinators

Edited by

David Fischer
Environmental Safety
Bayer CropScience LP
Research Triangle Park, North Carolina, USA

Thomas Moriarty
Office of Pesticide Programs
US Environmental Protection Agency
Washington DC, USA

from the SETAC Pellston Workshop on Pesticide Risk Assessment for Pollinators

January 15–21, 2011
Pensacola, Florida, USA

Coordinating Editor of SETAC Books
Lawrence Kapustka
LK Consultancy
Calgary, Alberta, Canada

WILEY Blackwell

Editorial offices: 1606 Golden Aspen Drive, Suites 103 and 104, Ames, Iowa 50010, USA
The Atrium, Southern Gate, Chichester, West Sussex, PO19 8SQ, UK
9600 Garsington Road, Oxford, OX4 2DQ, UK

For details of our global editorial offices, for customer services and for information about how to apply for permission to reuse the copyright material in this book please see our website at www.wiley.com/wiley-blackwell.

Contributions to this book by Thomas Steeger and Thomas Moriarty were made as part of their duties for the US Environmental Protection Agency (USEPA). However, the publication of this book has not been formally reviewed by the USEPA and does not necessarily reflect the views of the USEPA.

Library of Congress Cataloging-in-Publication Data

Pesticide risk assessment for pollinators / edited by David Fischer, Thomas Moriarty.
pages cm
Includes bibliographical references and index.
ISBN 978-1-118-85252-1 (cloth)
1. Bees–Effect of pesticides on. 2. Honeybee–Effect of pesticides on. 3. Bees–Health. 4. Pesticides–Environmental aspects. 5. Pesticides and wildlife. I. Fischer, David, 1955– II. Moriarty, Thomas.
SF538.5.P65P47 2014
638′.159–dc23

2013046761

A catalogue record for this book is available from the British Library.

Wiley also publishes its books in a variety of electronic formats. Some content that appears in print may not be available in electronic books.

Cover image: Leafcutting bee on blanket flower, photograph by Mace Vaughan (Xerces Society for Invertebrate Conservation).

Set in 10/12pt Times by Aptara Inc., New Delhi, India
Printed and bound in Malaysia by Vivar Printing Sdn Bhd

1 2014

SETAC Publications

Books published by the Society of Environmental Toxicology and Chemistry (SETAC) provide in-depth reviews and critical appraisals on scientific subjects relevant to understanding the impacts of chemicals and technology on the environment. The books explore topics reviewed and recommended by the Publications Advisory Committee and approved by the SETAC North America, Latin America, Asia/Pacific, or Africa Board of Directors; the SETAC Europe Council; or the SETAC World Council for their importance, timeliness, and contribution to multidisciplinary approaches to solving environmental problems. The diversity and breadth of subjects covered in the series reflect the wide range of disciplines encompassed by environmental toxicology, environmental chemistry, hazard and risk assessment, and life cycle assessment. SETAC books attempt to present the reader with authoritative coverage of the literature, as well as paradigms, methodologies, and controversies; research needs; and new developments specific to the featured topics. The books are generally peer reviewed for SETAC by acknowledged experts.

SETAC publications, which include Technical Issue Papers (TIPs), workshop summaries, newsletter (SETAC Globe), and journals (*Environmental Toxicology and Chemistry* and *Integrated Environmental Assessment and Management*), are useful to environmental scientists in research, research management, chemical manufacturing and regulation, risk assessment, and education, as well as to students considering or preparing for careers in these areas. The publications provide information for keeping abreast of recent developments in familiar subject areas and for rapid introduction to principles and approaches in new subject areas.

SETAC recognizes and thanks the past coordinating editors of SETAC books:

Joseph W. Gorsuch, Copper Development Association, Inc.
Webster, New York, USA

A.S. Green, International Zinc Association
Durham, North Carolina, USA

C.G. Ingersoll, Columbia Environmental Research Center
US Geological Survey, Columbia, Missouri, USA

T.W. La Point, Institute of Applied Sciences
University of North Texas, Denton, Texas, USA

B.T. Walton, US Environmental Protection Agency
Research Triangle Park, North Carolina, USA

C.H. Ward, Department of Environmental Sciences and Engineering
Rice University, Houston, Texas, USA

This book is dedicated to the memory of Dr. Peter Delorme of Health Canada's Pest Management Regulatory Agency. Dr. Delorme served as a member of the Steering Committee for the global SETAC Pellston Workshop on the Pesticide Risk Assessment for Pollinators, and is remembered for his contributions to this effort, and for his long service to protecting the environment.

Contents

Color plate section is located between pages 120 and 121.

List of Figures

List of Tables

Acknowledgments

We gratefully acknowledge the financial support for the workshop from the following organizations:

BASF Corporation
Bayer CropScience
Chemicals Regulation Directorate to the Health and Safety Executive
Dow AgroSciences
International Commission For Plant Bee Relationships
Julius Kuhn Institut
Monsanto Company
Pennsylvania State University
Pollinator Partnership
Project Apis m.
Syngenta
US Department of Agriculture
US Environmental Protection Agency
Valent U.S.A Corporation

About the Editors

Thomas Moriarty serves as a Risk Manager and Chemical Team Leader in the US Environmental Protection Agency, Office of Pesticide Programs (OPP), Pesticide Re-evaluation Division, and serves as the Chair of the USEPA Pollinator Protection Team. He has also worked as a Risk Assessor in the Human Health Effects Division of OPP, and served on the technical team that developed the USEPA Risk Assessment Framework for Bees that was presented to a Federal Insecticide, Fungicide, and Rodenticide Act (FIFRA) Scientific Advisory Panel in September 2013. He co-chairs the Risk Management workgroup of the Pesticide Effects on Insect Pollinators subgroup for the Organisation for Economic Co-operation and Development (OECD) Working Group on Pesticides, and was a member of the Steering Committee for the SETAC Pellston Workshop on Pollinator Risk Assessment. Tom has an MS in Public Policy from the University of Maryland and a BS in Political Science from Providence College.

David Fischer is Director and Head of the Environmental Toxicology and Risk Assessment group within the Development North America Department of Bayer CropScience LP. Fischer holds a BS degree in Zoology from the University of Massachusetts, an MS degree in Zoology from Western Illinois University, and a PhD in Zoology from Brigham Young University. His MS and PhD research projects were on the ecology of Bald Eagles and *Accipiter* Hawks. He has been working in the field of ecotoxicology and risk assessment since 1986, the last 26 years with Bayer CropScience and Legacy companies. Fischer has supervised the conduct of hundreds of laboratory and field toxicology studies of crop protection chemicals and animal pharmaceuticals, authored dozens of chemical risk assessments, and published more than 20 peer-reviewed scientific papers. His expertise is in the area of terrestrial ecotoxicology and risk assessment. A member of SETAC since 1988, he helped organize previous Pellston Workshops on *Wildlife Radiotelemetry Applications for Wildlife Toxicology Field Studies* and *Application of Uncertainty Analysis to Ecological Risks of Pesticides*. For the past decade or so, Fischer's research interests have focused on improving the testing and risk assessment process for honey bees and other pollinators.

Workshop Participants

Steering Committee

Dave Fischer, Bayer CropScience, USA
Tom Moriarty, US Environmental Protection Agency
Anne Alix, Ministry of Agriculture, Food, Fisheries, Rural Affairs and Spatial Planning, France
Mike Coulson, Syngenta Ltd., UK
Peter Delorme, Health Canada, Pest Management Regulatory Agency, Canada
Jim Frazier, Pennsylvania State University, USA
Christopher Lee-Steere, Australian Environment Agency Pty Ltd., Australia
Jeff Pettis, US Department of Agriculture
Jochen Pflugfelder, Swiss Bee Research Center, Switzerland
Thomas Steeger, US Environmental Protection Agency
Franz Streissl, European Food Safety Authority
Mace Vaughan, Xerces Society for Invertebrate Conservation, USA
Joseph Wisk, BASF Corporation, Crop Solutions, USA

Exposure Workgroup

Jens Pistorius, *co-chair*, Julius Kühn-Institut, Institute for Plant Protection in Field Crops and Grassland, Germany
Joseph Wisk, *co-chair*, BASF Corporation, Crop Solutions, USA
Mike Beevers, California Agricultural Research, Inc., USA
Richard Bireley, California Department of Pesticide Regulation, USA
Zac Browning, Browning's Honey Company, Inc., USA
Marie-Pierre Chauzat, French Agency for Food, Environmental and Occupational Health Safety, Sophia Antipolis, France
Alexander Nikolakis, Bayer CropScience AG, Development—Environmental Safety—Ecotoxicology—Bees, Germany
Jay Overmyer, Syngenta Crop Protection, LLC, USA
Robyn Rose, US Department of Agriculture Animal and Plant Health Inspection Service
Robert Sebastien, Health Canada, Pest Management Regulatory Agency, Canada
Bernard Vaissière, French National Institute for Agricultural Research, France
Mace Vaughan, Xerces Society for Invertebrate Conservation, USA

Hazard, Laboratory Workgroup

Jim Frazier, *co-chair*, Pennsylvania State University, USA
Jochen Pflugfelder, *co-chair*, Swiss Bee Research Center, Switzerland
Pierrick Aupinel, INRA, Centre Poitou-Charentes, UE d'entomologie, France
Pamela Bachman, Monsanto Company, USA
Axel Decourtye, ACTA, UMT PrADE, Germany
Axel Dinter, DuPont de Nemours (Deutschland) GmbH, Germany
Jamie Ellis, Honey Bee Research and Extension Laboratory, University of Florida, USA
Volker Grimm, Helmholtz Center for Environmental Research—UFZ, Leipzig, Germany

Zachary Huang, Michigan State University, USA
Roberta C.F. Nocelli, Center for Agricultural Science—UFSCar—Araras—SP, Brazil
Helen Thompson, Food and Environment Research Agency, UK
William Warren-Hicks, EcoStat, Inc., USA

Hazard, Semi-field and Field Workgroup

Ingo Tornier, *co-chair*, Eurofins Agroscience, Germany
Jeff Pettis, *co-chair*, US Department of Agriculture
Roland Becker, BASF Aktiengesellschaft, Germany
Mark Clook, Chemicals Regulation Directorate, Health and Safety Executive, UK
Mike Coulson, Syngenta Ltd., UK
Wayne Hou, Health Canada, Pest Management Regulatory Agency, Canada
Pascal Jourdan, ITSAP—Institut de l'abeille, France
Muo Kasina, Kenya Agricultural Research Institute, Kenya
Glynn Maynard, Office of the Chief Plant Protection Officer, Department of Agriculture, Fisheries and Forestry, Australia
Dick Rogers, Bayer CropScience, USA
Cynthia Scott-Dupree, School of Environmental Sciences, University of Guelph, Canada
Teodoro Stadler, Laboratorio de Toxicologia Ambiental, Instituto de Medicina y Biologia Experimental de Cuyo (IMBECU), Centro Cientifico Teconlogico CONICET, Argentina
Klaus Wallner, University of Hohenheim, Apiculture Institute, Germany
Bernard Vaissière, French National Institute for Agricultural Research, France

Risk Assessment Workgroup

Anne Alix, *co-chair*, Ministry of Agriculture, Food, Fisheries, Rural Affairs and Spatial Planning, France
Thomas Steeger, *co-chair*, US Environmental Protection Agency
Claire Brittain, Leuphana University of Lüneburg, Germany
Dave Fischer, Bayer CropScience, USA
Rolf Fischer, Federal Office of Consumer Protection and Food Safety, Germany
Michael Fry, American Bird Conservancy, USA
Erik Johansen, Washington State Department of Agriculture, USA
Reed Johnson, University of Nebraska—Lincoln, USA
Christopher Lee-Steere, Australian Environment Agency Party Ltd., Australia
Mark Miles, Dow AgroSciences, UK
Tom Moriarty, US Environmental Protection Agency
Franz Streissl, European Food Safety Authority

Non-*Apis* Workgroup

Mike Coulson, *co-chair*, Syngenta Ltd., UK
Mace Vaughan, *co-chair*, Xerces Society for Invertebrate Conservation, USA
Claire Brittain, Leuphana University of Lüneburg, Germany
Axel Dinter, DuPont de Nemours (Deutschland) GmbH, Germany
Erik Johansen, Washington State Department of Agriculture, USA
Muo Kasina, Kenya Agricultural Research Institute, Kenya

Glynn Maynard, Office of the Chief Plant Protection Officer, Department of Agriculture, Fisheries, and Forestry, Australia
Roberta C.F. Nocelli, Center for Agricultural Science—UFSCar—Araras—SP, Brazil
Cynthia Scott-Dupree, School of Environmental Sciences, University of Guelph, Canada
Helen Thompson, Food and Environment Research Agency, UK
Bernard Vaissière, French National Institute for Agricultural Research, France

Pellston Workshop Series

The workshop from which this book resulted, Potential Risks of Plant Protection Products to Pollinators, held in Pensacola Beach, Florida, USA, January 16–22, 2011, was part of the successful SETAC Pellston Workshop Series. Since 1977, Pellston Workshops have brought scientists together to evaluate current and prospective environmental issues. Each workshop has focused on a relevant environmental topic, and the proceedings of each have been published as peer-reviewed or informal reports. These documents have been widely distributed and are valued by environmental scientists, engineers, regulators, and managers for their technical basis and their comprehensive, state-of-the-science reviews. The workshops in the Pellston series are as follows:

- Estimating the Hazard of Chemical Substances to Aquatic Life. Pellston, Michigan, June 13–17, 1977. Published by the American Society for Testing and Materials, STP 657, 1978.
- Analyzing the Hazard Evaluation Process. Waterville Valley, New Hampshire, August 14–18, 1978. Published by the American Fisheries Society, 1979.
- Biotransformation and Fate of Chemicals in the Aquatic Environment. Pellston, Michigan, August 14–18, 1979. Published by the American Society of Microbiology, 1980.
- Modeling the Fate of Chemicals in the Aquatic Environment. Pellston, Michigan, August 16–21, 1981. Published by Ann Arbor Science, 1982.
- Environmental Hazard Assessment of Effluents. Cody, Wyoming, August 23–27, 1982. Published as a SETAC Special Publication by Pergamon Press, 1985.
- Fate and Effects of Sediment-Bound in Aquatic Systems. Florissant, Colorado, August 11–18, 1984. Published as a SETAC Special Publication by Pergamon Press, 1987.
- Research Priorities in Environmental Risk Assessment. Breckenridge, Colorado, August 16–21, 1987. Published by SETAC, 1987.
- Biomarkers: Biochemical, Physiological, and Histological Markers of Anthropogenic Stress. Keystone, Colorado, July 23–28, 1989. Published as a SETAC Special Publication by Lewis Publishers, 1992.
- Population Ecology and Wildlife Toxicology of Agricultural Pesticide Use: A Modeling Initiative for Avian Species. Kiawah Island, South Carolina, July 22–27, 1990. Published as a SETAC Special Publication by Lewis Publishers, 1994.
- A Technical Framework for [Product] Life-Cycle Assessments. Smuggler's Notch, Vermont, August 18–23, 1990. Published by SETAC, January 1991; 2nd printing September 1991; 3rd printing March 1994.
- Aquatic Microcosms for Ecological Assessment of Pesticides. Wintergreen, Virginia, October 7–11, 1991. Published by SETAC, 1992.
- A Conceptual Framework for Life-Cycle Assessment Impact Assessment. Sandestin, Florida, February 1–6, 1992. Published by SETAC, 1993.
- A Mechanistic Understanding of Bioavailability: Physical–Chemical Interactions. Pellston, Michigan, August 17–22, 1992. Published as a SETAC Special Publication by Lewis Publishers, 1994.
- Life-Cycle Assessment Data Quality Workshop. Wintergreen, Virginia, October 4–9, 1992. Published by SETAC, 1994.
- Avian Radio Telemetry in Support of Pesticide Field Studies. Pacific Grove, California, January 5–8, 1993. Published by SETAC, 1998.

- Sustainability-Based Environmental Management. Pellston, Michigan, August 25–31, 1993. Co-sponsored by the Ecological Society of America. Published by SETAC, 1998.
- Ecotoxicological Risk Assessment for Chlorinated Organic Chemicals. Alliston, Ontario, Canada, July 25–29, 1994. Published by SETAC, 1998.
- Application of Life-Cycle Assessment to Public Policy. Wintergreen, Virginia, August 14–19, 1994. Published by SETAC, 1997.
- Ecological Risk Assessment Decision Support System. Pellston, Michigan, August 23–28, 1994. Published by SETAC, 1998.
- Avian Toxicity Testing. Pensacola, Florida, December 4–7, 1994. Co-sponsored and published by Organisation for Economic Co-operation and Development (OECD), 1996.
- Chemical Ranking and Scoring (CRS): Guidelines for Developing and Implementing Tools for Relative Chemical Assessments. Sandestin, Florida, February 12–16, 1995. Published by SETAC, 1997.
- Ecological Risk Assessment of Contaminated Sediments. Pacific Grove, California, April 23–28, 1995. Published by SETAC, 1997.
- Ecotoxicology and Risk Assessment for Wetlands. Fairmont, Montana, 30 July–3 August 1995. Published by SETAC, 1999.
- Uncertainty in Ecological Risk Assessment. Pellston, Michigan, August 23–28, 1995. Published by SETAC, 1998.
- Whole-Effluent Toxicity Testing: An Evaluation of Methods and Prediction of Receiving System Impacts. Pellston, Michigan, September 16–21, 1995. Published by SETAC, 1996.
- Reproductive and Developmental Effects of Contaminants in Oviparous Vertebrates. Fairmont, Montana, July 13–18, 1997. Published by SETAC, 1999.
- Multiple Stressors in Ecological Risk Assessment. Pellston, Michigan, September 13–18, 1997. Published by SETAC, 1999.
- Re-evaluation of the State of the Science for Water Quality Criteria Development. Fairmont, Montana, June 25–30, 1998. Published by SETAC, 2003.
- Criteria for Persistence and Long-Range Transport of Chemicals in the Environment. Fairmont Hot Springs, British Columbia, Canada, July 14–19, 1998. Published by SETAC, 2000.
- Assessing Contaminated Soils: From Soil-Contaminant Interactions to Ecosystem Management. Pellston, Michigan, September 23–27, 1998. Published by SETAC, 2003.
- Endocrine Disruption in Invertebrates: Endocrinology, Testing, and Assessment (EDIETA). Amsterdam, The Netherlands, December 12–15, 1998. Published by SETAC, 1999.
- Assessing the Effects of Complex Stressors in Ecosystems. Pellston, Michigan, September 11–16, 1999. Published by SETAC, 2001.
- Environmental–Human Health Interconnections. Snowbird, Utah, June 10–15, 2000. Published by SETAC, 2002.
- Ecological Assessment of Aquatic Resources: Application, Implementation, and Communication. Pellston, Michigan, September 16–21, 2000. Published by SETAC, 2004.
- Toxicity Identification Evaluation/Toxicity Reduction Evaluation: What Works and What Doesn't. Pensacola, Florida, June 23–27, 2001. Proceedings published by SETAC in 2005.
- The Global Decline of Amphibian Populations: An Integrated Analysis of Multiple Stressors Effects. Wingspread, Racine, Wisconsin, August 18–23, 2001. Published by SETAC, 2003.
- Methods of Uncertainty Analysis for Pesticide Risks. Pensacola, Florida, 24 February–1 March 2002.
- The Role of Dietary Exposure in the Evaluation of Risk of Metals to Aquatic Organisms. Fairmont Hot Springs, British Columbia, Canada, 27 July–1 August 2002. Published by SETAC, 2005.

- Use of Sediment Quality Guidelines (SQGs) and Related Tools for the Assessment of Contaminated Sediments. Fairmont Hot Springs, Montana, August 17–22, 2002. Published by SETAC, 2005.
- Science for Assessment of the Impacts of Human Pharmaceuticals on Aquatic Ecosystem. Snowbird, Utah, June 3–8, 2003. Published by SETAC, 2005.
- Population-Level Ecological Risk Assessment. Roskilde, Denmark, August 23–27, 2003. Published by SETAC and CRC Press, 2007.
- Valuation of Ecological Resources: Integration of Ecological Risk Assessment and Socio-Economics to Support Environmental Decisions. Pensacola, Florida, October 4–9, 2003. Published by SETAC and CRC Press, 2007.
- Emerging Molecular and Computational Approaches for Cross-Species Extrapolations. Portland, Oregon, July 18–22, 2004. Published by SETAC and CRC Press, 2006.
- Veterinary Medicines in the Environment. Pensacola, Florida, February 12–16, 2006. Published by SETAC and CRC Press, 2008.
- Tissue Residue Approach for Toxicity Assessment: Invertebrates and Fish. Leavenworth, Washington, June 7–10, 2007. Published in SETAC Journal *Integrated Environmental Assessment and Management* (IEAM), 2011.
- Science-Based Guidance and Framework for the Evaluation and Identification of PBTs and POPs. Pensacola Beach, Florida, 27 January–1 February 2008.
- The Nexus between Ecological Risk Assessment (ERA) and Natural Resource Damage Assessment (NRDA). Fairmont, Montana, January 28–31, 2008. Published in SETAC Journal *Integrated Environmental Assessment and Management* (IEAM), 2009.
- Ecological Assessment of Selenium in the Aquatic Environment. Pensacola Beach, Florida, February 22–27, 2009. Published by SETAC and CRC Press, 2010.
- A Vision and Strategy for Predictive Ecotoxicology in the 21st Century: Defining Adverse Outcome Pathways Associated with Ecological Risk. Forest Grove, Oregon, April 19–23, 2009. Published in SETAC Journal *Environmental Toxicology and Chemistry* (ET&C), 2011.
- Problem Formulation for Ecological Risk Assessments. Pensacola Beach, Florida, April 18–24, 2010.
- Potential Risks of Plant Protection Products to Pollinators. Pensacola Beach, Florida, January 16–22, 2011.
- Life Cycle Assessment Database Global Guidance. Kanagawa, Japan, 30 January–4 February 2011. Published by the UNEP/SETAC Life Cycle Initiative, 2011.
- Influence of Global Climate Change on the Scientific Foundation and Application of Environmental Toxicology and Chemistry. Racine, Wisconsin, July 16–21, 2011. Published in ET&C, 2012.
- Guidance on Bioavailability/Bioaccessibility Measurements Using Passive Sampling Devices and Partitioning-Based Approaches for Management of Contaminated Sediments. Costa Mesa, California, November 7–9, 2012. To be published in ET&C and IEAM, 2013.

1 Introduction

CONTENTS

Worldwide declines in managed and non-managed pollinators have led to an increased global dialogue and focus concerning the potential factors that may be causing these declines. Although a number of factors have been hypothesized as potential contributors to pollinator declines, at this time, no single factor has been identified as the cause. The available science suggests that pollinator declines are a result of multiple factors which may be acting in various combinations. Research is being directed at identifying the individual and combined stressors that are most strongly associated with pollinator declines. Pesticide use is one of the factors under consideration.

In an effort to further the global dialogue, the Society of Environmental Toxicology and Chemistry (SETAC) held a Pellston Workshop[1] to explore the state of the science on pesticide risk assessment for pollinators. The proposal for this SETAC Workshop was developed by a steering committee (hereafter referred to as the Steering Committee) comprised of members from the government and nongovernmental organizations who were interested in advancing the science to understand the effect of pesticides on nontarget insects. Workshop participants were tasked to advance the current state of the science of pesticide risk assessment by more thoroughly vetting quantitative and qualitative measures of exposure and effects on the individual bee, and where appropriate, on the colony. In doing so, the Workshop aimed to synthesize the global understanding and work that has, thus far, taken place, and to move toward a harmonized process for evaluating and quantitatively characterizing risk to pollinators from exposure to pesticides; and to identify the data needed to inform that process. The Workshop focused on four major topics:

1. design and identify testing protocols to estimate potential exposure of bees to pesticide residues in pollen and nectar, as well as exposure through other routes;
2. design and identify testing protocols to measure the effects of pesticides on developing brood and adult honey bees at both the individual and the colony levels;
3. propose a tiered approach for characterizing the potential risk of pesticides to pollinators; and
4. explore the applicability of testing protocols, used for honey bees (*Apis* bees), to measure the effects of pesticides and pesticide risk to other non-*Apis* bee species.

[1] The first Pellston Workshop was held in 1977 to address the needs and means for assessing the hazards of chemicals to aquatic life. Since then, many workshops have been held to evaluate current and prospective environmental issues. Each has focused on a relevant environmental topic, and the proceedings of each have been published as a peer-reviewed or informal report. These documents have been widely distributed and are valued by environmental scientists, engineers, regulators, and managers because of their technical basis and their comprehensive, state-of-the-science reviews. The first four Pellston workshops were initiated before the Society of Environmental Toxicology and Chemistry (SETAC) was effectively functioning. Beginning with the 1982 workshop, however, SETAC has been the primary organizer and SETAC members (on a volunteer basis) have been instrumental in planning, conducting, and disseminating workshop results. Taken from http://www.setac.org/node/104

Pesticide Risk Assessment for Pollinators, First Edition. Edited by David Fischer and Thomas Moriarty.

Although the term "pollinators" encompasses a broad number of taxa, for the purposes of this SETAC Workshop and its proceedings, the term "pollinators" refers specifically to subspecies and strains of *Apis mellifera* that originated in Europe (i.e., the honey bee) and other (non-*Apis mellifera*) bees, for example, bumble bees, solitary bees, and stingless bees. The Workshop built upon the numerous efforts of different organizations, regulatory authorities, and individuals, both nationally and internationally, aiming to better understand the role and effects of pesticide products on honey bees[2] and other bee species.

1.1 WORKSHOP BALANCE AND COMPOSITION

Similar to other timely and relevant scientific issues addressed by SETAC Pellston Workshops, the issue of pollinator protection is of high interest to scientists employed by governments, business, academia, and nongovernmental organizations. For this reason, SETAC requires that its workshops be similarly balanced. The Workshop on Pesticide Risk Assessment for Pollinators represented an exceptionally diverse composition by both sector (employer) and geography. The 48 participants (35 panelists and 13 Steering Committee members) included individuals from industry, nongovernmental organizations, federal and state governments, the beekeeping community, and academia and represented five continents (South America, Europe, Australia, North America, and Africa) (see Acknowledgments).

This proceeding of the Workshop on Pesticide Risk Assessment for Pollinators has several sections:

- Chapters 2–6 provide background and overview of key elements such as bee biology, ecological risk assessment, and protection goals.
- Chapters 7–10 capture recommendations by the Workshop on the elements of exposure assessment, effects assessment (laboratory and field testing), and risk assessment.
- Chapters 11–14 capture discussion around statistical analysis, modeling, risk management, and research needs.

Pollinators, and the honey bee in particular, have been identified as a valued group of organisms because of the services they provide to agriculture and to ecosystem biodiversity. While both managed and unmanaged (*Apis* and non-*Apis*) bees contribute to crop pollination, most of the current knowledge of the side effects of agricultural pesticides on pollinators is in relation to the honey bee. Since it is not possible to test all species, regulatory authorities rely on one or several surrogate species to represent a wider range of species within a taxon. Unlike the North American process that uses the honey bee as a surrogate for other terrestrial invertebrates, the European process includes testing requirements for honey bees specifically (representing pollinating insects), and includes other surrogate test species for nontarget arthropods in general. The proposed process discussed herein relies mainly on the honey bee, but includes other species, such as bumble bees, for example, to represent the many different species of bees. Therefore, it is important to understand the ecology and biology of the *Apis* bee as a test organism, as well as that of non-*Apis* bees.

[2] USDA Technical Working Group Report on Honey Bee Toxicity Testing, July 8 and 9, 2009, http://www.aphis.usda.gov/plant_health/plant_pest_info/honey_bees/downloads/twg_report_july_2010.pdf; International Commission for Plant–Bee Relationships 10th International Symposium, 2009, http://www.uoguelph.ca/icpbr/pubs/2008%20ICPBR%20symposium%20archives%20Pesticides.pdf

2 Overview of the Honey Bee

J. Pettis

CONTENTS

A key goal of regulatory authorities charged with licensing the use of pesticide products is to protect nontarget organisms from the potential adverse effects from those pesticide products. As it is not possible to test all species, the pesticide risk assessment framework relies on surrogate species to represent major taxa, including insect pollinators. The European honey bee (*Apis mellifera*), among the many different bee species, is a desirable surrogate test species in that it is both commercially valued and is also adaptable to laboratory research. In many countries, such as Canada and the United States, the honey bee is used as a surrogate for insect pollinators and many other nontarget terrestrial insects. While honey bees may be subject to collateral effects from the use of pesticides in crop production, they are also the beneficiaries of pesticide applications, as beekeepers routinely employ registered pesticides to manage pest problems that occur in managed hives. The in-hive use of pesticides by beekeepers and the potential exposure of honey bees to environmental mixtures of pesticides used in agriculture coupled with the complex social organization and biology of honey bees can complicate pesticide risk assessment. While these are major challenges facing risk assessment, their resolution will require additional research efforts and so they are beyond the scope of this document and are not addressed further herein (see Chapter 14).

2.1 OVERVIEW OF HONEY BEE BIOLOGY

From a risk assessment perspective, there are several aspects of honey bee biology which are important to consider as they potentially influence the toxicity studies required as well as the approach for evaluating potential risks. Colony growth and survival are dependent on the collective actions of individuals that perform various critical tasks; therefore, honey bee colonies act collectively as a "superorganism." The different castes of bees within the hive structure have different functions which can result in differential exposure in terms of route, duration, magnitude, and mode (direct vs. indirect, secondary exposure). The survival of an individual bee may be of little consequence as colonies typically have a 10–30% reserve of workers, which reflects and accommodates the high turnover rate (of the individual) and flexibility of the colony to adapt to its environment. An examination of the roles of various castes within the hive and the implication for risk assessments follows.

A honey bee colony is made up of one queen, several drones, thousands of workers, and many immature bees in various stages of development (eggs, larvae, pupae). Worker bees are sexually undeveloped females and constitute the vast majority of the adults in a colony. All work, inside and outside the colony, is done by worker bees. Older workers forage outside the hive for pollen and nectar and thus are potentially more exposed to pesticides via contact during foraging (e.g., by direct overspray or by contact with pesticide residues on treated plant surfaces), as well as dietary exposure during collection or ingestion of pollen and nectar. Workers

Pesticide Risk Assessment for Pollinators, First Edition. Edited by David Fischer and Thomas Moriarty.

also are a medium by which environmental contaminants come back to the hive. Young workers clean cells and attend brood whereas middle-aged workers do a variety of tasks mainly within the hive. Both young and middle-aged workers can be exposed to pesticides through contaminated food brought back to the hive. Each colony has a single queen. Once she mates with drones, the queen returns to the hive to begin the task of egg laying; she will lay up to 1200 eggs per day for several years. The queen performs no other work in the hive and continues to be fed royal jelly throughout her lifespan. Drones are male bees whose sole function in the hive is to serve as sperm donors for new queens. Like younger and middle-aged workers, queens and drones can also be exposed to pesticides through contaminated food brought back to the hive or intentionally used in the colony by beekeepers.

Inputs by worker bees into the colony include pollen, nectar, water, and plant exudates (e.g., sap) used to make propolis. Pollen is used as the source of protein. It may be consumed directly, consumed and used to produce brood food or royal jelly, or stored and consumed later as bee bread. While larval bees may consume small quantities of raw pollen directly, they as well as the queen depend on processed secretions (brood food and royal jelly) produced by nurse bees. Availability and the quality of pollen can have a great influence on the health status of the colony. Nectar is used as a source of carbohydrates, it may be consumed directly or stored inside the hive converted to honey and consumed later.

From a risk assessment perspective, the large forage area of honey bees complicates the task of estimating potential exposure. Honey bees typically forage in the middle of the day for food within 2–3 km (1–2 mi) of the hive, but may forage 7 km (5 mi) or more if food of suitable quality is lacking nearby. The large forage range increases the potential that the pollen and nectar collected by the honey bee may contain pesticide residues used in the foraging vicinity. The time of day when foraging occurs in relation to pesticide application may also influence exposure and therefore the risk assessment. As will be discussed in the following chapters, numerous other factors should be considered in light of bee biology that can impact the design or interpretation of data intended to inform pesticide risk assessment with these organisms.

3 Overview of Non-*Apis* Bees

M. Vaughan, B.E. Vaissière, G. Maynard, M. Kasina, R.C.F. Nocelli, C. Scott-Dupree, E. Johansen, C. Brittain, M. Coulson, and A. Dinter

CONTENTS

3.1 INTRODUCTION

Honey bees (*Apis mellifera* L.) are frequently employed in pesticide toxicity testing either as a representative species (i.e., surrogate) for pollinating insects (such as in the European Union (EU)) or in other cases to represent other non-target terrestrial invertebrates (such as in North America). As with many surrogate test organisms, there are considerations or limitations to using *A. mellifera* as a representative species for pollinators and terrestrial invertebrates in general. For example, field tests with honey bees can be challenging because of their very long foraging range, the variability of their foraging area and the forage resources they utilize (Visscher and Seeley 1982). In semi-field tests, honey bees do not respond well to being kept in cages or indoor environments for a long period.

Uncertainties also exist regarding the extent to which pesticide toxicity data for honey bees can be considered protective for non-*Apis* bees. Studies have demonstrated variable and inconsistent toxicity among various bee groups (Johansen et al. 1983; Malaspina and Stort 1983; Torchio 1983; Macieira and Hebling-Beraldo 1989; Peach et al. 1995; Malone et al. 2000; Moraes et al. 2000; Scott-Dupree et al. 2009; Roessink et al. 2011; Biddinger et al. 2013). This variability results, in part, from the basic biological differences between the highly social honey bees and other non-eusocial species, as well as intrinsic differences in physiology, life cycle, and behavior between any two insect species (Thompson and Hunt 1999).

The need to thoroughly explore pesticide risk assessment for non-*Apis* pollinators is more important now than in the past as many areas around the world are seeing an increasing demand for insect pollination. The decline of available managed honey bees and the consequential rising costs for honey bee pollination services have left the needs of agriculture unmet. (Aizen and Harder 2009). As a result, across the globe many farmers are looking to other managed or wild (unmanaged) non-*Apis* bee species, and scientists are documenting

Pesticide Risk Assessment for Pollinators, First Edition. Edited by David Fischer and Thomas Moriarty.

that many crops are pollinated to a significant level by non-*Apis* bees (Garibaldi et al. 2013). For example, managed bumble bees (*Bombus* spp.) are increasingly being used to support agricultural and horticultural production. Over 1 million bumble bee colonies of different species were sold worldwide in 2006, primarily for greenhouse fruit and vegetable production (e.g., tomato, *Lycopersicon esculentum*), but also increasingly for commercial orchards and seed production (Velthuis and van Doorn 2006).

In the United States, many growers of alfalfa seed (*Medicago sativa*), almond (*Prunus dulcis*), apple (*Malus domestica*), blueberry (*Vaccinium* spp.), and sweet cherry (*Prunus avium*) are using managed solitary bees such as wood-nesting alfalfa leafcutting bees (*Megachile rotundata*), blue orchard bees (*Osmia lignaria*), and ground-nesting alkali bees (*Nomia melanderi*). In some places, the use of these non-*Apis* pollinators is already widespread or is becoming more common (Bosch and Kemp 2001). For example, in the United States, approximately 35 000 tons of alfalfa seeds are produced annually with pollination provided by alfalfa leafcutting bees from Canada (Mayer and Johansen 2003; Stephen 2003; Pitts-Singer 2008; James 2011; Pitts-Singer, personal communication, December 9, 2011). In Japan, the hornfaced bee (*Osmia cornifrons*) is managed to pollinate orchards of apple and pear (*Pyrus communis*) (Matsumoto et al. 2009), and in Brazil, the carpenter bee *Xylocopa frontalis* can be managed to pollinate the passion fruit (*Passiflora edulis*; Freitas and Oliveira Filho 2003). In Kenya, solitary bees have not yet been commercialized for pollination purposes, but efforts are underway to develop management protocols for solitary bees such as *Xylocopa calens*, *Xylocopa incostans*, and *Xylocopa flavorufa* for high-value greenhouse crops (Kasina, personal communication, October 5, 2011).

In the tropics, efforts are also underway to develop meliponiculture (stingless beekeeping) as a source of revenue from honey production, other hive products, and rentals for crop pollination. Meliponiculture is well established in countries such as Brazil and Mexico (Nogueira-Neto 1997; Villanueva-Gutiérrez et al. 2005). In Africa there are ongoing efforts to improve the management and expand the use of regionally native stingless bees, for example in Ghana (Kwapong et al. 2010) and in Kenya (Kasina, personal communication, October 5, 2011).

At the same time, across the world, there is a growing emphasis on the role of unmanaged or wild bees in agro-ecosystems among agriculture and conservation agencies. For example, in the United States this includes national-level ecosystem restoration efforts by the US Department of Agriculture's Natural Resources Conservation Service (USDA-NRCS), mandated under the Food, Conservation and Energy Act of 2008 (Vaughan and Skinner 2009). These conservation efforts are based upon general trends demonstrating declines in populations of wild bees in agricultural landscapes (Kremen et al. 2004; Biesmeijer et al. 2006; National Research Council 2007), as well as the increasingly large body of research demonstrating the significant role that unmanaged non-*Apis* bees may play in crop pollination (Kremen et al. 2002; Kremen et al. 2004; Njoroge et al. 2004; Winfree et al. 2007; Campos 2008; Winfree et al. 2008; Kasina et al. 2009; Isaacs and Kirk 2010; Vieira et al. 2010; Carvalheiro et al. 2011; Gariboldi et al. 2013). Furthermore, recent research highlights the importance of a diverse pollinator guild for optimal pollination (Klein et al. 2003; Höhn et al. 2008), as well as the benefits of the interaction between honey bees and wild bees to enhance the pollination effectiveness of honey bees (Greenleaf and Kremen 2006a, 2006b; Carvalheiro et al. 2011).

Non-*Apis* bees are often specialized for foraging on particular flower taxa, such as squash, berries, forage legumes, or orchard crops (Tepedino 1981; Bosch and Kemp 2001; Javorek et al. 2002; Brunet and Stewart 2010). This specialization is usually associated with more efficient pollination on an individual bee visit basis, which can lead to production of larger and more abundant fruit or seed from certain crops (Greenleaf and Kremen 2006a, 2006b; Klein et al. 2007, but see also Rader et al. 2009). In one study, researchers estimated that non-managed bees contribute an estimated US$3 billion worth of crop pollination annually to the US economy (Losey and Vaughan 2006). More recently, researchers estimated that in California alone, unmanaged non-*Apis* bees pollinated US$937 million to US$2.4 billion worth of crops (Chaplin-Kramer et al. 2011). In addition to their impact on agro-ecosystems, non-*Apis* pollinators are crucial to native flora. More than 85% of flowering plants benefit from animal pollinators (Ollerton et al. 2011), most of which are insects and the most important of which are bees (Apiformes).

Because of the recent increase in our understanding of the value of non-*Apis* bees for agriculture (Garibaldi et al. 2013) and the critical role they play in natural ecosystems, researchers have suggested that non-*Apis* bees could play a useful role in risk-assessment for pollinators (Biddinger et al. 2013). Specifically, they recommend that at least one solitary managed species, such as the wood-nesting alfalfa leafcutting bees (*M. rotundata*) or the blue orchard bees (*O. lignaria*) (Abbott et al. 2008; Ladurner et al. 2008), and one managed social non-*Apis* bee, such as bumble bees (e.g., *Bombus impatiens* or *Bombus terrestris*) in temperate climates (Thompson and Hunt 1999) or the highly social stingless bees (e.g., *Melipona* spp. or *Trigona* spp.) in the tropics (Valdovinos-Núñez et al. 2009) is incorporated into regulatory testing schemes. To develop appropriate toxicity tests and risk assessment protocols for non-*Apis* bees, however, it is important to understand more about non-*Apis* bees and the unique exposure pathways relevant for them.

3.2 NON-*APIS* BEE BIOLOGY AND DIVERSITY

Worldwide, there are over 20 000 recorded species of bees (Michener 2007; Ascher and Pickering 2011). They range in size from approximately 2 mm (1/12 inch) to more than 25 mm (1 inch), exhibit a wide variety of foraging and nesting strategies, vary from solitary to highly social, and exhibit other diverse life histories.

Bees use nectar mainly as a carbohydrate source and pollen as a source of protein, fatty acids, minerals, and vitamins. Some species also use other plant resources such as resins, leaves, plant hairs, oil, and fragrances to feed their larvae, build and protect nests, or attract mates (Michener 2007). Because they use plant products during all life cycle stages, they are vulnerable to plant protection products that are present or expressed in pollen and nectar, or that are found in or on other plant resources.

During their life cycle, bees undergo a complete metamorphosis where they develop through egg, larval, pupal, and adult stages. It is only the last of these stages, the adult, which most people see and recognize as a bee. During the first three stages, the bee is inside a brood cell of the nest. The length of each stage varies widely between species and is often defined by whether the bee is solitary or social (O'Toole and Raw 1999). In the case of solitary bees, each female works alone to create a brood cell, place a mixture of pollen and nectar into it, and then lay an egg on (or more rarely in) the food. Solitary bees may take a year to complete metamorphosis, although it can happen faster, that is, 4–6 weeks in those species that have two or three generations per year. Social bees, on the other hand, take only a few weeks to complete growth and emerge as adults.

The quantity of food provided at the time of egg-laying depends on whether the larvae are mass-provisioned (i.e., all of the bee's food is supplied in the cell at one time), or if the larvae are progressively fed (i.e., the food is delivered in small amounts over time). Most solitary bees mass provision their brood cells, as do most stingless bees, whereas honey bees and most bumble bees feed their brood progressively.

Female bees of most species have special morphological structures that enable them to carry pollen back to their nests. For example, the tibiae on the hind legs of honey bees, bumble bees, and stingless bees are modified into corbiculae (a flattened, shallowly depressed area margined with a narrow band of stiff hairs) into which the bee accumulates pollen wetted with nectar and packed into place. Other bee species have scopae to transport pollen. Scopae are fringes, tufts, or brushes of hair on their legs, their thorax, or the undersurface of the abdomen. Scopae are used to transport large amounts of pollen, usually in a dry state.

The wide range of life history traits of bees has implications for their exposure to pesticides (Brittain and Potts 2011) and so relevant aspects of their natural history is described below.

3.2.1 Generalist and Specialist Foragers

Bee species have several strategies for pollen collection. Certain species are considered generalist foragers (polylectic). Generalist foragers include species such as honey bees, stingless bees, and bumble bee species, which gather pollen from a wide range of flower species. Other species are considered specialist foragers, (oligolectic) and gather pollen from a narrow range of plant species that are usually related taxonomically. Specialist foragers, however, may gather nectar from a wider range of plants than from which they gather pollen. Examples of oligolectic bees include squash bees (*Xenoglossa* or *Peponapis* spp.), *Macropis* spp., and *Leioproctus* spp., which collect pollen from cucurbits (*Cucurbita* spp.), yellow loosestrife (*Lysimachia* spp.), and geebungs (*Persoonia* spp.), respectively. A third category of pollen collectors, of which there are very few species, are those bees which are monolectic. Monolectic foragers are those which feed on pollen from only a single species of plant, for example, *Hesperapis araria* which only visits flowers of the plant *Balduina angustifolia* (Asteraceae) on the coastal islands of the northern Gulf of Mexico (Cane et al. 1996). The life cycle of specialists (oligolectic and monolectic) are normally closely tied to their host plants, with the adult female bees emerging from their brood cells when their main pollen sources are flowering (O'Toole and Raw 1999).

3.2.2 Social and Solitary Behavior

Bees exhibit a wide range of social behaviors, but depending on their interdependency, bees can be broadly divided into two groups: social or solitary.

3.2.2.1 Social Bees

Social bees typically live as a colony in a nest with one queen (but occasionally can have more than one queen). The labor of building the nest, caring for offspring, protecting the colony, and foraging for resources is shared among female offspring with greatly reduced reproductive capacity. Only a few species of bees demonstrate highly social (eusocial) behavior. These eusocial species include all species of honey bees in the genus *Apis*, and approximately 400 stingless bee species in the tribe Meliponini. Eusocial bees are found primarily in the tropics and subtropics, with two species, *A. mellifera* and *Apis cerana*, living in temperate areas. Primitively social (or facultatively eusocial) bees exhibit lesser degrees of eusocial behavior (Michener 2007), where colonies are initiated by queens or dominant females on an annual basis (e.g., Halictidae (sweat bees)). Most remaining bee species, the vast majority, are solitary and while sometimes nest together in great numbers, these gregarious bees do not cooperate (Michener 2007; Cane 2008).

In the world's temperate zones, bumble bees are the best known non-*Apis* social bees. Bumble bees live in colonies, share the work of foraging and nest construction, and produce many overlapping generations throughout the year; and thus, they are eusocial. However, unlike honey bees, bumble bee colonies are seasonal. At the end of the summer, most of the bees in the colony die, leaving only a few fertilized queens to hibernate (usually underground) through the winter. In the spring, each surviving queen will start a new nest, which may eventually grow to include dozens to hundreds of workers, depending on the species. Apart from honey bees, bumble bees are often the first bees active in late winter (foraging at lower temperatures than honey bees) and the last bees active in the autumn (Kearns and Thomson 2001; Goulson 2003).

Most bumble bees are generalist foragers, visiting a wide diversity of flowers. Bumble bees can gather pollen by "buzzing" flowers—holding them tightly and vibrating their flight muscles (with an audible buzz), causing the poricidal anthers to release their pollen. Buzz pollinators are important for ensuring pollination in crops with poricidal anthers such as blueberries, cranberries, and other *Vaccinium* spp., as well as solanaceous plants including tomatoes and eggplants (*Solanum melongena*), but also others such as peppers (*Capsicum annuum*) and strawberry (*Fragaria x ananassa*).

Bumble bees need a suitable cavity in which to nest. Sometimes they build nests aboveground, under a tussock of grass or in hollow trees or walls, but generally they nest underground (Kearns and Thomson 2001). Abandoned rodent burrows are common nest sites, as this space is easily warmed and likely contains nesting and insulating materials, such as fur or dried grass. In this cavity, the queen creates the first few pot-like brood cells from wax secreted by her wax glands, lays eggs, and then forages to provide her brood with pollen and nectar (Goulson 2003). It will take about a month for her to raise this first brood. When this first brood emerges, these bees become workers. They take on the task of foraging and help the queen tend the growing number of brood cells through the summer. At the end of summer, new queens and drones emerge and mate. When the cooler weather of autumn arrives, most of the bees, including the old queen, will die, leaving only the newly mated queens to find appropriate sites in which to hibernate through the winter (Kearns and Thomson 2001).

Bumble bees mainly occur in temperate areas. However, as the pollination demand for greenhouse crops grows, there have been attempts to introduce bumble bee colonies in other non-native temperate zones. The threats of such introduction may include inbreeding with local bumble bee species, competition with the native bees for food resources, transfer of pathogens, (Oldroyd 1999; Thomson 2004; Stout and Morales 2009), which may result in a decline in the abundance and diversity of the native bee community, (Dafni et al. 2010) and disruption to the pollination of native plants. In temperate countries, the approach of winter controls the population of these bees through the death of all caste members except the newly mated queens. In warmer climates, weather may be more favorable year round and these bees may not diapause, increasing their numbers tremendously within a short duration of their introduction (Beekman et al. 1999; Dafni et al. 2010). Bumble bees, therefore, may not be appropriate for providing pollination services in the tropics and thus there is a need to study locally or regionally native stingless bees to provide pollination service for greenhouse crops in the tropics (Slaa et al. 2000; Del Sarto et al. 2005).

3.2.2.2 Social, Stingless Bees

Stingless bees live in the tropical and southern subtropical areas (Michener 2007). They live in colonies that number from a few dozen individuals to more than 25 000, and they are active year-round. The colony size and nest architecture are characteristic for each different species. Numerous species can be found in Central and South America. In the Yucatan Peninsula, for example, farming of stingless bees for honey and wax was so extensive that European honey bees were not introduced until the nineteenth century (Crane 1992; Vit and d'Albore 1994; Javier et al. 2001).

Stingless bees are generalist foragers, visiting a broad variety of flowers. However, individual colonies or populations may demonstrate a tendency to visit particular types of flowers or exhibit a temporary fidelity to specific plant species (Ramalho et al. 1994 1998, 2007). They are known to visit at least 90 crop species and are used to enhance pollination in some crops on a commercial to semi-commercial basis (Heard and Dollin 1998a; Heard 1999).

Most stingless bees nest in a cavity. Typically, these cavities are in trees or hollow logs; however, a few species will move into termite mounds, building walls, or even cavities underground. Nests are often located 2–30 m aboveground (Kajobe 2007). Stingless bees line their nest cavity with an envelope of batumen, a tough mixture of wax produced by the bees combined with resins, gums, plant material, and sometimes mud collected from around the nest. The nests are composed of many storage pots of honey and pollen and smaller brood cells. The pots (both storage and brood) are made of cerumen, a mixture of wax and plant resins.

Within the nest, each brood pot is mass provisioned with hypopharyngeal gland secretions, pollen, and honey. An egg is laid on top of these provisions and then the pot is sealed. The nests can have one to several queens depending on the species. Most species of stingless bees have brood cells of two different sizes; the large cells produce gynes (queens) while the small ones produce males and workers (Michener 1974). Caste determination is usually through food provisioning, with the quantity, not the quality, of food determining the caste. Thus gyne cells are provisioned with more food compared to the worker and male brood cells. This is in contrast to the honey bee caste determination where both quantity and quality of brood food are important.

New nests are initiated on a progressive basis. A virgin queen moves into a new cavity with some workers over a period of several weeks. They take materials from the old nest to create the new nest. Hence stingless bees are not capable of long-distance migration (Roubik 2006). However, with domestication, new colonies can be established through methods similar to splitting honey bee colonies. Young gynes are moved together with brood, workers, and males to another hive to establish a new colony (Nogueira-Neto 1997; Arzaluz et al. 2002; Villanueva-Gutiérrez et al. 2005; Kwapong et al. 2010).

3.2.2.3 Solitary Bees

The vast majority of bee species in the world are solitary. For these solitary species, the labor of nest construction and provisioning, foraging and egg-laying is all done by single, fertile female bees. A female solitary bee may lay twenty or thirty eggs in her life. For solitary species having one generation per year, 1–3 weeks after an egg is laid, it hatches and the larva emerges to feed on the combination of pollen and nectar ("bee bread") previously provided by the adult female. The larva grows rapidly for 6–8 weeks before pupating. The dormant prepupal or pupal stage typically lasts 8 or 9 months in temperate climates. When it emerges, the adult bee is fully grown and then needs food (primarily nectar) for egg maturation and energy. Most solitary bees have only one generation per year and have a fairly short season of adult activity. Some solitary species, such as some sweat bees in the genera *Halictus* and *Lasioglossum*, have two or three generations each year and so are present over a longer period of time.

Adult solitary bees are typically active for 3–6 weeks. Males usually emerge first from the nest, after which they typically loiter around a nesting area or a foraging site in search of a female to mate with. After a female bee emerges, she mates and then spends her time building and provisioning a nest in which to lay eggs (O'Toole and Raw 1999; Michener 2007; Cane 2008). The adults of a species emerge at roughly the same time each year, for example, early spring in the case of blue orchard bees (*O. lignaria*) or midsummer in the case of squash bees (*Peponapis pruinosa*). This emergence normally coincides with the flowering of forage plants, particularly if the bee is a specialist.

About 30% of solitary bee species are twig, or wood-nesting. Most species use hollow stems or abandoned beetle burrows or other tunnels in dead or dying standing trees, but some can chew out a nesting tunnel in the soft central pith of stems and twigs, or in a few cases they may bore their own tunnel in wood (Michener 2007). The other 70% nest in the ground, digging tunnels in bare or partially vegetated, well-drained soil (Potts et al. 2005). Each solitary bee nest will have one or more separate cells in which the female places all the provisions (pollen and nectar) required for the full development of her larvae. While some nests may have only a single cell, most have five or more. In the case of ground-nesting bees, females create a range of underground architecture, from simple tunnels to complex, branching systems with cells usually located 10 cm to 2 m underground. Wood-nesting bees on the other hand, usually stack cells in a single line inside their nest tunnels.

Most wood-nesting species separate individual brood cells with materials they collect, such as leaf pieces, leaf pulp, plant hairs, tree resin, or mud. For example, leafcutting bees (*Megachile*) use pieces of leaf or petal to create self-contained brood cells. Using their mandibles, they cut particular sizes and shapes to fit different parts of the brood cell, lining the entire cell. Most other wood-nesting bees, however, do not line the entire cell, but simply build dividing walls across the nesting tunnel, segmenting it into separate brood cells. Blue orchard bees (*Osmia*) make these walls with mud or leaf pulp. Large carpenter bees (*Xylocopa*) and small carpenter bees (*Ceratina*) use wood fibers scraped from the walls of the tunnel to form dividers of compacted sawdust. These bees seal the nest entrance when it is finished with the same materials they use to construct the inner partitions.

Rather than collecting materials from outside the nest with which to line their brood cells, many ground-nesting bee species smoothen the cell walls with their abdomens and then apply a waxy or oily substance produced from special glands near their mouths or on their abdomens to line the cells, thus stabilizing the soil and protecting their brood. The substance lining the cell usually soaks into the soil, making it look shiny and helping to exclude water and control microbes. Plasterer or polyester bees (*Colletes*), yellow-faced bees (*Hylaeus*), and other bees from the family Colletidae line each cell with a cellophane-like substance secreted from special glands to create a complete waterproof lining for their underground cells. A few species, such as tiny *Perdita* bees living in the southwestern deserts of the United States, leave their underground cells unlined.

3.2.3 Status of Toxicity Testing for Non-*Apis* Bees

In general, the research on pesticide toxicity and risk assessment for non-*Apis* bees lags behind that for honey bees (see Tables 8.2, 8.4, and 10.5 for examples of pesticide toxicity studies conducted on non-*Apis* bees). Except for bumble bees, most of the data referred to on non-*Apis* bees has been sourced from North America. The most commonly studied species are *M. rotundata* (the alfalfa leafcutting bee), *B. impatiens* (the eastern bumble bee), and *O. lignaria* (the blue orchard bee), all of which are managed species of economic importance. These species have been put through a range of lower and higher tier toxicity tests, but only for a handful of active ingredients, usually of regional importance. At present, the tests are not standardized.

Most of the non-*Apis* bee toxicity testing conducted in Europe has been on bumble bees, and in particular *B. terrestris*, which is the main species used for commercial pollination. Typically, bumble bee suppliers (e.g., Koppert Biological Systems, Biobest, and Syngenta Bioline) complete thorough higher tier testing of pesticide toxicity to ensure bumble bee safety in greenhouses when pesticides have to be applied. Lower tier toxicity tests (e.g., acute toxicity tests conducted in the laboratory) are somewhat limited, but comparative toxicities between *A. mellifera* and *Bombus* spp. have been reviewed by several authors (Thompson 2001; van der Steen et al. 2008). Comparison has been made both on a dose per bee level and a dose per gram of bee (factoring in the larger size of the bumble bee). The broad conclusions are that there is no consistent correlation between the toxicity for *Apis* and *Bombus* workers, but the general trends suggest that the toxicity to bumble bees is less on a per bee basis and similar on a per gram of bee basis (see also Figures 8.1 to 8.3).

Work on the comparative toxicity of pesticides to individuals or colonies of stingless bees in the subtropics and tropics is in its relative infancy. In part, this is because little is known of the biology of most stingless bee species and many species remain undiscovered or undescribed. However, because there is significant interest in the management in these species for the pollination of high value crops, the need to understand the effects of pesticides is growing. Already some toxicity work has been done using various species of Meliponini (*Melipona beecheii*, *Trigona nigra,* and *Nannotrigona perilampoides)* (Valdovinos-Núñez et al. 2009). Collaborations are underway between national regulatory authorities, national research institutions, and universities to develop toxicity testing protocols for non-*Apis* bees commonly used for field or greenhouse pollination in the tropics. Using OECD guidelines (OECD 1998) as a template protocol, these toxicity tests are being developed by partners in Brazil, Kenya, and the Netherlands to carry out comparative studies with native stingless bees, solitary bees, honey bees, and bumble bees (Roessink et al. 2011). Specifically, stingless bees in Kenya currently being studied include *Meliponula ferruginea* and *M. bocandei*, while in Brazil they include *Scaptotrigona postica* and *Melipona scutellaris*. The African honey bee (*Apis mellifera scutellata*) in Kenya and the Africanized honey bee (also *Apis mellifera scutellata*, but hybridized with European honey bees in the Americas) in Brazil are also study organisms. The results are expected to aid in understanding differences in sensitivity to various pesticides among stingless bees and honey bees in the tropics, compared to the western honey bee (*Apis mellifera mellifera*) and bumble bee (*B. terrestris*) found in the Netherlands.

In addition, tests will be performed on solitary bees in Brazil and Kenya (e.g., *Xylocopa* spp.) after optimizing procedures for their rearing to ensure enough individuals are available to meet the testing requirements.

3.3 OPPORTUNITIES FOR NON-*APIS* BEES TO INFORM POLLINATOR RISK ASSESSMENT

Specific life history traits of non-*Apis* bees lend themselves to providing useful information for risk assessors. For example, solitary non-*Apis* bees, such as *Osmia* and *Megachile* spp., have a more restricted foraging area than honey bees and use of these solitary species in field testing scenarios may provide more confidence that the test bees are foraging (receiving exposure) from the treated (test) crops (Maccagnani et al. 2003; Zurbuchen et al. 2010). In typical field test scenarios, it is only feasible to apply the product to a limited area (e.g., ≤2 ha.) of a bee-attractive crop. Compared to solitary species, honey bees forage over much larger areas (Visscher and Seeley 1982; Steffan-Dewenter and Kuhn 2003), which consequently can be a challenging variable to control in field test scenarios. In another example, managed non-*Apis* bees also lend themselves to semi-field experiments by virtue that they may be less stressed than honey bees in an enclosed cage or greenhouse setting, and thus behave more "naturally." Further research on the use of species available for toxicity testing and information on their management would also inform the use of *A. mellifera* as a surrogate for other non-*Apis* bees (Table 3.1). Available laboratory, semi-field, and field studies with representative groups of solitary and social non-*Apis* bee species can be found later in this book (Table 10.5).

Furthermore, because most non-*Apis* bees are solitary species, where single female bees build their nests, lay eggs, and forage for pollen and nectar to feed their offspring, the death of a foraging female or even her inability to provision her cells, can have a significant impact (Taséi 2002). Field kills of bumble bee queens early in the season can prevent the bumble bee colony from being established. When foragers (honey bee) are killed in the field, the loss of these workers may, to a certain extent, be compensated by the colony and may mask the event of the field kill. Because field kills have an immediate impact on non-*Apis* bees located near treatment areas, use of non-*Apis* species in semi-field and field studies may be advantageous by leading to more robust risk assessment.

3.4 CONCLUSIONS

It is clear that non-*Apis* bees play an important role in supporting diverse plant communities, and an increasingly important role in agriculture. They differ from honey bees in their biological characteristics, which consequently may make them subject to unique exposure routes (Tuell and Isaacs 2010; Brittain and Potts 2011), as well as unique challenges when it comes to risk management. Chapter 7 provides a very detailed discussion of specific biological, behavioral, or ecological traits—such as larval feeding behavior, foraging time and distance, and use of unique nesting materials—and how they affect exposure risk. Chapter 13 outlines suggested techniques for mitigating risk to non-*Apis* bees, in light of their unique biology. At the same time, some of these characteristics—such as their more limited foraging ranges and relative tolerance for foraging in enclosed areas—could be used to better assess the risks of pesticide applications for a wide range of pollinators, including honey bees. See Chapter 9 for additional details on semi-field and field studies, and Chapter 10, Table 10.5 for a list of examples.

For several reasons, it is important to consider non-*Apis* bees among its discussions on pesticide risk assessment for pollinators, including: (i) the increased understanding of the value of non-*Apis* bees in commercial agriculture; (ii) the critical role they play in natural ecosystems; (iii) increased research being conducted with them; and, (iv) the potential value they may add to the understanding of potential risks

TABLE 3.1
Potential Non-*Apis* Bee Species for Use in Laboratory, Semi-Field or Field Tests[a]

Species (common name)	Sociality	Region	References on Management
Megachile rotundata (Alfalfa leafcutting bee)	Solitary	Temperate North America, Asia	Mader et al. 2010
Osmia lignaria (Blue orchard bee)	Solitary	Temperate North America	Bosch and Kemp 2001; Mader et al. 2010
Osmia cornifrons (Japanese hornfaced bee)	Solitary	Temperate Asia, Europe	Sekita and Yamada 1993; Wilson and Abel 1996; White et al. 2009; Mader et al. 2010
Osmia rufa (Red orchard bee)	Solitary	Temperate Europe	Krunic et al. 1995; Biliñski and Teper 2004
Osmia cornuta (Hornfaced bee)	Solitary	Southern and Central Europe	Krunic et al. 1995; Maccagnani et al. 2003
Amegilla chlorocyanea (Blue-banded bee)	Solitary	Australia	Hogendoorn et al. 2006
Xylocopa spp. (Carpenter bees)	Solitary	Tropical (Brazil)	Freitas and Oliveira-Filho 2001; Freitas 2004
Bombus impatiens (Eastern bumble bee)	Social	Temperate (North America)	Readily available commercially. See also Evans et al. 2007; Mader et al. 2010
Bombus terrestris (European bumble bee)	Social	Temperate (Europe)	Readily available commercially. See also Evans et al. 2007; Mader et al. 2010
Melipona beecheii (stingless bee)	Social	Tropical (Central America)	González and de Araujo Freitas 2005; Villanueva-Gutiérrez et al. 2005; Quezada Euán 2005; Quezada Euán 2009
Trigona nigra (stingless bee)	Social	Tropical (Central America)	González & Medellín 1991a, 1991b
Nannotrigona perilampoides (stingless bee)	Social	Tropical (Central America)	González & Medellín 1991a, 1991b
Trigona carbonaria (stingless bee)	Social	Tropical (Australia)	Heard 1998; Heard and Dollin 1998b; Greco et al. 2011
Melipona subnitida (stingless bee)	Social	Tropical (Brazil)	de Oliveira Cruz et al. 2005
Meliponini tribe (stingless bees)	Social	Tropical (Brazil)	Nogueira-Neto 1997
Trigonini tribe (stingless bees)	Social	Tropical (Brazil)	Nogueira-Neto 1997
Meliponula bocandei (stingless bee)	Social	Tropical (Africa, Kenya)	Kwapong et al. 2010
Meliponula ferruginea (stingless bee)	Social	Tropical (Africa, Kenya)	Kwapong et al. 2010

[a] All of these species are either commercially available or they can be managed for crop pollination in various parts of the world. Analysis of data generated with these species would inform whether or which species may be an appropriate surrogate for other [non-*Apis*] bees or other pollinators in general, and whether their use in pesticide risk assessment would be sufficient to support regulatory decisions and attendant protection goals.

from pesticides to these taxa. For these reasons the Participants of the Workshop considered when and how non-*Apis* bee species may be incorporated and considered in a pesticide risk assessment for pollinators.

REFERENCES

Abbott VA, Nadeau JL, Higo HA, Winston ML. 2008. Lethal and sublethal effects of imidacloprid on *Osmia lignaria* and clothianidin on *Megachile rotundata* (Hymenoptera: Megachilidae). *J. Econ. Entomol.* 101(3):784–796.

Aizen MA, Harder LD. 2009. The global stock for domesticated honey bees is growing slower than agricultural demand for pollination. *Curr. Biol.* 19:915–918.

Arzaluz A, Obregón F, Jones R. 2002. Optimum brood size for artificial propagation of the stingless bee *Scaptotrigona mexicana*. *J. Apicult. Res.* 41:62–63.

Ascher JS, Pickering J. 2011. Discover Life bee species guide and world checklist (Hymenoptera: Apoidea: Anthophila). http://www.discoverlife.org/mp/20q?guide=Apoidea_species (accessed January 5, 2012).

Beekman M, van Stratum P, Veerman A. 1999. Selection for non-diapause in the bumblebee *Bombus terrestris*, with notes on the effect of inbreeding. *Entomol. Exp. Appl.* 93:69–75.

Biddinger D, Robertson J, Mullin C, Frazier J, Ashcraft S, Joshi NK, Vaughan M. 2013. Comparative toxicities and synergism of apple orchard pesticides to *Apis mellifera* (L.) and *Osmia cornifrons* (Radoszkowski). *PlosOne.* 8(9):1–6. e72587.

Biesmeijer JC, Roberts SPM, Reemer M, Ohlemüller R, Edwards M, Peeters T, Schaffers AP, Potts SG, Kleukers R, Thomas CD, Settele J, Kunin WE. 2006. Parallel declines in pollinators and insect-pollinated plants in Britain and the Netherlands. *Science* 313:351–354.

Biliñski M, Teper D. 2004. Rearing and utilization of the red mason bee—*Osmia rufa* L. (Hymenoptera, Megachilidae) for orchard pollination. *J. Apicult. Res.* 48:69–74.

Bosch J, Kemp W. 2001. *How to Manage the Blue Orchard Bee as an Orchard Pollinator*, Sustainable Agriculture Network, Beltsville, MD, pp. 88.

Brittain C, Potts SG. 2011. The potential impacts of insecticides on the life-history traits of bees and the consequences for pollination. *Basic Appl. Ecol.* 12(4):321–331.

Brunet J, Stewart C. 2010. Impact of bee species and plant density on alfalfa pollination and potential for gene flow. *Psyche* doi:10.1155/2010/201858

Campos MJO. 2008. Landscape management and pollinator richness in tomato (*Lycopersicum esculentum* Mill, Solanaceae) crops in southeastern Brazil. In: *Pollinator Management in Brazil*. Ministry of Environment, pp. 26–29.

Cane J. 2008. Bees (Hymenoptera: Apoidea: Apiformes). *Encyclopedia of Entomology*. Springer Verlag, 2:419–434.

Cane JH, Snelling RR, Kervin LJ. 1996. A new monolectic coastal bee, *Hesperapis oraria* Snelling and Stage (Hymenoptera: Melittidae), with a review of desert and neotropical disjunctives in the southeastern U.S. *J. Kansas. Entomol. Soc.* 69(4):238–247.

Carvalheiro LG, Veldtman R, Shenkute AG, Tesfay GB, Pirk CWW, Donaldson JS, Nicolson SW. 2011. Natural and within-farmland biodiversity enhances crop productivity. *Ecol. Lett.* 14:251–259.

Chaplin-Kramer R, Tuxen-Bettman K, Kremen C. 2011. Value of wildland habitat for supplying pollination services to California agriculture. *Rangelands* 34(3):33–41.

Crane E. 1992. The past and present status of beekeeping with stingless bees. *Bee World* 73:29–42.

Dafni A, Kevan P, Gross CL, Goka K. 2010. *Bombus terrestris*, pollinator, invasive and pest: an assessment of problems associated with its widespread introductions for commercial purposes. *Appl. Entomol. Zool.* 45(1):101–113.

de Oliveira Cruz D, Magalhães Freitas B, da Silva LA, Sarmento da Silva EM, Abrahão Bomfim IG. 2005. Pollination efficiency of the stingless bee *Melipona subnitida* on greenhouse sweet pepper. *Pesqui. Agropecu. Bras.* 40(12):1197–1201.

Del Sarto MCL, Peruquetti RC, Campos LAO. 2005. Evaluation of the neotropical stingless bee *Melipona quadrifasciata* (Hymenoptera: Apidae) as pollinator of greenhouse tomatoes. *J. Econ. Entomol.* 98: 260–266.

Evans E, Burns I, Spivak M. 2007. *Befriending Bumble Bees: A Practical Guide to Raising Local Bumble Bees*, Publication 08484. University of Minnesota Extension, St Paul, MN, pp. 65.

Freitas BM. 2004. *Solitary Bees: Conservation, Rearing and Management for Pollination*, Imprensa Universitaria, Fortaleza, pp. 285.

Freitas BM, Oliveira Filho JH. 2001. *Criação racional de mamangavas para polinização em áreas agrícolas*, Banco do Nordeste, Fortaleza, pp. 96.

Freitas BM, Oliveira Filho JH. 2003. Ninhos racionais para mamangava (*Xylocopa frontalis*) na polinização do maracujá-amarelo (*Passiflora edulis*). *Ciénc. Rural* 33:1135–1139.

Garibaldi LA, Steffan-Dewenter I, Winfree R, Aizen MA, Bommarco R, Cunningham SA, Kremen C, Carvalheiro LG, Harder LD, Afik O, Bartomeus I, Benjamin F, Boreux V, Cariveau D, Chacoff NP, Dudenhöffer JH, Freitas BM, Ghazoul J, Greenleaf S, Hipólito J, Holzschuh A, Howlett B, Isaacs R, Javorek SK, Kennedy CM, Krewenka KM, Krishnan S, Mandelik Y, Mayfield MM, Motzke I, Munyuli T, Nault BA, Otieno M, Petersen J, Pisanty G, Potts SG, Rader R, Ricketts TH, Rundlöf M, Seymour CL, Schüepp C, Szentgyörgyi H, Taki H, Tscharntke T, Vergara CH, Viana BF, Wanger TC, Westphal C, Williams N, Klein AM. 2013. Wild pollinators enhance fruit set of crops regardless of honey bee abundance. *Science*. 339:1608–1611.

González AJ, de Araujo Freitas JC. 2005. *Manual de Meliponicultura Mexicana*. Universidad Autónoma de Yucatán/Fundación Produce Guerrero. 46 pp. Edición de Impresos Gramma.

González AJ, Medellín MS. 1991a. La división artificial de la abeja Xunan Kab. Revista YIK'EL-KAB A.C.

González AJ, Medellín MS. 1991b. Manual práctico para criar abejas indígenas sin aguijón. YIK'EL-KAB.

Goulson D. 2003. *Bumblebees: Their Behaviour and Ecology*, Oxford University Press, Oxford, pp. 235.

Greco MK, Spooner-Hart RN, Beattie GAC, Barchia I, Holford P. 2011. Stocking rates of *Trigona carbonaria* for the pollination of greenhouse capsicums. *J. Apicult. Res.* 50(4):299–305.

Greenleaf S, Kremen C. 2006a. Wild bees enhance honey bees' pollination of hybrid sunflower. *Proc. Natl. Acad. Sci. USA* 103:13890–13895.

Greenleaf S, Kremen C. 2006b. Wild bee species increase tomato production and respond differently to surrounding land use in Northern California. *Biol. Conserv.* 133:81–87.

Heard TA. 1998. Propagation of hives of the stingless bee *Trigona carbonaria*. *J. Aust. Entomol. Soc.* 27:303–304.

Heard TA. 1999. The role of stingless bees in crop pollination. *Annu. Rev. Entomol.* 44:183–206.

Heard TA, Dollin A. 1998a. Crop pollination with native stingless bees. In: *Native Bees of Australia Series, Booklet 6*, Australian Native Bee Research Centre, Sydney.

Heard TA, Dollin A. 1998b. Keeping Australian stingless bees in a log or box. In: *Native Bees of Australia Series, Booklet 5*, Australian Native Bee Research Centre, Sydney, pp. 1–14.

Hogendoorn K, Gross CL, Sedgely M, Keller MA. 2006. Increased tomato yield through pollination by native Australian blue-banded bees (*Amegilla chlorocyanea* Cockerell). *J. Econ. Entomol.* 99:828–833.

Höhn P, Tscharntke T, Tylianakis JM, Steffan-Dewenter I. 2008. Functional group diversity of bee pollinators increases crop yield. *Proc. R. Soc. Lond. B Biol. Sci.* 275:2283–2291.

Isaacs R, Kirk AK. 2010. Pollination services provided to small and large highbush blueberry fields by wild and managed bees. *J. Appl. Ecol.* 47:841–849.

James R. 2011. Bee Importation, Bee Price Data, and Chalkbrood. *Proceedings of the Western Alfalfa Seed Growers Association (WASGA)*. Las Vegas, NV. January 23–25, 2011.

Javier J, Quezada-Euan JG, May-Itza WJ, Gonzalez-Acereto JA. 2001. Meliponiculture in Mexico: problems and perspective for development. *Bee World* 82:160–167.

Javorek SK, Mackenzie KE, Vander Kloet SP. 2002. Comparative pollination effectiveness among bees (Hymenoptera: Apoidea) on lowbush blueberry (Ericaceae: *Vaccinium angustifolium*). *Ann. Entomol. Soc. Am.* 95:345–351.

Johansen CA, Mayer DF, Eves JD, Kious CW. 1983. Pesticides and bees. *Environ. Entomol.* 12:1513–1518.

Kajobe R. 2007. Nesting biology of equatorial Afrotropical stingless bees (Apidae; Meliponini) in Bwindi Impenetrable National Park, Uganda. *J. Apicult. Res.* 46: 245–255.

Kasina M, Mburu J, Kraemer M, Holm-Müller K. 2009. Economic benefit of crop pollination by bees: a case of Kakamega small-holder farming in western Kenya. *J. Econ. Entomol.* 102(2):467–473.

Kearns CA, Thomson JD. 2001. *The Natural History of Bumblebees. A Sourcebook for Investigations*, University Press of Colorado, Boulder, CO, pp. 130.

Klein AM, Vaissiere BD, Cane JH, Steffan-Dewenter I, Cunningham SA, Kremen C, Tscharntke, T. 2007. Importance of pollinators in changing landscapes for world crops. *Proc. R. Soc. Lond. B Biol. Sci* 274:303–313.

Klein AM, Steffan-Dewenter I, Tscharntke T. 2003. Fruit set of highland coffee increases with the diversity of pollinating bees. *Proc. R. Soc. Lond. B. Biol. Sci.* 270:955–961.

Kremen C, Williams NM, Thorp RW. 2002. Crop pollination from native bees at risk from agricultural intensification. *Proc. Natl. Acad. Sci. USA* 99:16812–16816.

Kremen C, Williams NM, Bugg RL, Fay JP, Thorp RW. 2004. The area requirements of an ecosystem service: crop pollination by native bee communities in California. *Ecol. Lett.* 7:1109–1119.

Krunic M, Pinzauti M, Felicioli A, Stanisavljevic LJ. 1995. Further observations on *Osmia cornuta* Latr. and *O. rufa* L. as alternative fruits pollinators, domestication and utilization. *Arch. Biol. Sci. Belgrade*. 47:59–66.

Kwapong P, Aidoo K, Combey R, Karikari A. 2010. *Stingless Bees: Importance, Management and Utilization. A Training Manual for Stingless Beekeeping*, Unimax Macmillan Ltd, Ghana.

Ladurner E, Bosch J, Kemp WP, Maini S. 2008. Foraging and nesting behavior of *Osmia lignaria* (Hymenoptera: Megachilidae) in the presence of fungicides: cage studies. *J. Econ. Entomol.* 101:647–653.

Losey JE, Vaughan M. 2006. The economic value of ecological services provided by insects. *Bioscience* 56:311–323.

Maccagnani B, Ladurner E, Santi F, Burgio G. 2003. *Osmia cornuta* (Hymenoptera, Megachilidae) as a pollinator of pear (*Pyrus communis*): fruit- and seed-set. *Apidologie* 34:207–216.

Macieira OJD, Hebling-Beraldo MJA. 1989. Laboratory toxicity of insecticides to workers of *Trigona spinipes* (F. 1793) (Hymenoptera: Apidae). *J. Apicult. Res.* 28:3–6.

Mader E, Spivak M, Evans E. 2010. *Managing Alternative Pollinators: A Handbook for Beekeepers, Growers, and Conservationists*, SARE Handbook 11. USDA Sustainable Agriculture Research and Education (SARE), Beltsville, MD, pp. 162.

Malaspina O, Stort AC. 1983. Estudo da tolerância ao DDT e relação com outros caracteres em abelhas sociais. *Rev. Bras. Biol.* 43:327–330.

Malone L, Burgess E, Stefanovic D, Gatehouse H. 2000. Effects of four protease inhibitors on the survival of worker bumblebees, *Bombus terrestris* L. *Apidologie* 31:25–38.

Matsumoto S, Abe A, Maejima T. 2009. Foraging behavior of *Osmia cornifrons* in an apple orchard. *Sci. Hortic.* 121:73–79.

Mayer D, Johansen C. 2003. The rise and decline of *Nomia melanderi* (Hymenoptera: Halictidae) as a commercial pollinator for alfalfa seed. In: *For Nonnative Crops, Whence Pollinators of the Future*, Entomological Society of America, Lanham, MD, pp. 139–149.

Michener CD. 1974. *The Social Behaviour of Bees: A Comparative Study*. The Belknap Press of Harvard University Press, Cambridge, MA, pp. 404.

Michener CD. 2007. *The Bees of the World*, 2nd edn. John Hopkins University Press, Baltimore, MD, pp. 992.

Moraes SS, Bautista AR, Viana BF. 2000. Avaliação da toxicidade aguda (DL50 e CL50) de insecticidas para *Scaptotrigona tubida* (Smith) (Hymenoptera: Apidae): via de contacto. *Anais da Sociedade Entomológiva do Brasil.* 29:31–37.

National Research Council—Committee on Status of Pollinators in North America. 2007. *Status of Pollinators in North America*. The National Academies Press, Washington, DC.

Njoroge GN, Gemmill B, Bussmann R, Newton LE, Ngum VW. 2004. Pollination ecology of *Citrullus lanatus* at Yatta, Kenya. *Int. J. Trop. Insect. Sci.* 24(1):73–77.

Nogueira-Neto P. 1997. *Vida e criação de abelhas indígenas sem ferrão*, 1st edn. Nogueirapis, São Paulo.

OECD (Organization for Economic Co-operation and Development). 1998. Guidelines for the testing of chemicals. Honey Bees, Acute Oral and Contact Toxicity Test, n.213, n.214.

Oldroyd BP. 1999. Coevolution while you wait: *Varroa jacobsoni*, a new parasite of western honey bees. *Trends Ecol. Evol.* 14:312–315.

Ollerton J, Winfree R, Tarrant S. 2011. How many plants are pollinated by animals? *Oikos* 120:321–326.

O'Toole C, Raw A. 1999. *Bees of the World*. Blandford, London, pp. 192.

Peach ML, Alston DG, Tepedino VJ. 1995. Sublethal effects of carbaryl bran bait on nesting performance, parental investment, and offspring size and sex ratio of the alfalfa leafcutting bee (Hymenoptera: Megachilidae). *Environ. Entomol.* 24:34–39.

Pitts-Singer T. 2008. Past and present management of alfalfa bees. In: Pittsinger TL, James R (eds), *Bee Pollination in Agricultural Ecosystems*. Oxford Press, New York, pp. 105–123.

Potts SG, Vulliamy B, Roberts S, O'Toole C, Dafni A, Ne'eman G, Willmer PG. 2005. Role of nesting resources in organizing diverse bee communities in a Mediterranean landscape. *Ecol. Entomol.* 30:78–85.

Quezada Euán JJG. 2005. Biología y uso del las abejas nativas sin aguijón de la peninsula de Yucatán, Mexico (Hymenoptera: Meliponini) Tratados 16, ediciones de la Universidad Autónoma de Yucatán).

Quezada Euán JJG. 2009. Potencial de las abejs nativas en la polinización de cultivos. *Acta boil. Colom.* 14:169–172.

Rader R, Howlett BG, Cunningham SA, Westcott DA, Newstrom-Lloyd LE, Walker MK, Teulon DAJ, Edwards W. 2009. Alternative pollinator taxa are equally efficient but not as effective as the honeybee in a mass flowering crop. *J. Appl. Ecol.* 46: 1080–1087.

Ramalho M, Giannini TC, Malagodi-Braga KS, Imperatriz-Fonseca VL. 1994. Pollen harvest by stingless bees foragers (Hymenoptera, Apidae, Meliponinie). *Grana* 33:239–244.

Ramalho M, Imperatriz-Fonseca VL, Giannini TC. 1998. Within-colony size variation of foragers and pollen load capacity in the stingless bee *Melipona quadrifsasciata anthidioides* Lepeletier (Apidae, Hymenoptera). *Apidologie* 29:221–228.

Ramalho M, Silva MD, Carvalho CAL. 2007. Dinâmica de uso de fontes de pólen por *Melipona scutellaris* Latreille (Hymenoptera: Apidae): uma análise comparativa com *Apis mellifera* L. (Hymenoptera: Apidae), no Domínio Tropical Atlântico. *Neotrop. Entomol.* 36(1):38–45.

Roessink I, van der Steen JJM, Kasina M, Gikungu M, Nocelli RCF. 2011. Is the European honey bee (*Apis mellifera mellifera*) a good representative for other pollinator species? SETAC Europe 21st Annual Meeting: Ecosystem Protection in a Sustainable World: A Challenge for Science and Regulation. Milan, Italy, May 15–19, 2011.

Roubik DW. 2006. Stingless bee nesting biology. *Apidologie* 37:124–137.

Scott-Dupree CD, Conroy L, Harris CR. 2009. Impact of currently used or potentially useful insecticides for canola agroecosystems on *Bombus impatiens* (Hymenoptera: Apidae), *Megachile rotundata* (Hymentoptera: Megachilidae), and *Osmia lignaria* (Hymenoptera: Megachilidae). *J. Econ. Entomol.* 102(1):177–182.

Sekita N, Yamada M. 1993. Use of *Osmia cornifrons* for pollination of apples in Aomori Prefecture, Japan. *Jpn. Agr. Res. Q.* 26(4):264–270.

Slaa JE, Sanchez LA, Sandi M, Salazar W. 2000. A scientific note on the use of stingless bees for commercial pollination in enclosures. *Apidologie* 31:141–142.

Steffan-Dewenter I, Kuhn A. 2003. Honeybee foraging in differentially structured landscapes. *Proc. R. Soc. Lond. B. Biol. Sci.* 270:569–575.

Stephen WP. 2003. Solitary bees in North American agriculture: a perspective. In: *For Nonnative Crops, Whence Pollinators of the Future*. Entomological Society of America, Lanham, MD, pp. 41–66.

Stout J, Morales C. 2009. Ecological impacts of invasive alien species on bees. *Apidologie* 40:388–409.

Taséi JN. 2002. Impact of agrochemicals on non-*Apis* bees. In: Devillers J, Pham-Delègue MH (eds), *Honey Bees: Estimating the Environmental Impact of Chemicals*. Taylor & Francis, London, pp. 101–131, 332 pp.

Tepedino VJ. 1981. The pollination efficiency of the squash bee (*Peponapis pruinosa*) and the honey bee (*Apis mellifera*) on summer squash (*Cucurbita pepo*). *J. Kansas. Entomol. Soc.* 54:359–377.

Thompson HM, Hunt LV. 1999. Extrapolating from honeybees to bumblebees in pesticide risk assessment. *Ecotoxicology* 8:147–166.

Thompson HM. 2001. Assessing the exposure and toxicity of pesticides to bumblebees (*Bombus* sp.). *Apidologie* 32:305–321.

Thomson D. 2004. Competitive interactions between the invasive European honey bee and native bumble bees. *Ecology* 85:458–470.

Torchio P. 1983. The effects of field applications of naled and trichlorfon on the alfalfa leafcutting bee, *Megachile rotundata* (Fabricius). *J. Kansas. Entomol. Soc.* 56:62–68.

Tuell JK, Isaacs R. 2010. Community and species-specific responses of wild bees to insect pest control programs applied to a pollinator-dependent crop. *J. Econ. Entomol.* 103:668–675.

Valdovinos-Núñez GR, Quezada-Euán JJG, Ancona-Xiu P, Moo-Valle H, Carmona A, Sánchez ER. 2009. Comparative toxicity of pesticides to stingless bees (Hymenoptera: Apidae: Meliponini). *J. Econ. Entomol.* 102(5):1737–1742.

van der Steen JJM, Bortolloti L, Chauzat MP. 2008. Can pesticide acute toxicity for bumblebees be derived from honeybee LD50 values? *Hazards of pesticides to bees—10th International Symposium of the ICP-BR Bee Protection Group.* October 8–10, Bucharest (Romania).

Vaughan M, Skinner M. 2009. *Using Farm Bill programs for pollinator conservation.* United States Department of Agriculture Natural Resources Conservation Service. Pollinator Technical Note No. 78.

Velthuis HHW, van Doorn A. 2006. A century of advances in bumblebee domestication and the economic and environmental aspects of its commercialization for pollination. *Apidologie* 37:421–451.

Vieira PFSP, Cruz DO, Gomes MFM, Campos LAO, Lima JE. 2010. Valor econômico da polinização por abelhas mamangavas no cultivo do maracujá-amarelo. *Revista de la rede Iberoamericana de Economia Ecológica* 15:43–53.

Villanueva-Gutiérrez R, Buchmann S, Donovan AJ, Roubik D. 2005. *Crianza y Manejo de la Abeja Xunan Cab en la Península Yucatán*. ECOSUR-University of Arizona, pp. 35.

Visscher PK, Seeley TD. 1982. Foraging strategy of honey bee colonies in a temperate deciduous forest. *Ecology* 63:1790–1801.

Vit P, d'Albore R. 1994. Melissopalynology of stingless bees (Apidae: Meliponinae) from Venezuela. *J. Apicult. Res.* 33: 145–154.

Wilson RL, Abel CA. 1996. Storage conditions for maintaining *Osmia cornifrons* (Hymenoptera: Megachilidae) for use in germplasm pollination. *J. Kansas. Entomol. Soc.* 69(3):270–271.

White J, Youngsoo S, Park Y. 2009. Temperature-Dependent Emergence of *Osmia cornifrons* (Hymenoptera: Megachilidae) Adults. *J. Econ. Entomol.* 102(6):2026–2032.

Winfree R, Williams NM, Dushoff J, Kremen C. 2007. Native bees provide insurance against ongoing honey bee losses. *Ecol. Lett.* 10:1105–1113.

Winfree R, Williams NM, Gaines H, Ascher JS, Kremen C. 2008. Wild bee pollinators provide the majority of crop visitation across land-use gradients in New Jersey and Pennsylvania, USA. *J. Appl. Ecol.* 45:793–802.

Zurbuchen A, Landert L, Klaiber J, Müller A, Hein S, Dorn S. 2010. Maximum foraging ranges in solitary bees: only few individuals have the capability to cover long foraging distances. *Biol. Conserv.* 143:669–676.

4 Overview of Protection Goals for Pollinators

T. Moriarty, A. Alix, and M. Miles

CONTENTS

4.1 INTRODUCTION

The decades-long evolution of cropping systems management is a response to higher demands for food and other products (e.g., fiber, fuel). Along with this has come an increased need to control pest populations and diseases. Pesticides have become an integral part of commercial production. Regulatory authorities serve a critical function in assessing and balancing the benefits of pesticides with other potential consequences of their use in order to maximize overall benefits to the societies they serve. Authorities articulate the objectives of their efforts in broad terms, such as "protecting human health and the environment" as a guide (EFSA 2010). At this level, multiple factors, in addition to estimated risk, are considered when guiding the actions of a regulatory authority and may be economic, legal, or political. Together, all the variables are considered and balanced in a way that produces an assessment that is consistent with the protection goals of a regulatory authority.

Regulatory authorities base their interest in assessing the potential impact of pesticides to a specific organism or taxon in different factors such as:

- the market value or the role an organism (or taxon) plays in ecosystem services, both in natural and cultivated systems;
- the estimated or measured pesticide exposure concentrations derived from modeling, testing, or monitoring efforts;
- information on potential or actual effects of a pesticide on a taxon derived from experimental data, incident reports, or surveys; and
- the relevance of the species or taxon to a regulatory authority's protection goals.

Protection goals therefore reflect a certain level of information and certain values of a society. Regulatory authorities, in turn, use risk assessment tools to determine whether the use of a pesticide is consistent with

Pesticide Risk Assessment for Pollinators, First Edition. Edited by David Fischer and Thomas Moriarty.

its general goal(s). A risk assessment process must be designed to provide clear information for the risk assessor and risk manager to determine whether the proposed use of a pesticide product would or would not be consistent with the protection goals of a regulatory authority. General protection goals, however, do not necessarily inform or provide adequate guidance at the risk assessment level. Therefore, more specific protection goals may need to be considered which would be more appropriate for use at the risk assessment level. Specific protection goals, however, must be linked to the general protection goals. In this way, protection goals of a risk assessment (e.g., for a particular taxon or nontarget species) are consistent with and support the general protection goal of "protecting human health and the environment." Over time, entities such as the Organization for Economic Co-operation and Development (OECD), the US Environmental Protection Agency (USEPA), and the European and Mediterranean Plant Protection Organisation (EPPO) have developed a number of documents to guide the risk assessment processes to be used in decision making with respect to registering pesticides.

The participants came to the Workshop with an understanding of the value of honey bees and of the current science on potential exposure and effects of pesticides on bees. Participants spent time discussing specific protection goals for pollinators such that a pesticide risk assessment process for pollinators would be consistent with general protection goals of regulatory authorities.

From this discussion developed surrogate protection goals that served the Workshop participants as they developed recommendations for a pesticide risk assessment process, and for the data to inform that process. However, the participants of the Workshop were aware that, since protection goals reflect a range of considerations (including legal, societal, and resource considerations) that are specific to a government or authority, it was not within the scope of this effort to define the protection goal of any one country or protection authority.

4.2 ELEMENTS AND PROPOSED PROTECTION GOALS

During the Workshop, participants discussed the longstanding global importance of *Apis* and non-*Apis* bees in terms of both commercial and ecological significance. Participants of the Workshop agreed that the maintenance of the pollinating function is a critical ecological service of pollinators (bees in particular) that needs to be protected. The goal would be to ensure adequate pollination (sufficient frequency of floral visits) to support healthy crop survival and yield. While such a protection goal is relevant for commercial agricultural production, it may not be relevant at a larger scale, that is, the landscape, where the role of non-*Apis* species is more relevant as these species pollinate adjacent cropland or the non-cropped landscape. For this to be taken into account, non-*Apis* (i.e., non-managed) pollinating insect species would need to be considered with their interactions in the larger landscape. While pollination ensures a healthy and ecologically diverse landscape, and remains the critical function of these species, consideration of non-*Apis* species and their contribution to landscape ecology reflects the role that ecological diversity plays in supporting a healthy environment. Protection of the pollinator community at the landscape level ensures pollination services and also contributes to the diversity of the species associated with pollination services within that landscape. Finally, participants identified honey and other hive products as a potentially specific goal to be protected as well as a measure of honey bee health. Model (surrogate) protection goals upon which to build a risk assessment framework were then defined as:

- protection of pollination services provided by both *Apis* and non-*Apis* species;
- protection of pollinator biodiversity (i.e., protection of adequate number and diversity of bee species that contribute to the health of the environment); and
- protection of honey production and other hive products.

4.3 LINKING PROTECTION GOALS WITH ASSESSMENT ENDPOINTS

With possible protection goals defined, they then had to be linked to risk assessment endpoints, and further linked to specific endpoints measured in either exposure or effects studies (i.e., measurement endpoints). Assessment endpoints are attributes of an entity (e.g., an organism or environmental component) that are essential for its continued survival. In ecological risk assessments for wildlife, assessment endpoints have traditionally been defined as the growth, reproduction, and survival of an organism. These same assessments can be applied to the honey bee, but it must be recognized that the honey bee functions as a superorganism and therefore the attributes of growth, reproduction, and survival apply to the colony, not the individual bee.

A risk assessment (e.g., for a particular taxon) is based upon data from one or several studies that provide sufficient information for the risk assessor to determine whether the intended use of a pesticide will have a significant adverse effect on one or more of the assessment endpoints. Data provided by specific studies should inform one or more of the assessment endpoints in either a direct fashion (e.g., treatment related mortality) or an indirect fashion (e.g., reduced foraging activity). Both exposure studies and effects studies produce measurement endpoints (e.g., pesticide residue levels in pollen, body length, adult bee longevity, or mortality of different castes or stages) informing the risk assessor whether the intended use of a pesticide results in a significant exposure or reduction in an organism's ability to either grow, reproduce, or survive. When measurement endpoints are appropriately linked to assessment endpoints and specific protection goals, they then support generic protection goals (Figure 4.1; see Table 10.1 for specific examples of protection goals, assessment endpoints, and measurement endpoints).

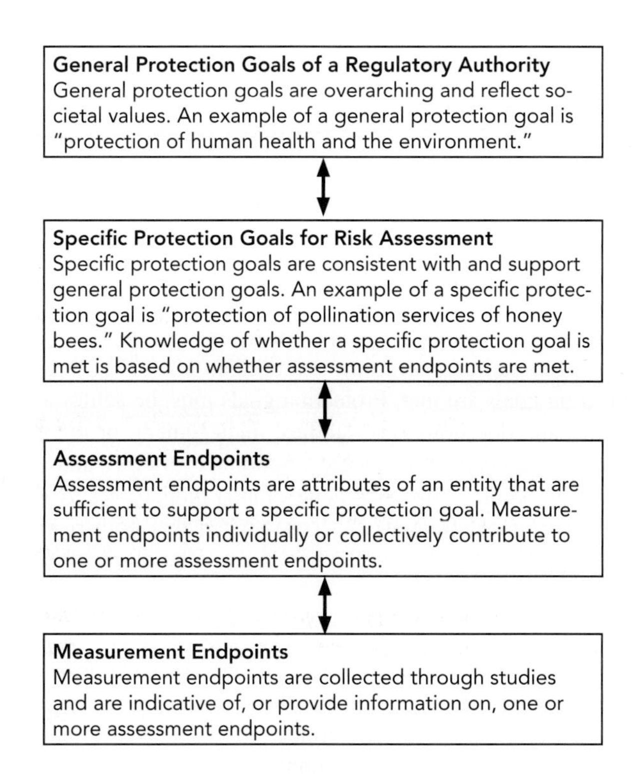

FIGURE 4.1 Relationship between measurement endpoints to generic protection goals, used in assessing ecological risks.

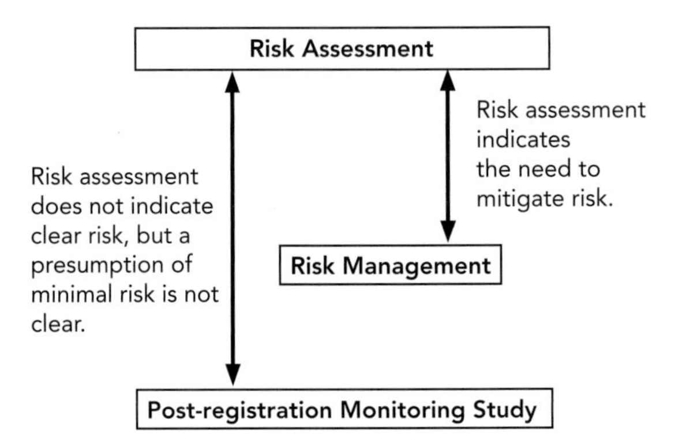

FIGURE 4.2 Post-registration monitoring studies in a risk assessment framework.

4.4 PROTECTION GOALS AND MONITORING

The risk assessment process proposed by the Workshop participants is designed to support the protection goals articulated at the Workshop. The process also provides an avenue for additional feedback information to continue to inform the assessment of risk. Confirmatory information, such as incident or monitoring data, provides direct feedback on whether the regulatory decisions are effective and whether specific and generic and protection goals are being achieved. However, field monitoring studies can be complex since they often reflect natural events or scenarios that impact bees, such as disease, predation, and competition. Thus, it is important that when defining protection goals, consideration is given to the risk assessment parameters and potential monitoring parameters in a way that makes the relationship between them clear (Figure 4.2).

4.5 CONCLUSION

Well-defined protection goals guide a risk assessment by providing criteria for decisions within the paradigm of risk assessment (study design and interpretation), risk mitigation, or post-registration monitoring actions to determine whether protection goals are met. Protection goals must be achievable and sustainable through appropriate scientific analysis and decisions (i.e., studies, management, or monitoring). During the Workshop, participants discussed the longstanding global importance of *Apis* and non-*Apis* bees in terms of both commercial and ecological realms. Participants developed model (or surrogate) protection goals suitable as the basis for a risk assessment framework. It was noted that both risk assessment and risk management are complementary options to meet protection goals. Therefore, suitable protection goals were defined as:

- protection of pollination services provided by *Apis* and non-*Apis* species,
- protection of honey production and other hive products
- protection of pollinator biodiversity, that is, protection of adequate number and diversity of bee species that contribute to the health of the environment (primarily non-*Apis* bees).

REFERENCE

EFSA Panel on Plant Protection Products and their Residues (PPR). 2010. Scientific Opinion on the development of specific protection goal options for environmental risk assessment of pesticides, in particular in relation to the revision of the Guidance Documents on Aquatic and Terrestrial Ecotoxicology (SANCO/3268/2001 and SANCO/10329/2002). *EFSA J.* 8(10):1821. doi:10.2903/j.efsa.2010.1821

5 Overview of the Pesticide Risk Assessment and the Regulatory Process

C. Lee-Steere and T. Steeger

CONTENTS

5.1 INTRODUCTION

As discussed earlier, regulatory authorities have the responsibility to evaluate pesticides and the potential risks associated with their use. They have developed tools and methods to do this in a consistent manner with respect to different taxons. However, with the introduction of new plant protection products, changes in agricultural practices, and advances in the understanding of honey bee health and ecology, the ability to accurately characterize potential risks to insect pollinators with the existing tool set has been seen as a challenge. While many countries share the same broad risk-based environmental assessment approach, differences between approaches exist that account for national conditions, such as policies, legal requirements, or preferences.

The Workshop considered a generic, tiered risk assessment methods, and worked to develop a process that included three phases: (1) problem formulation, (2) exposure, and (3) effects assessment, risk characterization. In Phase 1 (i.e., problem formulation), measurement endpoints, derived from studies, are selected with an understanding of how they relate to assessment endpoints (and ultimately specific protection goals and generic protection goals); a conceptual model is prepared that describes a risk hypothesis; and an analysis plan to test that hypothesis is described. In Phase 2 (i.e., analysis), measures of exposure and effects are evaluated. In Phase 3 (i.e., risk characterization), measures of exposure and measures of effect are integrated to develop risk estimates, and uncertainties are discussed.

Analysis is carried out in a tiered manner, where a Tier 1 analysis is intended to be a conservative screen that efficiently separates compounds that are not anticipated to present a potential risk from those compounds that may. Higher tiers are intended to refine the estimates or measures of potential exposure, effects, and the resulting characterization of risk. Risk assessors and risk managers proceed through the risk assessment process (i.e., ascending through higher tiers of analysis) to determine whether the intended use of a compound is consistent with defined protection goals. If the estimate of risk indicates that proposed use is not consistent

Pesticide Risk Assessment for Pollinators, First Edition. Edited by David Fischer and Thomas Moriarty.

with the protection goals, then risk mitigation techniques may be implemented proactively to resolve concerns. During the Workshop, risk mitigation was briefly discussed as it is a component of the overall regulatory management of plant protection products (see Chapter 13).

5.2 CURRENT APPROACH FOR ASSESSING EFFECTS OF PESTICIDE PRODUCTS TO POLLINATORS

In the United States, the first tier of toxicity testing with honey bees consists of an acute contact toxicity test (USEPA 2012a) with adult honey bees that provides a median Lethal Dose (LD50), that is, the dose that causes death to 50% of the exposed organisms from a single dose of the test compound, along with any sublethal effects that may have occurred as a result of chemical exposure. The acute LD50 is assessed after 24 and 48 hours, but depending upon the outcome of the test, its duration can be extended up to a maximum of 96 hours, if necessary. Based upon the outcome of the acute LD50 toxicity test, pesticides are classified as practically non-toxic, moderately toxic, or highly toxic to bees on an acute exposure basis. If the LD50 is less than 11 μg/bee, additional testing may be required in the form of a foliar residue study (USEPA 2012b) to determine the duration over which field-weathered foliar residues remain toxic to honey bees. On a case-by-case basis, additional higher-tiered studies such as field pollinator studies with honey bees (USEPA 2012c) (i.e., hive studies) may be necessary if the data from toxicity studies indicate potential chronic effects or adverse effects on colonies.

In the European Union (EU), risk to honey bees from exposure to pesticides is based on the European and Mediterranean Plant Protection Organization (EPPO) process and includes a three-tiered progression of testing (2010)[1]. Guidelines describe laboratory tests, (OECD 1998a, 1998b), as well as semi-field (cage or tunnel) tests, and field tests for evaluating the lethal and sub-lethal effects of pesticides on adult honey bees (OECD 2007; EPPO 2010). The testing approach in the EU is similar to that of the United States and Canada in that it consists of a tiered approach, where laboratory studies are considered Tier 1 tests, and semi-field and field tests are considered higher tiers. In contrast to the United States, the EU and Canada require the acute oral toxicity (LD50) on adult workers (OECD 1998a) in addition to the acute contact toxicity (OECD 1998b). In the EU, it is also standard practice to conduct both acute oral and acute contact LD50 studies on formulated end-use products, in cases where either exposure to the end use product itself is possible, or in the case where products have more than one active component, as well as the technical grade (relatively pure) active substance.

In addition to guideline toxicity test requirements, regulatory authorities around the world also make use of published open literature and dedicated studies for nontarget arthropods to evaluate the potential effects of pesticides on terrestrial invertebrates, or as a line of evidence to require higher tiered testing. Along with guideline and open literature studies, adverse effect (e.g., bee kill incident) reports, and monitoring studies are considered in order to gauge the effects of pesticides on nontarget organisms.

5.2.1 RISK ASSESSMENT FOR SYSTEMIC COMPOUNDS

Many who are familiar with pesticide risk assessment recognize that the techniques and assessment schemes employed for foliar application products (where exposure may be primarily through surface contact) are not well adapted to assess potential risk from compounds with systemic properties. With better understanding of the ability of these chemicals to be present in pollen and nectar during flowering, there has followed a

[1] Risk Assessment: PP 3/10 (2) (OEPP/EPPO), Test Methodologies: Guideline No. 170 (OEPP/EPPO); OECD 75.

better understanding of how systemic compounds present potential for both oral and contact exposure and, therefore, need to be considered.

The EPPO has recently put forward a risk assessment scheme (Alix et al. 2009) for systemic compounds that includes the same tiered testing system, but replaces the hazard quotient (HQ) calculation with a toxicity exposure ratio (TER), where TER = PNEC/PEC. The PNEC is the Predicted No Effect Concentration, while the PEC is the Predicted Exposure Concentration. The PEC is determined from estimated or measured residue concentrations in the whole plant, flowers, pollen or nectar. The dose that individual bees might ingest is then calculated for different categories of honey bees (e.g., larvae, queen, foragers) depending on the amount of contaminated pollen and nectar they may consume. PNECs are derived from acute, sublethal, and chronic toxicity data and may also include a factor to account for uncertainty. These factors range from 1 to 10 depending on whether the toxicity endpoint is assessed in a laboratory (Tier 1) or in a semi-field or field test, that is, uncertainty decreases as toxicity data become more representative of how the pesticide will be used.

5.2.2 Trigger Criterion and Levels of Concern

A "trigger criterion" is a value, a threshold, used to define the limit of risk that is consistent with protection goals. A trigger criterion or level of concern (LOC) is compared to a quantitative risk estimate (e.g., HQ employed in Europe, or a risk quotient (RQ) employed in North America (USEPA 1998)) to determine if the estimated risk is acceptable or not. If the comparison between an LOC and an estimated risk indicates that the use of a compound is inconsistent with defined protection goals, then it may be appropriate to either further refine the risk with additional data, or seek action to mitigate potential risk (In Europe, for example, when assessing a spray formula, the trigger criterion at the screening level is where HQ ≥50; such that when HQ ≥50, either higher tier data, or risk mitigation may be sought (Alix et al. 2009; EPPO 2010.). In the United States, estimates of risk (i.e., RQ) are compared against the LOC to determine whether further refinement is needed. Participants of the Workshop noted that while levels of concern promote efficiency in decision making, risk assessment is an iterative process between risk assessors and risk managers, and is composed of multiple lines of evidence in order to determine whether the use of a compound on a specific crop is consistent with protection goals. Ultimately, trigger criterion and levels of concern are policy tools and, as such, they are outside the purview of the SETAC Pellston Workshop and remain the right and responsibility of respective regulatory authorities to define.

REFERENCES

Alix A, Chauzat MP, Duchard S, Lewis G, Maus C, Miles MJ, Pilling E, Thompson HM, Wallner K. 2009. Environmental risk assessment scheme for plant protection products. In: Oomen PA, Thompson HM (eds), *10th International Symposium: Hazards of Pesticides to Bees*, October 8–10, Bucharest (Romania). Julius–Kühn Archives. pp. 27–33.

EPPO. 2010. Environmental risk assessment scheme for plant protection products. *EPPO Bull.* 40:323–331.

OECD. 1998a. OECD Guidelines for the Testing of Chemicals. Test Number 213: Honey Bees, Acute Oral Toxicity Test. http://www.oecd-ilibrary.org/environment/test-no-213-honey bees-acute-oral-toxicity-test_9789264070165-en;jsessionid=5p2ngklfmv8p4.epsilon

OECD. 1998b. OECD Guidelines for the Testing of Chemicals. Test Number 214: Acute Contact Toxicity Test. http://www.oecd-ilibrary.org/environment/test-no-214-honey bees-acute-contact-toxicity-test_9789264070189-en;jsessionid=43gvto47wnue9.delta

OECD. 2007. Guidance Document on the Honey Bee (*Apis mellifera*) Brood Test under Semi-Field Conditions. OECD Environment, Health and Safety Publications. Series on Testing and Assessment No. 75. ENV/JM/MONO (2007) 22.

USEPA. 1998. Guidelines for Ecological Risk Assessment. Published on May 14, 1998, Federal Register 63(93):26846–26924. http://www.epa.gov/raf/publications/pdfs/ECOTXTBX.PDF
USEPA. 2012a. Ecological Effects Test Guidelines. OCSPP 850.3020: Honey Bee Acute Contact Toxicity [EPA 712-C-019], January 2012. http://www.regulations.gov/#!documentDetail;D=EPA-HQ-OPPT-2009-0154-0016
USEPA. 2012b. Ecological Effects Test Guidelines. OCSPP 850.3030: Honey Bee Toxicity of Residues on Foliage [EPA 712-C-018], January 2012. http://www.regulations.gov/#!documentDetail;D=EPA-HQ-OPPT-2009-0154-0017
USEPA. 2012c. Ecological Effects Test Guidelines. OCSPP 850.3040: Field Testing for Pollinators [EPA 712-C-017], January 2012. http://www.regulations.gov/#!documentDetail;D=EPA-HQ-OPPT-2009-0154-0018

6 Problem Formulation for an Assessment of Risk to Honey Bees from Applications of Plant Protection Products to Agricultural Crops

D. Fischer, A. Alix, M. Coulson, P. Delorme, T. Moriarty, J. Pettis, T. Steeger, and J.D. Wisk

CONTENTS

Pesticide Risk Assessment for Pollinators, First Edition. Edited by David Fischer and Thomas Moriarty.

As mentioned in Chapter 5, Phase 1 of the risk assessment process is problem formulation[1] (PF), where measurement endpoints are selected; a conceptual model is prepared that describes a risk hypothesis; and an analysis plan that articulates what data is needed and how it will be used to test the stated hypothesis is described. The PF is intended to provide a foundation for the risk assessment by articulating the purpose of the assessment, defining the nature of the problem (i.e., potential for adverse effects given the nature of the chemical stressor and its existing or proposed use), and establishing the plan for analyzing available data and characterizing risk. Participants of the Workshop discussed the generic principles of PF and developed PFs for the assessment of risk of honey bees for two types of pesticide use scenarios: (1) application of a systemic chemical to the soil or seeds planted into the soil, and (2) application of a non-systemic chemical as a foliar spray. It should be noted that there are other possible scenarios such as foliar spray application of a systemic chemical, which may require a separate PF because both contact and oral exposure routes may be important. Likewise, some modification of the PF examples presented herein by the Workshop will likely be needed to apply them to non-*Apis* species in order to account for differences in behavior and life history. The goal here is to illustrate the process for developing a PF for assessment of pesticide risk to honey bees and other insect pollinators by providing some relevant examples.

6.1 WHAT IS PROBLEM FORMULATION?

PF is the first step of an ecological risk assessment (Figure 6.1). The objective of PF is to develop a working risk hypothesis including a conceptual model that describes the potential exposure to and resulting effects of a stressor (e.g., a pesticide) on an ecological receptors of concern (e.g., honey bees). During PF, objectives of the anticipated risk assessment are identified and underlying uncertainties and assumptions (constraints) regarding data are described. During PF, initial scoping and integration of available information begins, and data or information gaps are identified. Within the context of a PF for a pesticide risk assessment, the active ingredient is identified as the stressor. To better define the stressor, use information is considered such as: label information, formulations, application parameters (rates, methods, and timing), crop types, or information on target pests (see Box 6.1).

[1] PF is a widely utilized generic process for framing and developing an ecological risk assessment. This process is not necessarily employed by all regulatory authorities, nor employed in the same manner by those regulatory authorities that do employ the PF process.

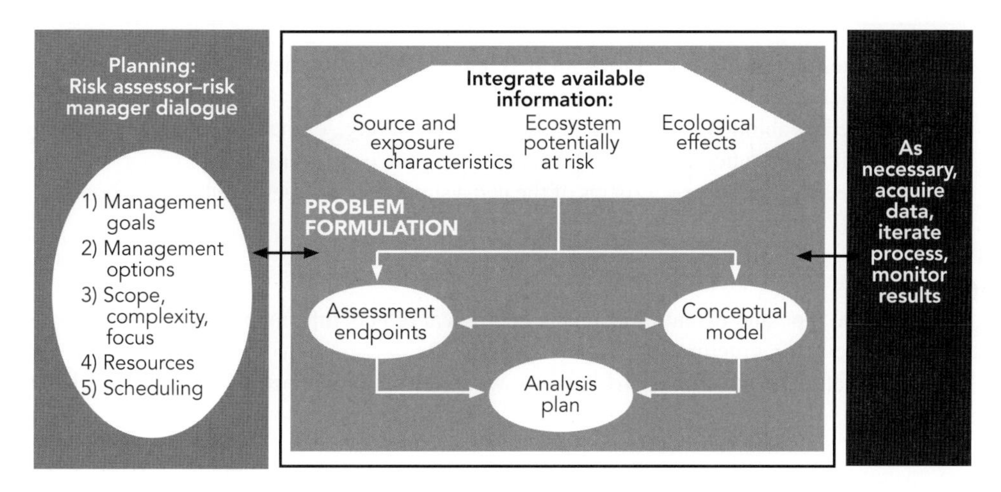

FIGURE 6.1 Scheme depicting problem formulation phase of the ecological risk assessment process. (Taken from USEPA 1998).

BOX 6.1 PROBLEM FORMULATION QUESTIONS: ASSESSING AVAILABLE INFORMATION

Source and Stressor Characteristics

- What is the source of the stressor (anthropogenic, natural, point source, etc.)?
- What type of stressor is it (chemical, physical, or biological)?
- What is the intensity of the stressor (the dose or concentration, the magnitude, or extent of the disruptions)?
- What is the mode of action? How does the stressor act on organisms or ecosystem functions?

Exposure Characteristics

- With what frequency does the stressor event occur (is it isolated, episodic, continuous)?
- What is the duration of the exposure? How long does it persist in the environment? (half-life, does it bioaccumulate, does it alter habitat, does it reproduce, or proliferate)
- What is the timing of exposure? When does it occur in relation to critical organism life cycles or ecosystem events?
- What is the spatial scale of exposure? Is the extent or influence of the stressor local, regional, global, habitat-specific, or ecosystem-wide?
- What is the distribution? How does the stressor move through the environment?

Ecosystems Potentially at Risk

- In what habitat is the stressor present?
- How do these characteristics influence the susceptibility (sensitivity and likelihood of exposure) of the ecosystem to the stressors?
- Are there unique features that are particularly valued (i.e., the last representative of an ecosystem type)?
- What is the landscape context within which the ecosystem occurs?

- What are the geographic boundaries of the endpoint? How do they relate to the functional characteristics of the ecosystem or endpoint?
- What are the key abiotic factors influencing the endpoint (e.g., climatic, geology, hydrology)?
- Where and how are functional characteristics driving the ecosystem?
- What are the structural characteristics of the ecosystem (e.g., species number and abundance, trophic relationships)?

Ecological Effects

- What are the type and extent of available ecological effects information (e.g., field surveys, laboratory tests, or structure–activity relationships)?
- Given the nature of the stressor (if known), which effects are expected to be elicited by the stressor?
- Under what circumstances will effects occur?

PF has three deliverables (see middle box of Figure 6.1):

1. Risk assessment endpoints that reflect management or protection goals, and the ecosystem they represent;
2. Conceptual models that describe key relationships between a stressor and assessment endpoint; and
3. An analysis plan.

A critical component of PF is planning dialog (left box of Figure 6.1) where risk assessors and risk managers identify and agree on management objectives and identify issues associated with the chemical. PF is intended to be iterative and is informed by existing data (including open literature, existing data, or incident information). As more data become available, the risk hypothesis may change to reflect a more refined understanding of potential risks. The PF identifies available data and information gaps and enables risk managers to convey potential limitations to registrants (chemical manufacturers who support labels) who may be able to provide information to address uncertainties.

Components of PFs include:

1. A description of the nature of the chemical stressor (typically a single technical grade active ingredient, but may include formulations, inerts or degradates of the active ingredient based on the availability of data);
2. A broad overview of pesticide existing or proposed uses;
3. A description of assessment endpoints—valued entities (biological receptors) and their attributes—characteristics which are to be protected (survival, growth, and reproduction), relevant to management or protection goals;
4. A conceptual model that identifies the relationship between ecological entities and the chemical stressor under consideration. The conceptual model has two components, the risk hypothesis and conceptual diagram.
   a. The risk hypothesis describes the predicted relationships among the chemical stressor, exposure and assessment endpoint responses along with a rationale to support it.
   b. The conceptual model diagram illustrates the relationships presented in the risk hypothesis and is typically represented by a flow diagram depicting the source (use), stressor, receptor, and change in [endpoint] attribute.

5. An analysis plan is then presented to identify how the risk hypothesis will be assessed. It identifies data needs and methods for conducting the assessment and what measurements, for example, model-estimated environmental concentrations, no-observed-adverse-effect concentrations (NOAEC) and attribute changes, such as foraging behavior, will be used.

6.1.1 Selecting Assessment Endpoints

Assessment endpoints are explicit expressions of the actual environmental value that is to be protected. Selection of assessment endpoints begins to structure the assessment toward addressing management concerns. Assessment endpoints must be measurable ecosystem characteristics that represent protection goals. Selection of ecological characteristics to protect then becomes the basis for defining assessment endpoints, which connects broad protection goals with specific measures in risk assessment.

The element or characteristic of an ecosystem to be valued or protected must

1. have ecological relevance;
2. be susceptible to known or potential stressors; and
3. be relevant to protection goals and societal values.

6.1.2 Ecological Relevance

Ecologically relevant endpoints reflect important characteristics of the system and may be defined at any level of organization (e.g., individual, community, population, ecosystem, landscape). Ecologically relevant endpoints often help sustain the natural structure, function, and biodiversity of a system or its components.

Ecologically valuable endpoints are those that, when changed, cause multiple or widespread effects (i.e., are upstream of other effects in the ecosystem).

6.1.3 Susceptibility to Known or Potential Stressors

An ecological resource is susceptible when it is sensitive to a stressor, that is, it is affected by the stressor such as through a mode of action. The sensitivity of an ecological resource may be relative to timing—a life stage of an organism (or system)—or may be affected by the presence of other stressors or natural disturbances. Measures of sensitivity may include mortality, behavioral abnormalities, loss of offspring, habitat alteration, community structural change, or other factors. Susceptibility (of an ecological resource) requires exposure such as through co-occurrence or contact. Typically, the amount and conditions of exposure directly influence how an ecological resource will respond to a stressor. Thus, the timing of exposure, timing of effects, presence or absence of other stressors, and other variables add complexity to evaluations of sensitivity or susceptibility.

6.1.4 Defining and Relation of Assessment Endpoints to Protection Goals

As noted earlier, measurement endpoints, assessment endpoints, specific protection goals, and generic protection goals must all be related. Protection goals must be appropriately scaled in order to be represented by assessment endpoints. Assessment endpoints should remain neutral and specific, whereas protection goals represent a desired achievement (i.e., a goal). As such, assessment endpoints do not contain words like "protect," "maintain," or "restore," or indicate a direction for change such as "loss," or "increase." Instead, assessment endpoints are ecological values defined for specific entities and their measurable attributes, providing a framework for measuring stress–response relationships.

Risk assessors and risk managers should share their professional judgment when selecting and defining potential endpoints. Assessment endpoints themselves must be: (i) scientifically valid, (ii) important to the public, and (iii) valued by risk managers (i.e., reflect statutory obligations) in order for them to be relevant. Once ecological values are selected as potential endpoints (attribute changes), they must then be operationally defined. Two elements are required for operational definition:

1. identification of the specific valued ecological entity, such as a species, or a functional group of species, or a community or ecosystem or specific habitat or unique place; and
2. the characteristics (attributes) of the entity that are important to protect.

For practical reasons, it may be helpful to use assessment endpoints that have well-developed test methods, field measurement techniques, and predictive models. However, this is not necessary, as appropriate measures for an assessment endpoint are identified during the development of the conceptual model and further specified in the analysis plan. The number and type of measurement endpoints depend upon the specificity of the questions being asked through the risk assessment and the complexity of the ecological entity being examined. Final assessment endpoint selection is an important risk manager–assessor checkpoint during PF. Risk assessors and risk managers should agree that selected assessment endpoints effectively represent the protection goals.

Common problems in selecting assessment endpoints are:

- the endpoint is a goal
- the endpoint is vague
- the ecological entity is better suited as a measure rather than an endpoint
- the ecological entity may not be sensitive to the stressor
- the ecological entity is irrelevant to the assessment
- the attribute is not sufficiently sensitive for detecting important effects (e.g., survival compared with recruitment for endangered species)

6.1.5 Conceptual Models

Conceptual models provide a written and visual representation of predictive relationships between ecological entities and the stressors and may describe primary, secondary, or tertiary exposure pathways, co-occurrences, ecological effects, or ecological receptors that are reflective of valued attribute changes in these receptors. Multiple conceptual models may be developed to address several issues in a given risk assessment. When conceptual models are used to describe pathways of individual stressors and assessment endpoints and the interaction of multiple and diverse stressors and endpoints, more complex models and sub-models will often be needed.

Conceptual models are flexible and can be modified to accommodate new or additional data. For example, conceptual models can start out as broad and identify as many potential relationships as possible, then narrow as information is acquired. The complexity of a risk hypothesis is commensurate with the complexity of the risk assessment.

Conceptual models consist of two principal components:

1. a set of risk scenario that describe predicted relationships among stressor, exposure, and endpoint response; and
2. a diagram that illustrates the relationships presented in the risk scenario.

Diagrams are typically flow diagrams with boxes and arrows. Elements considered for inclusion in the diagram include the number of relationships depicted; the comprehensiveness of the information; data abundance or scarcity; or the relative certainty of the pathways. Several smaller diagrams may be more effective than a single diagram that contains too much detail. Diagrams should reflect or document the "state-of-the-science" and degree of certainty regarding its components, and should be discussed with risk managers to ensure that they reflect and communicate the regulatory agency's policies and protection goals prior to analysis.

6.2 CASE 1: PROBLEM FORMULATION FOR A SYSTEMIC CHEMICAL APPLIED TO THE SOIL, OR AS A SEED-DRESSING

6.2.1 Stressor Description

Workshop participants developed a risk assessment process through two case examples that were representative of two general types of pesticide delivery modes, systemic and foliar. Briefly outlined next is an example of a PF for the pesticide risk assessment for pollinators first for a systemic compound, and then for a foliar applied compound.

The stressor of concern is a systemic plant protection product (insecticide or acaricide) applied to the soil of field and orchard crops such as cotton, maize, oil-seed rape, wheat, barley, potatoes, sugar beets, cucurbits (e.g., melons), citrus, and pome fruit, or as a coating on seeds of field crops (cotton, maize, oil-seed rape, wheat, barley). Crop plants absorb the chemical through the roots and translocate it into aboveground tissues of the plant. Magnitude of residue studies demonstrate that the parent compound, *per se*, comprises the residues found in treated plants. Use of the product provides effective control of several economically important chewing and sucking pest insects such as aphids, psyllids, and whiteflies. Application timing is at planting or during transplant of field crops and after flowering of orchard crops.

The above paragraph covers the first two components of a PF, which were listed as (1) a description of the nature of the chemical stressor, and (2) a broad overview of pesticide existing or proposed uses. The third component of a PF is a description of assessment endpoints, that is, valued entities (biological receptors) and their attributes, that is characteristics to be protected (e.g., survival, growth, and reproduction), which are relevant to protection goals.

6.2.2 Protection Goals

As discussed, protection goals are policy decisions set by government agencies and other organizations that represent the interests of the societies they serve. In the absence of specific protection goals, the participants used those developed during the Workshop, which included

- protection of pollination services provided by *Apis* and non-*Apis* species;
- protection of honey production and other hive products; and
- protection of pollinator biodiversity.

The first and third of these goals are applicable to pollinators in general (*Apis* and non-*Apis*). The second statement is applicable to managed pollinators (*Apis*).

6.2.3 Assessment Endpoints

For honey bees, logical assessment endpoints are colony strength (population size and demographics) and colony survival (persistence). Bumble bees too can be measured against colony strength (larval ejection, number of offspring, or colony weight) and colony survival (persistence). As a colony, loss simply represents the situation when colony strength is minimal, it could be argued that *colony survival* is not needed as a separate assessment endpoint. Various measures of colony strength are often made when beehives are rented and placed at agricultural crops. Rental fees are greater for strong colonies than weak colonies because colony strength is expected to be related to the quality of pollination service provided by the colony. Colony strength will likely be significantly impacted if queen viability, brood development, or general worker bee health is adversely affected for an extended period of time. There are many known cases where pesticide exposure has caused effects on colony strength, which meets the criteria for an assessment endpoint which includes:

1. the affected organism has ecological relevance;
2. the affected organism is sensitive, or susceptible to known or potential stressors; and
3. the affected organism is relevant to the management or protection goals and societal values associated with maintenance of pollination services.

For solitary bees, possible assessment endpoints may include adult survival, adult fecundity, larval survival, and larval development time. Populations will be significantly impacted by decreased adult or larval survival and adult fecundity. Increased time for larval development, for example, could impact (be delaying) individual bee emergence time and reduce the number of generations per year in multivoltine species, or cause bees to enter diapauses too late which could ultimately relate to fecundity.

6.2.4 Conceptual Model

The fourth component of PF is the conceptual model that identifies the relationship between ecological entities and the chemical stressor under consideration. The conceptual model has two components: the risk hypothesis and the conceptual diagram.

6.2.4.1 Risk Hypothesis

For a systemic pesticide applied to the soil or as a seed dressing, the risk hypothesis may involve the following steps describing how exposure most likely occurs and results in effects on an assessment endpoint (e.g., colony strength). The hypothesis is:

1. the use of the systemic plant protection product results in concentrations in nectar, pollen, or other parts of plants visited by honey bees;
2. forager honey bees collect the contaminated nectar and pollen and transport it back to the hive where it is incorporated into the food stores of the colony;
3. foragers, hive bees, bee brood, and the queen are exposed to concentrations of the chemical mainly via ingestion;
4. if the exposure concentration is high enough, toxic effects on forager bees, hive bees, bee brood, or the queen result in reduced queen fecundity, brood development success, or survival of adult bees; and
5. colony strength is affected as a result of reduced fecundity, brood success, or adult survival.

The duration of exposure of forager bees depends on the persistence of the chemical in the soil and within the treated plants, the duration of bloom, and the chronology of application (planting of treated seeds or application to the soil) of the chemical to agricultural fields within the landscape around the hive. Based on the risk hypothesis, key questions that need to be answered during risk analysis are:

1. To what extent do foraging honey bees visit treated plants and collect materials (pollen, nectar, resins, honey) that may contain residues of the chemical being assessed?
2. At what level is the parent compound and the toxic metabolite present in materials (pollen, nectar, etc.) collected by honey bees?
3. How do the subject concentrations change over time when stored in the hive?
4. What concentrations in pollen and nectar when fed to a bee colony result in a significant decrease in queen fecundity, brood success, adult survival, and ultimately, colony strength?

6.2.4.2 Conceptual Model Diagram

The conceptual model diagram (Figure 6.2) illustrates the relationships presented in the risk hypothesis for the assessment of risk of a systemic pesticide applied to the soil or as a seed dressing.

The source of exposure is application of the systemic plant protection product to the soil or as a coating to seeds planted in the soil. The primary routes of exposure are assumed to be via residues in pollen and nectar (yellow boxes); however, other routes of exposure such as ingestion of residues in surface water, plant exudates (e.g., guttation fluid), and abraded seed dust are also included. Primary routes of residue transfer are indicated by thick arrows and lesser routes by thin arrows. Forager worker bees may be exposed by both contact and oral ingestion; however, since the chemical is applied to the soil, the potential for contact exposure is assumed to be limited. The attendees of the Workshop believe that the main route of exposure for worker bees is the oral route, particularly the ingestion of nectar, since nectar is the primary food consumed by forager worker bees. Pollen is also collected on hairs on the forager worker bees' bodies, or in small pouches (pollen baskets) on their hind legs. The nectar and pollen collected by worker bees are brought back to the hive where they are incorporated into the food stores, consumed by hive bees, and in turn used to produce food for the queen and the developing brood. If the pesticide concentration is high enough, toxic effects on forager bees, hive bees, bee brood or the queen may result in reduced queen fecundity, brood development success, or survival of adult bees. If these effects are severe enough or last long enough, a significant effect on colony strength may result.

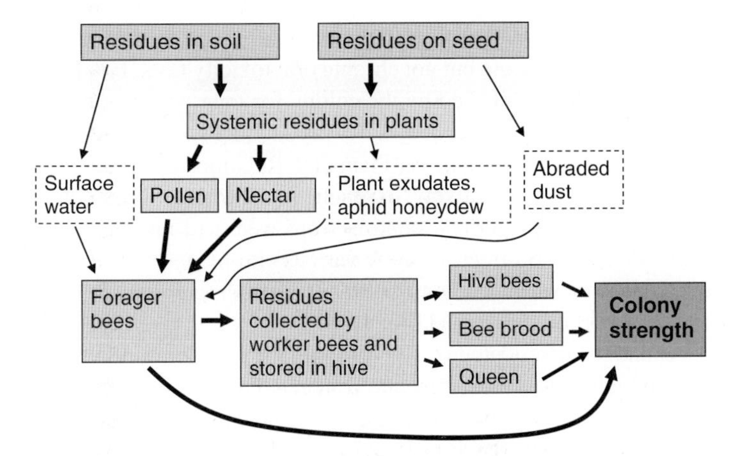

FIGURE 6.2 Depiction of stressor source, potential routes of exposure, receptors and attribute changes for a systemic pesticide applied to the soil or as a seed dressing. (For a color version, see the color plate section.)

6.2.5 Analysis Plan

The final component of the PF is the analysis plan, which identifies how the risk hypothesis will be assessed. The analysis plan identifies the data needs and the methods for conducting the assessment. The analysis plan describes the measures of exposure (e.g., estimated environmental concentrations, monitoring data) and measures of effects (e.g., NOAEC) that will be used. In the case of this example, the analysis plan may generally discuss the attribute changes that will be used for assessing risk to pollinators, including individual bee mortality and colony strength (such as percent coverage of hive frames by adult bees, percent open brood, or percent capped brood).

6.2.6 Data Needs for Exposure Characterization

While it may be possible to develop a computer model to predict residues of systemic chemicals in various plant tissues, such models are not currently available and direct measurements are obtained through field studies. For the purposes of this PF, let us assume that field studies have been conducted to measure residue levels of the parent compound and the toxic degradates in pollen and nectar. These measurements can be used to determine the median (50th percentile) and high end (defined here as the 95th percentile) concentrations expected in the pollen and nectar following an application. Estimated daily intake rates for pollen and nectar by various castes of honey bees (Table 1) of Rortais et al. (2005) may be used to convert food concentrations (μg chemical/g of food) to a daily dose (μg chemical/individual bee/day). Some toxicity endpoints are expressed in units of a test concentration (e.g., μg chemical/kg test matrix = parts per billion or ppb); or as a dose (e.g., μg chemical/individual bee). The units of the measure of exposure must match the units of the measure of toxicity in order for a valid risk estimate to be calculated.

6.2.7 Data Needs for Effects Characterization

As described briefly in Chapter 8, the progression of effects data development begins with standard laboratory assays and then, if necessary, continues on to higher-tier studies which may consist of specialized laboratory, semi-field or field tests. In this sort of testing sequence, the results of higher-tier studies are used to refine the overall conclusions about risk.

Because the main route of exposure expected for systemic chemicals is oral ingestion, toxicity testing of the oral route of exposure is needed to characterize potential effects of residues in bee foods. Standard protocols are available for conducting acute but not chronic oral toxicity tests. Food with residues of systemic compounds may be stored in the hive and used by the colony for long periods of time. The development of a standardized chronic feeding test may be needed. A 10-day feeding test of individual adult honey bees has been proposed by the International Commission on Plant-Bee Relationships (Alix et al. 2009) as a means to provide a chronic toxicity measure. Alternatively, experiments in which whole colonies are fed prescribed concentrations of the test chemical for periods ranging from weeks to months have been performed with some systemic chemicals. Measures of effects of these various chronic tests have included the median lethal concentration and the NOAEC for various colony attributes, including colony strength (e.g., percent frame coverage with adult bees, open brood, or capped brood).

If unacceptable risks cannot be discounted on the basis of simple laboratory test results, and conservative exposure assumptions, then higher-tier studies may be conducted to determine the likelihood and severity of risks under conditions simulating actual agricultural use. Semi-field and field studies may have the advantage of evaluating all routes of exposure simultaneously under conditions reasonably similar to actual field use, whereas laboratory studies are generally limited to evaluation of a single route of exposure under artificial conditions.

6.2.8 Risk Characterization Approach

Most assessments of ecological risks of pesticides use a conventional risk quotient (RQ) or toxicity exposure ratio (TER) approach that compares point estimates of exposure (e.g., typical and high end estimates of residue levels in various food types) to estimated thresholds of toxicity (i.e., median lethal concentration or NOAEC). The RQ equals the exposure point estimate divided by the toxicity point estimate. Although RQ values are dimensionless numbers, the greater the RQ, the greater is the presumed risk. TERs are the reciprocal of the RQ, so the greater the TER, the lower the risk. Regulatory agencies compare the RQ or TER to an established level of concern (LOC) that is presumed to represent a threshold between minimal and non-minimal risk. If the RQ is less than the LOC, or the TER is greater than the LOC, the risk may be presumed to be minimal and further testing is unnecessary provided the constituent elements of the RQ are considered to be sufficiently inclusive. Risk assessment is iterative with screening-level point estimates of exposure and toxicity often used in initial assessments. If the RQ of a screening-level assessment exceeds the LOC, it can be concluded that the risk is potentially not minimal, and further testing may be appropriate to clarify the risk. If semi-field or field tests are performed, these results may be incorporated into the risk characterization (provided the studies are of sufficient quality) using a weight-of-evidence approach.

6.3 CASE 2: PROBLEM FORMULATION FOR A CONTACT CHEMICAL APPLIED AS A FOLIAR SPRAY

6.3.1 Stressor Description

The stressor of concern is a "knock-down" insecticide product applied as a spray to field and orchard crops such as cotton, maize, vegetables, citrus, and pome fruit to control pest insects that feed on stems, leaves, inflorescences, and fruit. In this model, the pesticide does not penetrate treated plant surfaces and so it is not translocated systemically throughout the plant (note, however, that certain pesticides that have systemic properties may be foliarly applied). For the purposes of this example, it is assumed that residues on plant foliage dissipate fairly rapidly, with a foliar dissipation half-life of 2–3 days. Because of the short residual toxicity, several applications may be necessary to protect plants during critical phases of the growing season. Based on their chemical structure, none of the chemical's major breakdown products are expected to exhibit significant toxicity to insects. The product label recommends application rates that vary from 20 to 30 g active ingredient (a.i.) per hectare (ha), depending on crop and growth stage.

6.3.2 Management Goals

As discussed earlier, protection goals are policy decisions that are set by government agencies and other organizations that represent the interests of the societies they serve. In the absence of specific protection goals, the participants used those developed during the Workshop, which included

- protection of pollination services provided by *Apis* and non-*Apis* species;
- protection of honey production and other hive products; and
- protection of pollinator biodiversity.

6.3.3 Assessment Endpoints

For honey bees, logical assessment endpoints include colony strength (population size and demographics) and colony survival (persistence). Bumble bees too can be measured against colony strength (larval ejection, number of offspring, or colony weight) and colony survival (persistence). Since a colony loss simply represents the situation when colony strength is minimal, it could be argued that *colony survival* is not needed as a separate assessment endpoint. Various measures of colony strength are often made when beehives are rented and placed in agricultural crops. Rental fees are greater for strong colonies than weak colonies because colony strength is expected to be related to the quality of pollination service provided by the colony. Colony strength will likely be significantly impacted if queen viability, brood development, or general worker bee health is negatively impacted for an extended period of time. There are many known cases where pesticide exposure has caused effects on colony strength. Colony strength appears to meet very well the identified criteria for an assessment endpoint. Colony strength

1. has ecological relevance;
2. is susceptible to known or potential stressors; and
3. is relevant to protection goals and societal values.

As previously said, for solitary bees, assessment endpoints may include adult survival, adult fecundity, larval survival, and larval development time. Populations will be significantly impacted by decreased adult or larval survival and adult fecundity. Increased time for larval development could impact individual bee emergence time and reduce the number of generations per year in multivoltine species, or by causing bees to enter diapauses too late and, ultimately relate to fecundity or a sign that larvae will not emerge as healthy adults. There are known cases where pesticide exposure has affected these endpoints. These endpoints also fulfill the identified criteria for an assessment endpoint (see (1), (2), and (3) above).

6.3.4 Conceptual Model

The fourth component of PF listed previously is the conceptual model, which identifies the relationship between ecological entities and the chemical stressor under consideration. The conceptual model has two components, that is, the risk hypothesis and conceptual diagram.

6.3.4.1 Risk Hypothesis

The risk hypothesis describes the predicted relationships among the chemical stressor, exposure, and assessment endpoint responses along with a rationale to support the hypothesis.

For a nonsystemic pesticide applied as a foliar spray, the risk hypothesis involves the following steps describing how exposure most likely occurs and results in effects on the assessment endpoint (colony strength). The hypothesis is:

1. Residues in spray droplets may (1) contact bees directly (i.e., bees hit directly by the spray), (2) be deposited on plant surfaces visited by honey bees, and (3) contaminate standing water (e.g., puddles) from which bees drink.
2. Spray deposits hitting open flowers may contaminate nectar and pollen sources for a short period of time post-application (until these flowers are replaced by others that were not open during spray).
3. Forager honey bees may ingest contaminated water or contaminated nectar, and may collect and transport contaminated nectar and pollen back to the hive where these materials are processed, then incorporated into the food stores of the colony.
4. If the exposure concentration is high enough, toxic effects on forager bees, hive bees, bee brood or the queen may result in reduced survival of adult bees, brood development, or queen fecundity.
5. Colony strength is affected as a result of reduced fecundity, brood development, or adult survival if these effects are severe enough or last long enough.

6. As the chemical is knock-down insecticide with short residual time on foliage, the primary effect expected may be direct mortality of forager bees shortly after spraying (i.e., a bee kill event).

The duration of exposure of forager bees will depend on the persistence of the chemical on plant surfaces, and the persistence (duration of bloom) of individual flowers that were hit by the application. As new blooms replace old ones, the potential for exposure may rapidly decrease. Thus, the main concern for foliar spray applications has traditionally been acute exposure of forager worker bees that results in a discrete bee kill event. However the possibility of residues in bee-collected pollen and nectar being brought to, processed, and stored in the hive should be considered since this scenario may lead to chronic exposure of the hive bees, queen, and bee brood.

Based on the risk hypothesis, key questions that need to be answered during risk analysis are:

1. To what extent are forager honey bees active when spray applications are made? (Or, what is the relation between the application and the flowering of that crop?)
2. If forager bees incur contact exposure during or shortly after application, are the levels of exposure great enough to cause "knock-down" intoxication?
3. If spray deposits represent an initial lethal hazard to honey bees, how long does this situation last?
4. To what extent do foraging honey bees visit sprayed plants and water sources and collect materials (e.g., pollen, nectar, resins, water) that may contain residues of the chemical?
5. What levels of the chemical are present in materials (e.g., pollen, nectar, resins, water) collected by honey bees and brought back to the hive?
6. How do the above concentrations change over time, including changes in concentrations in hive-stored pollen and nectar?
7. What concentrations in pollen, nectar, or bee bread when fed to a bee colony result in a significant decrease in queen fecundity, brood development, adult survival, and ultimately, colony strength?

6.3.4.2 Conceptual Model Diagram

The conceptual model diagram depicted in Figure 6.3 illustrates the relationships presented in the risk hypothesis for the assessment of risk of a nonsystemic chemical applied as a foliar spray.

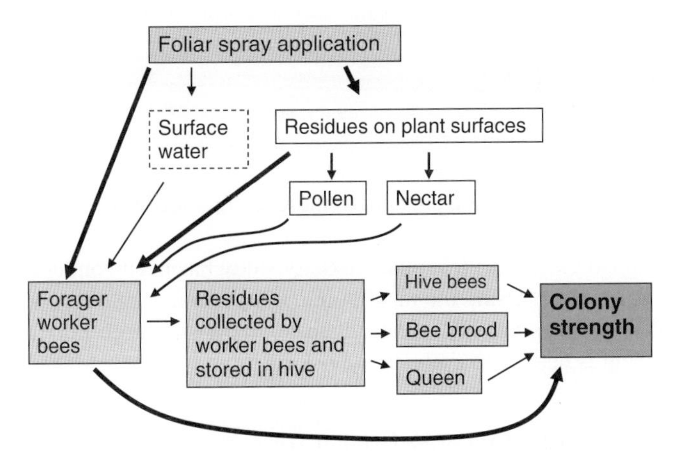

FIGURE 6.3 Depiction of stressor source, potential routes of exposure, receptors and attribute changes for a nonsystemic pesticide applied as a foliar spray. (For a color version, see the color plate section.)

The source of exposure is foliar spray application of the nonsystemic plant protection product to crop plants. The primary routes of exposure are assumed to be via contact of foraging bees with spray as it is applied or with freshly deposited residues on plant surfaces. For flowers open during spraying, residues may occur in pollen and nectar, and these materials may be brought back into the hive, processed and stored as food that is later utilized by hive bees, bee brood, and the queen. Another possible route of exposure is via surface water (e.g., puddles) that are oversprayed and used by bees as a source of drinking water. Primary routes of residue transfer are indicated by thick arrows, lesser routes by thin arrows. Greatest exposure is expected for forager bees that may be exposed via contact with spray droplets and residues on plant surfaces, and via ingestion of residues in water and nectar. If the exposure level is sufficient enough, then forager bees may be killed to the extent that colony strength is reduced (e.g., large bee kill event).

Bees in the hive could also be exposed, but the exposure levels are not expected to be as great as for forager bees unless the hive is inadvertently sprayed (overspray) during application. However, if pesticide residues in the forage area are high, then other bees may be exposed to these high residues during social grooming. In addition, if concentrations in pollen and nectar brought into the hive are high enough, toxic effects on hive bees, bee brood, or the queen may result. If these effects are severe enough or last long enough, a significant adverse effect on colony strength may result.

6.3.5 Analysis Plan

The final component of the PF is the analysis plan. The analysis plan identifies how the risk hypothesis will be assessed. It identifies data needs and methods for conducting the assessment and what measures of exposure (e.g., estimated environmental concentrations) and measures of effects (e.g., NOAEC) and attribute changes (e.g., colony strength attributes might include estimates of the percent coverage of hive frames by adult bees, open brood, and capped brood) will be used.

6.3.6 Screening Assessment

A simple hazard quotient (HQ) approach is currently used in Europe to predict whether foliar applications of plant protection products have the potential to cause observable bee kills of adult foragers. This screen has been validated by comparing predictions to results of field studies and incident monitoring programs (see Mineau et al. 2008).

The HQ calculation is made as follows:

HQ = application rate (g a.i./ha) / LD50 (μg/bee)
If HQ <50, a minimal risk may be presumed
If HQ >50, a potential risk concern may be presumed (more testing needed)

For example, it is assumed that an acute contact toxicity study has been conducted and the LD50 for the chemical in question is 0.1 μg/bee. Using the maximum application rate of 30 g a.i./ha, the HQ calculation would be 30/0.1 = 300. Since this value is greater than 50, the risk of bee kills cannot be discounted as minimal. Further assessment is needed to evaluate the risk.

6.3.7 Data Needs for Refined Exposure Characterization

A label statement prohibiting application to crops during bloom until the evening or night time hours could go a long way toward eliminating the possibility that foraging bees will be hit by the spray droplets as they are applied to the crop. A key piece of information needed is how long residues on sprayed vegetation remain toxic to visiting honey bees. This could be estimated from field studies that measure the magnitude and dissipation of residues on sprayed vegetation. It may be simpler to determine this using a standard EPA Tier 2 bioassay, with honey bees i.e., toxicity of residues on foliage (USEPA 2012) (discussed in greater detail in Chapter 7). Another key piece of information is to determine the residue levels in plant materials (mainly pollen and nectar) collected by forager bees and brought in to the hive. It may be necessary to conduct field studies to obtain direct measurements. Such measurements can be used to determine the median (50th percentile) and high-end (e.g., 95th percentile) concentrations expected to be present in pollen and nectar following an application. Estimated daily intake rates for pollen and nectar by various castes of honey bees listed in Table 1 of Rortais et al. (2005) may be used to convert food concentrations (μg chemical/g of food) to a daily dose (μg chemical/individual bee/day). Some toxicity endpoints are expressed in units of a test concentration (e.g., μg chemical/kg test matrix = ppb); others as a dose (e.g., μg chemical/individual bee). The units of the measure of exposure must match the units of the measure of toxicity in order for a valid risk estimate to be calculated.

6.3.8 Data Needs for Effects Characterization

The logical progression of effects data development is to begin with standard laboratory assays and, if necessary to conduct higher-tier studies that may consist of specialized laboratory, semi-field or field tests. In this sort of testing sequence, the results of higher-tier studies are used to refine the assessment and are weighted more heavily in reaching overall conclusions about the risk.

Because the main route of exposure for forager bees is expected to be contact, the standard EPA Tier 2 bioassay with honey bees (i.e., toxicity of residues on foliage [USEPA 2012]) may be appropriate. In this test, groups of honey bees are exposed via contact to vegetation which was sprayed in the field and then collected for testing after prescribed time intervals. For example, a common protocol is to evaluate the contact toxicity of vegetation at 2, 8, and 24 hours post application. In the case of this chemical, let us assume it was found that a high level of mortality occurred in bees exposed to 2-hour-old foliar residues, but that normal honey bee survival was noted when bees were exposed to foliar residues collected 8 and 24 hours after application. Because this is a laboratory-based study, results such as these would indicate that there is a window of acute hazard from contact that exists for 2–8 hours after application of the subject pesticide.

To assess the significance of residues in pollen and nectar that may be brought into and stored in the hive, oral toxicity testing is needed. As a minimum, an acute oral toxicity test can be used to establish oral dose levels that are potentially lethal to adult bees. If there are indications that significant residues will be contained in stored food (pollen, honey, bee bread), then a chronic feeding study may be needed to identify the NOAEC. A 10-day feeding test of individual adult honey bees has been proposed by the International Committee on Plant-Bee Relationships (ICPBR) as a means to provide a chronic toxicity measure to adult bees. Various kinds of larval feeding tests have been developed to establish dose levels that affect larval survival and development. Alternatively, experiments in which whole colonies are fed prescribed concentrations of the test chemical for periods ranging from weeks to months have been performed with some chemicals. Measures of effects directly related to colony strength can be obtained from such studies.

If adverse effects cannot be discounted on the basis of simple laboratory test results, higher-tier studies may be conducted to determine the likelihood and severity of effects under conditions simulating actual agricultural use. Semi-field (tunnel) and field studies may have the advantage of evaluating all routes of exposure simultaneously under conditions reasonably similar to actual field use, whereas laboratory studies are generally limited to the evaluation of a single route of exposure under artificial conditions.

6.3.9 Risk Characterization Approach

Calculation of the screening assessment HQ represents an initial risk characterization of the chemical. If the HQ <50, there is a presumption of minimal acute risk in the EU, based on historical investigations of bee kill incidents (Mineau et al. 2008). Based upon the results of the acute toxicity test and the use pattern, higher-tier tests may be required by the USEPA, which may provide some insight into whether the label statement requiring applications be made in late afternoon or evening will mitigate the potential risk. Since, in this example, a study showed residual toxicity lasting less than 8 hours, residues from applications made in the late afternoon or evening should not pose an acute hazard to bees that begin foraging the following day. An RQ or TER calculation could be calculated to assess the risk posed by residues in pollen and nectar. The RQ or TER calculation would compare the concentration measured in these matrices or dose taken in by various castes of bees to available toxicity endpoints (LD50, NOAEC, etc.). Finally, well-designed semi-field or field studies may provide the more reliable information regarding the level of risk actually occurring under field use conditions. A weight-of-evidence approach may be taken to integrate the various lines of evidence.

REFERENCES

Alix A, Chauzat M-P, Duchard S, Lewis G, Maus C, Miles MJ, Pilling E, Thompson HM, Wallner K. 2009. Guidance for the Assessment of Risks to Bees from the Use of Plant Protection Products Applied as Seed Coating and Soil Applications—Conclusions of the ICPBR Dedicated Working Group. Julius-Kühn-Archiv 423, pp. 15–27.

Mineau P, Harding KM, Whiteside M, Fletcher MR, Garthwaite D, Knopper LD. 2008. Using reports of bee mortality in the field to calibrate laboratory-derived pesticide risk indices. *Environ. Entomol.* 37(2):546–554.

Rortais A, Arnold G, Halm M-P, Touffet-Briens F. 2005. Modes of honey bees' exposure to systemic insecticides: estimated amounts of contaminated pollen and nectar consumed by different categories of bees. *Apidologie* 36, 71–83.

USEPA. 1998. Guidelines for Ecological Risk Assessment. Published on May 14, 1998, Federal Register 63(93):26846–26924. http://www.epa.gov/raf/publications/pdfs/ECOTXTBX.PDF (accessed December 15, 2013).

USEPA. 2012. Ecological Effects Test Guidelines. OCSPP 850.3030: Honey Bee Toxicity of Residues on Foliage [EPA 712-C-018], January 2012. http://www.regulations.gov/#!documentDetail;D=EPA-HQ-OPPT-2009-0154-0017

7 Assessing Exposure of Pesticides to Bees

J.D. Wisk, J. Pistorius, M. Beevers, R. Bireley, Z. Browning, M.P. Chauzat, A. Nikolakis, J. Overmyer, R. Rose, R. Sebastien, B.E. Vaissière, G. Maynard, M. Kasina, R.C.F. Nocelli, C. Scott-Dupree, E. Johansen, C. Brittain, M. Coulson, A. Dinter, and M. Vaughan

CONTENTS

Pesticide Risk Assessment for Pollinators, First Edition. Edited by David Fischer and Thomas Moriarty.

7.1 INTRODUCTION

An essential component of an ecological risk assessment is a prediction of exposure of the organisms being assessed. This chapter outlines exposure pathways for the different pesticide delivery methods, both nonsystemic and systemic, and discusses methods used to predict pesticide exposure to honey bees and non-*Apis* bees. This chapter also provides an outline of techniques employed to measure pesticide residues in relevant matrices and discusses higher tier field study designs that are used to refine bee exposure assessments for specific products. Finally, this chapter presents perspectives regarding pesticide application technologies that can be employed to mitigate bee exposure, as well as future research needs to further refine exposure assessments for this taxa.

7.1.1 Potential Exposure to Foraging Bees

7.1.1.1 Sprayed Compounds

Honey bees can be exposed to direct spray or through contact with the crop to which a pesticide is applied. Bees can be exposed to pesticides that drift to plants on the edges of the treated field, potentially leading to either contact or oral exposure, as well as water sources near the treated field that may contain residues either from drift or surface run-off. Pesticide drift can also reach hives directly if the hives are located in or near a treated field. When foliar applications are made directly onto flowers, oral exposure can occur through the collection of contaminated pollen, nectar, or honeydew or by contact exposure if the product is directly sprayed on foraging bees or the plant parts that they can come in contact with during foraging.

7.1.1.2 Microencapsulated Compounds

Microencapsulation technology is designed to increase adhesion of the formulated pesticide to the plant surface or soil and to control exposure by slowing the release of the product. Bees can potentially be exposed to certain microencapsulated pesticides if the microcapsules are similar in size to pollen. Bees may inadvertently collect the microcapsules and bring them back to the hive. If the microcapsules are collected by bees and mixed into the bee bread, the exposure may affect the whole colony as the pesticide will thus be fed to the larvae. Such incidents have been reported following the use of Pencap-M, a microencapsulated formulation of methyl-parathion (Mason 1986).

7.1.1.3 Dust

Abraded dust that is contaminated with pesticide can be released from treated seed during planting operations involving pesticide-treated seed (Alix et al. 2009c). The exposure can be oral or contact from bees foraging on flowers upon which abraded dust falls. Bees may also be exposed if they fly through the dust or vapors released during planting operations (Forster 2009; Pistorius et al. 2009; Alix et al. 2009c) or, may receive exposure if they forage on weeds and flowers (i.e., understory or in material that is adjacent to the target site) covered with contaminated dusts.

7.1.1.4 Compounds with Systemic Properties

Pesticides that have systemic properties will move within the plant and may be expressed in the pollen and nectar. Pollen and nectar of plants treated with systemic compounds (such as treated seed, soil applications, ground drench, or chemigation applications) may contain pesticide residues. These residues may be collected by foragers and brought back to the hive to be stored, processed, and fed to adults and larvae.

Bees may be exposed to pesticide residues that may occur in rotational crops or alternative forage (understory or adjacent areas) that may take up and express pesticide residues applied at an earlier date. Even if target crops are not attractive to bees, compounds that are persistent may represent a potential source of exposure through soil, or through residues in the nectar and pollen of the succeeding (rotational) crop or associated weeds. The presence of pesticide residues in a succeeding crop may be influenced by the type of crop, treatment pattern, the physicochemical properties, and of course the environmental fate of the compound.

Other potential routes of exposure for foraging bees include inhalation (Seiber and McChesney 1987; Seiber et al. 1991), and consumption of aphid honeydew, guttation water (Girolami et al. 2009; Schenke et al. 2010), or chemigation water from soil treatments.

7.1.2 Potential Exposure to Non-foraging Bees From Beeswax

All members of a honey bee colony may be potentially exposed to contaminants through the wax that composes the hive. Larvae are reared in cells made of beeswax, and as adults they are in constant contact with the wax while they are in the hive. After pupation, bees chew through the wax coating on the brood capping and emerge as an adult. During colony development, worker bees continuously modify the wax cell structure (e.g., converting male cells into worker cells, cleaning brood cells to stock honey and vice versa). Pesticides that are lipophilic tend to accumulate in wax (Tremolada et al. 2004) and if the beeswax contains pesticide residues, members of the colony, especially larvae, may be subject to contact exposure, depending upon the bioavailability of the pesticide (Chauzat et al. 2007).

7.1.2.1 Nurse Bees

For the first 1–3 weeks after emergence adult honey bee workers remain in the hive to perform many duties including, but not limited to, feeding and cleaning larvae, cleaning cells, building new cells, processing nectar and storing honey, packing pollen, and capping cells. Nurse bees may be potentially exposed to higher levels of pesticide residues by virtue of their duties. Nurse bees process pollen and nectar into bee bread and honey, respectively, and also produce larval jelly. Nurse bees are the only caste or life-stage of honey bees that consume significant amounts of raw pollen, which is regurgitated and processed into bee bread. Bee bread is then stored in the hive until it is processed by nurse bees into brood food and fed to larvae. In addition, nurse bees can potentially be exposed to pesticides through water brought back to the hive for cooling and brood rearing. Nurse bees may also be exposed as they process nectar into honey within beeswax cells as well as through contact with wax while moving through the hive. Pesticides applied directly to the hive for *Varroa* sp. control and other pests are a direct route of exposure to nurse bees (Martel et al. 2007). Nurse bees can potentially be exposed to pesticides during all of these activities if residues are present in the hive.

7.1.2.2 Drones

Upon emergence as adults, honey bee drones receive food from worker bees or eat stored honey. As larvae, drones receive more food than worker larvae, but the composition of that food is similar (Free 1977). Like larvae and nurse bees, drones may be exposed to pesticides through food or residues within the hive.

7.1.2.3 Queens

Larvae that are fed only royal jelly beyond 3 days after hatching develop into queens (Free 1977). A queen may live within the hive from 6 months to several years. Therefore, the queen may be exposed to multiple pesticides and residues within the hive over a relatively long period of time. Feeding on royal jelly and contact with residues in the hive are the potential routes of contaminant exposure for queens.

7.1.2.4 Honey Bee Larvae

Honey bee larvae can be exposed to pesticides through ingestion of contaminated food including pollen, bee bread, honey, and larval jelly. Larval worker bees are fed royal jelly (also referred to as worker jelly or larval jelly) for 3 days after egg hatch. Royal jelly is a glandular secretion from the hypopharyngeal glands of nurse bees, and consists of some white components (mostly lipids) and a clear secretion (Free 1977). Honey bees exposed to some pesticides can potentially produce contaminated larval jelly (Tremolada et al. 2004) that could be fed to the queen, workers, and the larvae. From 4–6 days after egg hatch, worker larvae are fed bee bread, which is largely processed pollen, but also includes some larval jelly, honey, and pollen (Free 1977). The bee bread can be contaminated if processed with contaminated pollen (Orantes Bermejo et al. 2010).

Water is brought back to the hive and used to cool the hive, dilute stored honey, and prepare larval food. If pesticide residues are present in the water that is brought back to the hive, larvae may be exposed through direct contact or through ingestion of food prepared with it. Larvae may also be exposed via contact exposure to pesticides that accumulate in wax or from residues on foraging bees. Additionally, larvae, as well as adults, may be exposed to insecticides or miticides applied directly to the hive by the beekeeper for *Varroa* control or fungicides, bactericides, or any other active substance applied for disease control.

7.1.3 Residue Movement Through the Hive

Pesticides can be transferred into the hive environment from foraging honey bees that bring residues back to the hive in contaminated pollen and nectar. Pesticide residues can also move throughout the hive as workers

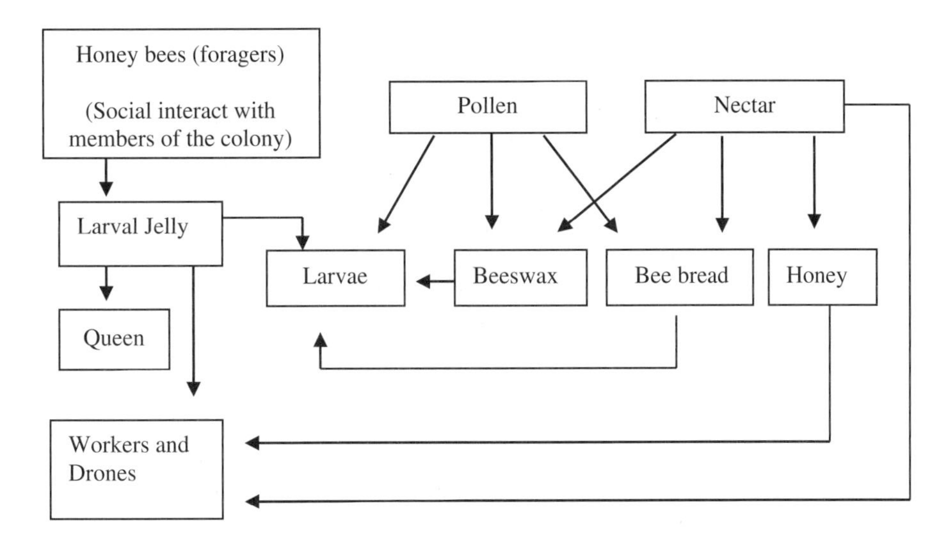

FIGURE 7.1 Conceptual model showing how contaminants may potentially reach various matrices within honey bee colonies. Pollen and nectar are the main sources of in-hive contamination. Arrows show potential major contamination transfer routes. For minor routes, please refer to the text.

pass food (especially nectar and diluted honey) among themselves as it is processed, stored, or consumed. All potential pesticide transfer to, and movement within in a hive is highly dependent on the use pattern of the pesticide product, as well as the physical and chemical properties of the contaminants. Some chemicals may persist in the hive, resulting in prolonged exposures, while others dissipate or degrade into metabolites. Some pesticide metabolites can also be toxic to honey bees (Suchail et al. 1999; Martel and Lair 2011). Therefore, while research continues to shed light on the fate and movement of a compound in a hive, it is important to understand and consider these properties of a compound in assessing potential exposure (Figure 7.1).

7.2 POTENTIAL ROUTES OF EXPOSURE FOR NON-*APIS* BEES

Most routes of exposure that have been examined for honey bees are valid for non-*Apis* bees as well. However, because of their diverse and often different biology (see also Chapter 3), non-*Apis* bees may be prone to other routes of pesticide exposure. Understanding different exposure routes is important because it is not feasible to conduct tests on the more than 20 000 species of non-*Apis* bees worldwide (Michener 2007). A risk assessment for non-*Apis* bees can be based mainly on the exposure routes reviewed for honey bees and tailored for different non-*Apis* species groups. If more specific exposure information is required for risk assessment refinements, actual measures of unique exposure pathways may be adapted from tests conducted on some key non-*Apis* species (see Section 7.11). Because of the large diversity of non-*Apis* biological features, this section will be structured around some broad features of non-*Apis* bee ecology.

7.2.1 NESTING SITES AND NESTING MATERIALS FOR NON-*APIS* SPECIES

Social non-*Apis* bees, such as stingless bees nest in cavities that are usually located aboveground. In addition, plant resins used for nest construction may be contaminated by pesticide applications (Romaniuk et al. 2003),

FIGURE 7.2 Leafcutter bee on blanket flower, photo by Mace Vaughan (Xerces Society for Invertebrate Conservation). (For a color version, see the color plate section.)

FIGURE 7.3 Micropipetting nectar samples, photo by Mike Beevers. (For a color version, see the color plate section.)

FIGURE 7.4 Hand collecting pollen by removing flower anthers, photo by Mike Beevers. (For a color version, see the color plate section.)

and while honey bees also use resin in nest construction, certain non-*Apis* species employ resins to a greater extent in nest building (Murphy and Breed 2008; Roubik 1989). Most bumble bee species (e.g., *Bombus terrestris*, *Bombus lapidaries,* and *Bombus subterraneus*), nest underground in abandoned nests of rodents and, therefore, are protected from direct spray applications. However, other non-*Apis* species nest above ground in cavities (e.g., *Melipona* spp. and *Trigona* spp.) or under patches of grasses and vines (e.g., *Bombus pascuorum* and *Bombus ruderarius)* where there is greater potential exposure to drift, or direct pesticide

FIGURE 7.5 Honey bee semi-field study with *Phacelia*, photo provided by BASF SE. (For a color version, see the color plate section.)

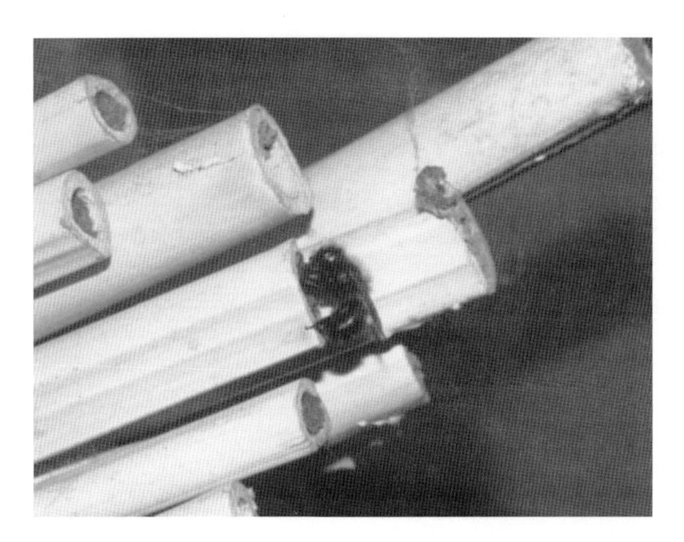

FIGURE 7.6 Mason bee, photo by Mace Vaughan (Xerces Society for Invertebrate Conservation). (For a color version, see the color plate section.)

applications (Pouvreau 1984; Thompson 2001). Stingless bees and bumble bees mainly use wax to build their nests, but, unlike honey bees, they also commonly mix it with pieces of grass, leaves, and various substrates (Pouvreau 1984; Roubik 1989), that may also be a source of exposure to contaminants.

Among solitary bees, the location of the nests as well as the material used to build them can vary considerably. Most solitary bees nest in tunnels underground. The gregarious ground-nesting species can occur in large aggregations of several thousand individuals in natural sites (e.g., Potts and Willmer 1998) or in man-made bee beds such as for *Nomia melanderi* (Cane 2008). In addition, ground-nesting bees can be found along the border of fields planted with annual crops, but also in the soil within such fields (Vaissière et al. 1985; Shuler et al. 2005; Kim et al. 2006). Therefore, the dissipation rate of pesticides in soil is a key factor affecting potential exposure to species that nest in the field.

The second largest group of solitary bees consists of species that nest in pre-existing cavities (mostly tunnels) in dead wood, hollow twigs and bamboo (Figure 7.6), or pithy stems such as elderberry (*Sambucus* spp.). These include most bees in the genera *Osmia* and *Megachile* (Cane et al. 2007). Among the "tunnel nesters," leafcutter bees (Figure 7.2) (Megachilidae, especially *Megachile* spp.) use excised leaf or petal pieces, as their common name suggests, to line their burrows and seal each cell once their egg has been laid on a ball of pollen and nectar. These leaf pieces are collected from a large array of plants, such as alfalfa and rose bushes. Other species, such as carpenter bees (*Ceratina* spp., *Lithurgus* spp., and *Xylocopa* spp.) drill their nest tunnels in soft wood or the soft pith of some plant stems.

Other bees build their nests with flower petals (e.g., *Hoplitis* spp.), or plant hairs (e.g., wool-carder bees such as *Anthidium manucatum*) (Gibbs and Sheffield 2009), and many mason bees, *Osmia* spp., use mud to build partitions between the different cells of their nests (e.g., Bosch and Kemp 2001; Mader et al. 2010); exposure to pesticides may occur from these materials if contaminated (Waller 1969; Johansen and Mayer 1990). The increasing use of systemic insecticides, not only in commercial agriculture but also in residential or recreational scenarios, may result in exposure of certain species (Vera Krischik, personal communication), especially some species of *Osmia* that chew up pieces of leaves to create walls of pulp to separate brood cells. This, however, requires further study to better understand.

7.2.1.1 Exposure at Immature Stages of Non-*Apis* Species

As stated previously, honey bee worker and drone larvae feed on food that has been processed, which may result in modifications (e.g., degradation) of pesticide active ingredients in food stores. However, this differs from scenarios of solitary non-*Apis* bees whose larvae feed directly on raw pollen and nectar in either a mass provisioning manner or sequential mass provisioning manner (i.e., brood cells are provisioned over various timeframes). As such, exposure via food may differ between *Apis* and non-*Apis* species feeding on mostly unprocessed pollen, nectar, and other floral resources (O'Toole and Raw 1999; Pereboom 2000). Therefore, exposure estimates based on stored honey bee pollen which is converted to royal jelly may not be predictive of the chemical residues fed to the non-*Apis* bee brood (Konrad et al. 2008). In addition, with bees that mass provision their cells (i.e., most non-*Apis* bees), the eggs and larvae are in direct contact with the pollen and nectar provision during the early life stages (i.e., the egg and first instar). Honey bees, on the other hand, are isolated in their cells and are fed progressively by nurse honey bees, and therefore, have a very different exposure profile (Winston 1987).

7.2.1.2 Foraging Time and Mating

Peak foraging time for honey bees is generally during warm, non-overcast conditions (Johansen and Mayer 1990; Tew 1997). However, this is not the case for many non-*Apis* bee species, such as bumble bees and mason bees (*Osmia* spp.), which are known to forage during cool, inclement weather, as well as earlier and later in the day and earlier and later in the season than honey bees (Thompson and Hunt 1999; Vicens and Bosch 2000; Bosch and Kemp 2001; Thompson 2001). Similarly, squash bees (*Peponapis,* and *Xenoglossa* spp.) are active in the early predawn hours (Sampson et al. 2007). Male solitary non-*Apis* bees are also the first to emerge from the nest, followed a few days later by females. Males may therefore be present when not expected. In addition, males of many non-*Apis* bees often spend the night in flowers or hanging from plants, potentially leading to higher exposures (Sapir et al. 2005). However, male squash bees that spend the night in closed squash blossoms may receive some level of protection from night time pesticide applications because the blossoms close tightly around them.

7.2.1.3 Food Sources

Honey bees are extreme generalists in that a colony will forage for nectar and pollen on a large array of plant species (polylecty). This is not so for most non-*Apis* bees, especially for the 80% or more which are solitary. These species often gather their pollen on a few species of taxonomically related plant species (oligolecty) and sometimes on a single species. Indeed, non-*Apis* bees may also forage, and even specialize, on plants not readily visited by honey bees (e.g., potato, many legumes, and some ornamentals). As a result, pesticide exposure (to generalists) may be "diluted" from various floral resources across a wide landscape. For example, tomato and potato flowers do not produce nectar but will release their pollen through buzz pollination (sonication). Although, it is possible that pollen from flowers of this type could be shielded from foliar pesticide applications (because of the unique plant morphology), and considered safe for honey bees, because honey bees visit these flowers infrequently, they remain a potential exposure scenario for non-*Apis* bees.

7.2.1.4 Size

Another factor affecting foraging and exposure in non-*Apis* bees is the size of some non-*Apis* bees, and the relationship between foraging distance and species size. Some non-*Apis* bees are much smaller than honey bees (e.g., small bees of the genera *Perdita* or *Dialictus* in the United States and *Nomioides* in Europe), and therefore are subject to relatively greater exposure because of the higher surface area to volume ratio of smaller bodies (i.e., µg of pesticide that contacts the body/mg body weight). Indeed, even intraspecific tests of pesticide toxicity to bumble bees have confirmed that smaller bees may be more effected than larger bees for a specific dose (Van der Steen 1994; Thompson and Hunt 1999; Malone et al. 2000).

A second size-related factor affecting potential exposure of non-*Apis* bees is the relationship between size and foraging distance. Whereas large bees, such as honey bees, bumble bees, or carpenter bees (*Xylocopa* spp.) easily forage over several kilometers from their nest (Beekman and Ratnieks 2000; Goulson and Stout 2001; Pasquet et al. 2008); small bees may only fly a few hundred meters from their nest site (Greenleaf et al. 2007). This factor may potentially result in higher exposure to small bees, compared to larger species, that are attracted to blooming crops, where their limited foraging range necessitates nearby nesting, and ongoing exposure to pesticide applications throughout the growing season. In some landscapes (e.g., New Jersey, USA), small bees (e.g., *Halictus* and *Lasioglossum* spp.) perform a significant amount of crop pollination (Winfree et al. 2007, 2008).

Somewhat related to foraging distance is the tendency of certain solitary bees to collect pollen from one area, and often from only one or a few plant species, whereas honey bees forage on a wide variety of plant species across a large landscape. Honey bee foraging areas and sources of nectar and pollen can vary considerably from one day to the next (Visscher and Seeley 1982). Therefore, due to the foraging behavior, the pesticide residues on one crop may be diluted in a honey bee colony diet, but not so in the nest of a non-*Apis* species.

7.3 METHODS AND MODELS FOR ESTIMATING EXPOSURE OF BEES TO PESTICIDES

Currently, there are no globally accepted approaches for estimating exposure of pesticides to bees for screening-level risk assessments. Participants of the Workshop reviewed current methodologies employed in the United States and European Union, and evaluated information that can be used or developed to establish exposure estimates for screening-level risk assessments for both honey bees and non-*Apis* bees.

7.3.1 Screening Level Exposure Estimates

Atkins et al. (1981) conducted laboratory contact toxicity studies and corresponding field studies with 65 pesticides. The field hazards were studied in a large number of commercial fields during bloom using crops that were highly attractive to honey bees. Data developed by Atkins et al. (1981) indicated that, for foliar-applied products, the median lethal dose (LD50) as measured in micrograms of active ingredient per bee (µg a.i./bee) can be coverted and expressed as the equivalent number of kilograms of chemical per hectare (kg a.i./ha) (that would yield an LD50) by multiplying by 1.12. For example, an acute contact LD50 of 1 µg a.i./bee (highly toxic according to Atkins et al. (1981) classification scheme) would equate to an application rate of 1.12 kg a.i./ha, (or pound per acre). In the European Union, the hazard quotient (HQ) approach is used as a screening-level assessment to distinguish between compounds with either potentially low or high risk of acute poisoning from foliar pesticide applications. The HQ relates the application rate of a product with laboratory oral and contact LD50 values.

$$HQ = \text{Application rate (g a.i./ha)}/\text{Contact or Oral LD50 } (\mu\text{g a.i./bee})^{1}$$

[1] See Chapter 8 for a discussion on acute (dermal or oral) toxicity tests.

7.3.1.1 Environmental Protection Agency Residue Unit Dose (TREX), Comparison of Lab Contact Toxicity Data with Residue Data From TREX

Environmental Protection Agency (EPA) has typically employed the terrestrial residue exposure model (TREX) when investigating foliar-applied pesticides. This model is used to predict residues on food items (e.g., vegetation, seeds, insects) for birds and mammals, and is based on a nomogram developed by Hoeger and Kenaga (1972). The contact exposure to a bee (which to this point has only been done for endangered species analysis) is calculated by multiplying the residue predicted for broadleaf plants or small insects by the assumed weight of a foraging honey bee (0.128 g) (Mayer and Johansen 1990) to establish a dose per bee (μg a.i./bee).

Although the TREX method could potentially be useful for developing a screening-level exposure estimate for bees in a risk assessment process, the values developed by Hoeger and Kenaga (1972) are not based on residue data for insects but rather on plants or plant parts of similar size (Fletcher et al. 1994). Data from Hart and Thompson (2001) indicate that the 95th percentile value for an insect residue per unit dose (RUD) is 24 mg/kg compared to 135 mg/kg for broadleaf plants (EPA's surrogate for small insects) which is approximately six-fold higher. Data from additional studies (Brewer et al. 1997; Fischer and Bowers 1997) also suggest that the insect residue estimates developed by Hoeger and Kenaga (1972) are greatly overestimated.

7.3.1.2 ICPBR (EPPO) Proposal for Seed Treatment or Soil-Applied Systemic Compounds

The main route of exposure of bees to residues from systemic compounds (such as those applied as a seed treatment or soil application) is through the translocation of the compound into nectar and pollen. Data on measured residue levels in different plant parts have been compiled and analyzed by Alix et al. (2009a). Residue levels in plant parts were measured after treatment with systemic insecticides for the purpose of developing Tier 1 exposure assessments. The compiled residue database considered residue values as close as possible to flowering. Based on their analysis, a default maximum residue value of 1 mg a.i./kg plant matrix has been proposed as a peak value for the screening-level exposure estimate for systemic compounds used as seed treatments or applied to soil (Alix et al. 2009a; Alix and Lewis 2010). In the event the Tier 1 risk assessment based on this worst-case estimate indicates a potential risk, actual measured residues from higher tier studies can be used for a refined risk assessment. If there is a need to transform the Tier 1 predicted concentrations in pollen and nectar into predicted doses for honey bees, it is recommended to follow the proposals as outlined by International Commission for Plant–Bee Relationships (ICPBR) (Alix et al. 2009a), which uses pollen and nectar consumption rates by different castes of honey bees (Rortais et al. 2005). The published consumption rates are provided later in this chapter (see Section 7.7).

7.4 PHYSICAL AND CHEMICAL PROPERTIES OF PESTICIDE ACTIVE INGREDIENTS WHICH AFFECT EXPOSURE

The physicochemical properties of the pesticide active ingredient determine its fate in soil and in hive matrices which can affect the exposure of the various life stages of both *Apis* and non-*Apis* species to these chemicals.

1. Fate in soil—systemic products
   Systemic products applied to soil can be taken up by the plant and translocated into plant foliage, floral nectar, and pollen. Persistent systemic products that remain in the soil for over an year could potentially be translocated into the nectar and pollen of rotational crops planted in succeeding years. The dissipation time (DT50) is used to characterize the persistence of pesticides in soil.

Physicochemical properties of the pesticide active ingredient that can affect persistence in soil include water solubility, the octanol–water partition coefficient (K_{ow}), dissociation constant (K_a), the soil adsorption coefficient (K_d), and the organic carbon partition coefficient (K_{oc}). Pesticides with high water solubility and low K_{oc} values (e.g., <50) have a higher potential for mobility, do not strongly adsorb to soil particles and can be prone to leaching depending on soil conditions, weather, and persistence of the compound. The log of the K_{ow} (log K_{ow} or log P) is the measure of a chemical's propensity to bioaccumulate. Pesticides with a high log P (e.g., >3) usually have low water solubility and are not highly mobile in soil. The log of the dissociation constant (pK_a) is a measure of the extent to which a substance ionizes in equilibrium with water. The p*K*a of a pesticide indicates the ratio of the forms (ionized or undissociated) in which the chemical will exist in environments of various pH values, and the extent of its potential involvement in ion-exchange binding processes in soils or sediments. The form of a pesticide (anion or cation) can influence its mobility and hence persistence in soil. Soil type and meteorology (amount of rainfall, temperature) can also influence the persistence of a pesticide in soil.

Specific criteria to classify compounds as being persistent in soil have been identified by the European Union (EEC 2006) and other regulatory agencies to trigger the requirement of rotational crop residue studies (used to inform human health risk assessment). It has been proposed that similar criteria be used to require assessment for the risk of residues in pollen and nectar for succeeding crops (Alix and Lewis 2010).

2. Fate in hive matrices—systemic and nonsystemic products
   Physicochemical properties including water solubility, log P, and the pK_a can influence the fate of the active ingredient in the hive. Compounds with a high log P that are hydrophobic (i.e., tending to be insoluble in water) may accumulate in wax, pollen, and bee bread, which contain lipids. Compounds with a high solubility in water (hydrophilic) can partition to nectar and honey which contain water. If the compound dissociates, the dissociation constant may be used to indicate the fate in acidic matrices such as honey.

7.5 INFORMATION NEEDED TO DEVELOP REFINED PREDICTIVE EXPOSURE MODELS

As stated earlier, there are no defined predictive models currently used for estimating the exposure levels in bees or bee matrices for use in a screening-level ecological risk assessment. The procedures described here that have been previously used by the European Union and Canada for example, and employ values for potential exposure, have been effective in screening-out compounds that have low potential risk to adult worker bees from foliar-applied products. However, for crop protection products where potential risk cannot be excluded based on current Tier 1 screening analysis, the current method to refine assessments consists of higher tier effects or exposure assessment studies (e.g., EPA Tier 2 foliar residue study, EPPO tunnel test).

Optimally, there should be methods to predict residue levels in relevant matrices (e.g., bees, pollen, nectar). These predicted exposure concentrations could then be used to compare with laboratory toxicity data, such as acute contact LD50 values for adult bees, and acute and chronic dietary toxicity data for adult bees and larvae to estimate risk to both foraging bees and other castes and life-stages in the hive, including larvae.

7.6 PREDICTED CONTACT EXPOSURE FOR FOLIAR-APPLIED PRODUCTS

For foliar-applied products, the prediction of residues on foraging bees due to contact exposure (i.e., direct spray on foraging bees or bees contacting residues post spray) can be estimated. The US EPA has proposed using predicted concentrations in insects based on estimates in their TREX wildlife exposure model. However, as noted earlier, there are some inherent uncertainties with using this approach. In this approach, values from TREX Version 1.4.1, which relies on residue estimations developed by Hoeger and Kenaga (1972) for plants, fruits, and seeds, would be used as surrogate data to estimate contact exposure for insects. However, actual field residue data are available for honey bees (Koch and Weißer 1997) and a variety of flying, soil-dwelling, and leaf-dwelling arthropods (Schabacker et al. 2005) that can be used for estimating contact exposure to bees. In a multiyear study by Koch and Weißer (1997), the fluorescent tracer sodium fluorescein was applied to flowering apple orchards or flowering *Phacelia* fields while honey bees were actively foraging, to determine contact doses in individual honey bees. After applications of 20 g sodium fluorescein/ha, doses in honey bees ranged from 1.62 to 20.84 ng/bee, and 6.34 to 35.77 ng/bee for honey bees foraging in apples and *Phacelia*, respectively. If the maximum detected residue in this study (35.77 ng/bee after an application of 20 g/ha) was used as a point estimate for a screening-level exposure assessment, a predicted environmental dose due to contact exposure (PEDc) in adult honey bees after an application of 1 kg/ha (1000 g/ha) would be 1789 ng/bee or 1.79 μg/bee. The assumption here is that there will be a linear relationship between application rate and contact dose of foraging bees, which is an area of uncertainty.

In the report by Schabacker et al. (2005), maximum residues in flying, ground-dwelling, and foliage-dwelling arthropods from a number of field trials were compiled and residue unit doses (RUDs) were calculated. The mean and 90th percentile RUDs in mg/kg after application of pesticides at a rate of 1 kg a.s./ha were calculated (Table 7.1).

When residue data for flying insects are used to develop a screening-level point estimate for contact exposure of foraging bees, a 90th percentile PEDc after an application of 1 kg a.i./ha is calculated to be 0.84 μg/bee. This is derived by multiplying the 90th percentile concentration in flying insects (6.6 mg/kg) by the weight of an adult foraging honey bee (128 mg) (Mayer and Johansen 1990). This point estimate (0.84 μg/bee) is close to the exposure value calculated using the data of Koch and Weißer (1.79 μg/bee), and is consistent with the data developed by Atkins et al. (1981), where a dose of 1 μg/bee represents an application rate of 1 lb a.i./acre. Therefore, according to the Atkins method, an application of 1 kg a.i./ha is equivalent to an exposure value of 0.89 μg/bee. Based on this information, a worst-case estimate PEDc to honey bees after an application of 1 kg a.i./ha is 1.79 μg/bee.

TABLE 7.1
Predicted Concentrations (in mg/kg) After Foliar Application of 1 kg/ha

Arthropod Classification	Mean Predicted Concentration (mg/kg)	90th Percentile Predicted Concentration (mg/kg)
Flying insects	1.4	6.6
Ground dwellers (orchard/vines, grasslands, late growth stages of leafy crops and cereals (insecticides and fungicides))	3.6	9.8
Ground dwellers (orchard/vines (herbicides), early growth stages of leafy crops and cereals (all pesticides)	6.7	15.6
Leaf dwellers	9.5	47.8

Source: Data from Schabacker et al. (2005).

TABLE 7.2

Comparison of Hazard Quotient (HQ), Toxicity/Exposure Ratios (TER) and Risk Quotients (RQ) Assuming a Predicted Contact Exposure Dose (PEDc) of 1.79 μg a.i./bee After an Application of 1 kg a.i./ha

Use Rate (kg/ha)	PEDc (μg/bee)	Contact LD50 (μg/bee)	HQ	TER	RQ
0.1	0.179	1	100	5.6	0.18
0.1	0.179	2	50	11	0.09
0.1	0.179	5	20	28	0.036
0.1	0.179	20	5	112	0.009

To evaluate the sensitivity of the proposed point estimate of exposure for honey bees, a generic data set (LD50 values) can be used to calculate HQs and toxicity/exposure ratio (TERs),[2] along with the value of 1.79 μg/bee after an application of 1 kg a.i./ha. Using a generic data set with an application rate of 100 g a.i./ha, the corresponding HQ, TER, and RQ values were calculated (Table 7.2).

According to Annex VI of the EU Uniform Principles, a TER of ≥10, designed to cover potential variabilities (such as interspecies), typically indicates acceptable risk for terrestrial organisms, and has been recommended as an appropriate assessment factor for oral exposure to systemic insecticides by ICPBR (Alix et al. 2009a, 2009b; Alix and Lewis 2010). US EPA on the other hand uses a level of concern (LOC) RQ of 0.1 for non-listed threatened or endangered aquatic or avian species. Based on this analysis, the screening-level risk assessment based on a PEDc of 0.179 μg/bee is in line with the current European Union screening HQ of 50.

Although the published field trial data (Koch and Weißer 1997) for residues on honey bees are most appropriate for developing exposure estimates for honey bees, it might be more appropriate to use the data for leaf-dwelling and soil-dwelling arthropods developed by Schabacker et al. (2005) to address exposure to leaf-dwelling and soil-nesting non-*Apis* bee species, respectively. Therefore, for the initial conservative point estimate of contact exposure, the 90th percentile predicted concentration for leaf-dwelling arthropods (47.8 mg/kg) can be used to develop a PEDc for leaf-dwelling species, while the 90th percentile predicted concentration for soil-dwelling arthropods (15.6 mg/kg) can be used to develop a PEDc for soil-nesting species. However, in order to complete this analysis and develop recommended PEDc values for leaf-dwelling and soil-nesting non-*Apis* bees, focal species need to be identified. For leaf-dwelling species, the leafcutter bee (e.g., *Megachile rotundata*) is recommended as a surface dwelling non-*Apis* reference species, while the bumble bee (*Bombus* spp.), which typically nests on or underground, or the mason bee (*Osmia* spp.), which collects mud for nest construction, is recommended for soil-nesting (gregarious) focal species. Ideally, ground-nesting solitary bees, such as sweat bees (e.g., *Halictus* or *Lasioglossum* spp.), squash bees (*Peponapis* or *Xenoglossa* spp.), or alkali bees (e.g., *Nomia melanderi*) could also be considered a representative soil-nesting species, for these insects dig nests underground. However, at least in North America, only *Nomia melanderi* is currently managed successfully on a larger scale. With the identification of focal species, the typical body weights of the species can be used to convert predicted exposure concentrations in mg/kg to PEDc values in μg/bee for direct comparison to laboratory toxicity data.

Prior to adopting this proposed methodology into a formal regulatory assessment paradigm for bees, the method should be used to calculate toxicity/exposure ratios for some representative compounds to ensure

[2] TER = LD50 in μg a.i./bee/PEDc in μg a.i./bee; and, Risk Quotients (RQ) = PEDc/LD50.

that the exposure assessment methodology is sensitive enough to predict an acute risk to compounds that are highly toxic to non-*Apis* bees (e.g., pyrethroid insecticides), while not predicting a high risk for compounds that are known to have low inherent toxicity and present a low risk to non-*Apis* bees. Such an exercise would provide some feedback that the proposed methodology would not potentially be inconsistent with protection goals.

7.7 PREDICTED DIETARY EXPOSURE FOR FOLIAR-APPLIED PRODUCTS

For assessing acute or chronic dietary risk to adults or larvae, predicted concentrations in relevant food items (e.g., pollen, nectar, bee bread, honey, and larval jelly) should be used as the dietary exposure estimate. Currently, models to predict residues in these items from foliar-applied pesticide products do not exist. Although the results from survey-style analysis indicate that agricultural pesticides are entering managed honey bee colonies through contaminated pollen (Chauzat et al. 2010; Mullin et al. 2010), there are limited published data from controlled studies that relate foliar application rates to measured pesticide levels in pollen and nectar or in any processed hive food.

In a study by Choudhary and Sharma (2008), residues of three foliar-applied pesticides were determined in nectar and pollen following applications to flowering mustard. Pesticides evaluated in this two-year study were endosulfan, lambda-cyhalothrin, and spiromesifen. Mean measured residues in pollen and nectar, and predicted concentrations after application of 1 kg a.i./ha were calculated (Table 7.3).

In a study by Wallner (2009), residues of the fungicides boscalid and prothioconazole were determined in pollen and nectar samples from foraging bees following applications to oilseed rape (canola). Mean measured residues in pollen and nectar and predicted concentrations after application of 1 kg a.i./ha are summarized in Table 7.4.

Finally, in a study by Dinter et al. (2009), concentrations of the insecticide chlorantraniliprole in pollen and nectar collected from foraging bees following applications to *Phacelia* in a semi-field study were determined. The maximum concentrations in pollen and nectar, 1 day after treatment were calculated (Table 7.5).

TABLE 7.3
Day 0 Measured Concentrations of Three Foliar Applied Pesticides in Pollen and Nectar After Application to Flowering Mustard

Compound	Application Rate (g a.i./ha)	Mean Measured Residues Nectar[a] (mg/kg)	Mean Measured Residues Pollen[a] (mg/kg)	Mean Predicted Nectar Residues (mg/kg) After Application of 1 kg/ha	Mean Predicted Pollen Residues (mg/kg) After Application of 1 kg/ha
Endosulfan	525	1.725 ± 0.031	2.126 ± 0.088	3.15	3.99
		1.583 ± 0.006	2.068 ± 0.048		
Lambda-cyhalothrin	75	0.858 ± 0.038	1.607 ± 0.004	10.6	21.2
		0.728 ± 0.022	1.577 ± 0.018		
Spiromesifen	225	1.541 ± 0.078	2.003 ± 0.040	6.54	8.45
		1.401 ± 0.016	1.799 ± 0.033		

Source: Data from Choudhary and Sharma (2008).
[a]Mean measured residues from two successive application and sampling years.

TABLE 7.4
Day 0 Measured Concentrations of Two Foliar Applied Fungicides in Pollen and Nectar Collected from Honey Bees After Application to Flowering Oilseed Rape

Compound	Application Rate (g a.i./ha)	Mean Measured Residues Nectar (mg/kg)	Mean Measured Residues Pollen (mg/kg)	Mean Predicted Nectar Residues (mg/kg) After Application of 1 kg/ha	Mean Predicted Pollen Residues (mg/kg) After Application of 1 kg/ha
Boscalid	500	1.43	26.2[a]	2.86	52.4
Prothioconazole	250	0.69	nd (LOQ = 0.001)	2.76	

Source: Data from Wallner (2009).
[a]Concentrations 1 day after treatment, which were higher than day 0 values.

It is difficult to draw any firm conclusions based on these limited data. For instance, there is not a linear relationship between application rate and measured concentration in pollen and nectar across the different compounds. Therefore, the predicted concentrations after applications of 1 kg/ha (i.e., PEDc's) may be greatly exaggerated for some compounds. It is likely that the variation in residue levels seen between these two studies (Dinter et al. 2009 and Wallner 2009) is a result of different factors such as sampling, extraction methods, fate properties of the different compounds, or product formulation.

Although limited published data are available for maximum residue levels in nectar and pollen after controlled applications of foliar products, there is likely to be a significant amount of data that have been developed by pesticide manufacturers for individual products. Therefore, the participants of the workshop proposed that nectar and pollen residue data from semi-field exposure studies conducted according to EPPO guidelines be compiled and analyzed. These data should represent maximum residues in bee food items in a bee-attractive crop, and developing models around these data would likely provide realistic, worst-case predicted residues for a screening-level risk assessment.

Once these data are compiled, a conservative estimate for residues on or in pollen and nectar (e.g., 90th percentile RUDs) can be used to calculate TER or RQ values. These screening-level predicted values would represent a conservative estimate of dietary exposure for honey bees from foliar application of pesticide

TABLE 7.5
Day 1 Measured Concentrations of Chlorantraniliprole in Pollen and Nectar Collected from Honey Bees After Application to Flowering *Phacelia*

Compound	Application Rate (g a.i./ha)	Maximum Measured Residues Nectar (mg/kg)	Maximum Measured Residues Pollen (mg/kg)	Maximum Predicted Nectar Residues (mg/kg) After Application of 1 kg/ha	Maximum Predicted Pollen Residues (mg/kg) After Application of 1 kg/ha
Chlorantraniliprole	60	0.033	2.60	0.55	43.3

products. For a dietary risk assessment, the predicted concentration of residues in food items can be directly compared with the results from dietary toxicity studies with adult bees and bee larvae, if the results from the studies are expressed as exposure concentrations (i.e., LC50, NOEC). However, if the toxicity results are expressed as a dose (i.e., LD50 in μg/bee), the predicted dose can be calculated based on predicted concentrations on food items and consumption rates by different castes of bees. Honey bee consumption data, based on complete life-stages, have been reported by Rortais et al. (2005), and are summarized as follows.

Nectar foragers: 224–898.8 mg sugar
Pollen foragers: 72.8–109.2 mg sugar
Nurse bees: 65 mg pollen
Worker larvae: 59.4 mg sugar + 5.4 mg pollen
Drone larvae: 98.2 mg sugar

The following daily consumption rates for the different honey bee castes were calculated by Thompson (2007):

Nectar foragers: 32–128.4 mg sugar/bee/day
Pollen foragers: 10.4–15.6 mg sugar/bee/day
Nurse bees: 6.5 mg pollen/bee/day
Worker larvae: 11.9 mg sugar + 1.1 mg pollen/bee/day
Drone larvae: 15.1 mg sugar/bee/day

For dietary exposure estimates, it will be important to choose the appropriate consumption rate with respect to life stage, that is, the daily consumption rate should be compared with acute oral toxicity data to estimate acute risks, while life-stage consumption data should be compared with chronic toxicity data to estimate chronic risk.

7.8 PREDICTED EXPOSURE FOR SOIL AND SEED TREATMENT SYSTEMIC COMPOUNDS

For soil-applied or seed treatment systemic products, the current ICPBR proposal recommends using a default maximum exposure value of **1 mg/kg for pollen and nectar**, which is based on the analysis of existing residue data (Alix et al. 2009a). Currently, the number of standardized exposure studies evaluating residues in pollen and nectar for systemic pesticides is limited to a few compounds for the same class of chemistry (i.e., neonicotinoids) (Alix et al. 2009b). Therefore, there may not be enough data to develop a predictive exposure model applicable to all soil-applied or seed treatment systemic compounds. In the case of systemic compounds, it appears that residues in pollen and nectar are not only influenced by the physical and chemical properties of the compound (e.g., K_{oc}, soil DT50, K_d, pollen and nectar uptake and dissipation), but also by soil properties, crop, weather, and application timing versus the time of bloom. Therefore, as pollen and nectar residue data for other classes of systemic compounds are developed, the additional variables should be considered. As more residue data are developed for systemic compounds (both neonicotinic and other classes), the concept of developing a predictive screening-level exposure model should be explored further. In the interim, the default value of 1 mg/kg is recommended as the point estimate for exposure in Tier 1 risk assessment for dietary exposure to systemic compounds, as it represents a current worst-case estimate of residues in matrices that are consumed by bees (i.e., pollen and nectar). However, as more data are developed for systemic compounds, the value of 1 mg/kg should be re-evaluated to ensure that it is sufficiently conservative for use in a screening-level risk assessment.

7.9 PREDICTED EXPOSURE FOR TREE-INJECTED COMPOUNDS

Certain insecticides can be directly injected into tree trunks for control of wood boring insects. The chemical enters the xylem and is systemically transported to all parts of the tree including nectar (if produced) and pollen, and potentially propolis, which is not consumed, but is used by bees in the construction and maintenance of nests and hives. There is a scarcity of data on residues of pesticides resulting from tree injections. Until more data are developed or collected, it is unclear if the residue value of 1 mg/kg, as proposed by ICPBR for soil and seed treatments, is appropriate as a maximum default residue for a screening-level risk assessment for tree injection.

7.10 MEASURING PESTICIDES IN MATRICES RELEVANT FOR ASSESSING EXPOSURE TO BEES

When quantification of pesticide residues in bees or bee food is required to refine an exposure assessment, it must be determined whether the goal is to assess exposure of adult forager bees or other members of the hive (queen, nurse bees, drones, and larvae). To determine exposure of foragers from foliar applications, analysis of bees collected from the sprayed crop can be conducted. For exposure of forager bees from oral sources, samples of nectar and pollen can be collected by hand from flowers or from foraging bees on the crop. Bees may be sampled by drawing nectar from the honey stomach and pollen can be removed from the pollen baskets. Whether it is more time- and cost-effective to use bees to collect samples or to do it by hand sampling is dependent on the type of crop flower being sampled.

Where collection of nectar from the target crop is possible by hand, this can be done by inserting a microcapillary tube or pipette into the nectary and extracting the nectar (Figure 7.3). Collection of pollen by hand can be done by shaking flowers or using scissors to remove anthers followed by separation of the pollen from the anthers either in the field (Figure 7.4) or after transportation to a laboratory. Flowers from several crops have very little, if any, nectar and pollen, making hand collection impractical. In these instances, bees can be used to collect the samples. Obtaining nectar samples using bees can be done by collecting the bees that are actively foraging on flowers in the crop of interest (such as by vacuuming, which, in certain cases may be impractical). Another way to sample bees is by collecting them at the hive entrance. In either scenario, verification of exposure from the crop of interest should be done by identifying pollen brought back to the hive or by confining the bees during the exposure portion of the study using a semi-field study design. To obtain the nectar sample from honey bees, the honey stomach can be dissected from the bee and the contents drained into a vial or be pierced with a syringe or micropipette and the nectar extracted. Pollen can be obtained from bees collected from flowers or at the hive entrance by removing the pollen from the pollen baskets. Pollen samples can also be collected in pollen traps attached to the hive entrance. If either pollen or nectar cannot be efficiently collected in large enough quantities for residue analysis, whole flower samples could also be analyzed for possible use as a surrogate (pending further collection and analysis of these data).

Samples from the hive can be drawn for potential exposure to residues in stored pollen, nectar, and larval jelly. Stored pollen can be sampled by identifying frames where fresh pollen is being stored and removing this pollen with a spatula from individual cells. Adding an empty comb can ensure that the pollen and nectar is freshly collected. Nectar can be sampled by identifying the frame where fresh nectar is being stored, removing the frame from the hive, and shaking the frame into a large pan to release the nectar. The released nectar can then be transferred to a vial using a pipette, or pouring if the volume allows. Alternatively, fresh nectar can be identified and extracted from individual cells using a syringe or pipette and transferred to a vial. Larval jelly can be identified on the frames and collected either by extracting it from the cells with a capillary tube or pipette, or by removing the larvae and scooping out the jelly with a spatula and transferring it to a vial.

All samples collected in the field should be kept on ice until received by the analytical laboratory. At the laboratory, samples should be stored frozen (−20°C) and protected from light until analysis. Experience shows that plastic storage containers should be used with caution because some pesticides can sorb to plastic. Standardized procedures for sampling, including appropriate storage and transport, should be established in order to avoid contamination, and provide adequate sample size. Specific, statistically valid plans for sample size and number also should be established in the study protocol. Dedicated coolers, chain of custody, records of transport and storage conditions, and other appropriate good laboratory practice procedures should be used and documented to ensure sample integrity. The quantity of samples needed for analysis of pesticide residues should be determined prior to sampling and might vary based on the limits of detection and limits of quantification for each pesticide in the individual matrices. Use of spiked samples, to accompany samples collected from the field, can be used to ensure sample integrity (as well as sample stability). Analytical methods also need to be properly validated to ensure that extraction methods are adequate and the residues of interest are accurately identified.

At the present time, it is recommended that collection of nectar and pollen directly from the flowers, or collecting and removing pollen and nectar from foraging bees would be the most conservative and most relevant estimates of exposure for bees outside the hive. For larvae, nurse bees, drones, and the queen in the hive, sampling freshly deposited nectar and pollen from the combs would be the most conservative dietary exposure estimate; considering additional processing of these materials by bees may result in lower concentrations in other hive food sources. To further refine these estimates, data on the comparative residue levels in flowers, nectar, pollen, and hive products (such as stored pollen, nectar, honey, larval jelly, and bee bread) can to be generated to determine worst-case oral exposure estimates for either foraging bees or hive bees.

7.11 HIGHER TIER STUDIES TO ASSESS EXPOSURE OF PESTICIDES TO BEES

7.11.1 Higher Tier Study to Evaluate Contact Exposure to Honey Bees

In the United States, if a compound is classified as toxic to honey bees by contact exposure (i.e., LD50 <11 μg/bee), a Tier 2 contact residue study is required. In this study, a bee attractive plant (typically alfalfa) is sprayed with formulated product at the maximum application rate. Groups of worker bees are caged over the treated crop at various time points after application (typically, 0, 4, 8, and 24 hours), to evaluate the bioavailability and persistence of pesticide residue. These data are used to determine the length of time between application and when bees can be safely exposed to a treated crop. From this test, a residual toxicity time (RT) is established indicating where the pesticide residue is lethal to 25% of the test population, referred to as the RT_{25}.

7.11.2 Higher Tier Exposure Studies Using Honey Bee Colonies

Since it is not economical to conduct exposure studies in every crop, realistic worst-case model crops should be used for assessing exposure of bees under field-relevant use conditions in semi-field and field trials. Choosing a realistic worst-case model crop should include the following considerations:

- attractive to bees
- provides both nectar and pollen
- provides sufficient flower density and sufficient duration of flowering

EPPO PP 1/170 (OEPP/EPPO 2001) proposes *Phacelia*, oilseed rape (canola), and mustard. Buckwheat (*Fagopyrum esculentum*) may also be used. Application parameters (i.e., rate, interval, formulation) used in any higher tier study should be those that are expected to produce the greatest potential exposure that is prescribed by the product label being assessed.

For a worst-case assessment of exposure, semi-field, or tunnel studies can be conducted. In these studies, colonies are placed within a tent or mesh tunnel (Figure 7.5) and exposed to the treated crop during or immediately after application. Using a highly bee-attractive crop would simulate a worst-case exposure to residues in pollen and nectar. Because of the controlled nature of semi-field studies for foliar-applied products, the location of the study is not as important as it is for a field study. Therefore, data from semi-field studies may be useful in risk assessments beyond the country in which it was performed, assuming that maximum application rates are assessed. However, in some instances, soil type and weather can influence nectar production. See Chapter 8 for additional discussion on effects measurements through semi-field studies.

7.11.3 Studies to Evaluate Exposure from Seed Treatments and Soil Applications of Systemic Compounds

Regarding seed treatments and soil applications with systemic compounds, specific semi-field or field studies can be designed to measure residues in nectar and pollen in order to refine a screening-level risk assessment for systemic compounds. If the purpose of the study is to measure residue data only, the actual crop of interest should be used. If higher tier studies are conducted with a foliar-applied compound and the aim is to concurrently assess residues and potential effects, preferably a crop with the highest application rate and highest attractiveness to bees should be used. If such an effort is undertaken with a systemic compound, then the target crop per se, should be considered first as the test crop, utilizing the maximum application rate for that use scenario. If the target crop is not feasible for conduct of either semi-field or field studies, the use of a surrogate crop is recommended but must be scientifically justified (e.g., supported by plant metabolism data, measured residue levels in nectar and pollen). Data on the uptake and decline of pesticide residues in pollen and nectar after systemic pesticide applications to the test crop should be evaluated prior to initiating field testing with honey bees. (Certain residue chemistry information, typically used for human health assessments may be useful in these cases.) In reviews of reports for two compounds submitted to the State of California (Bireley 2008; Omer 2008; Papathakis 2008; Bireley 2009), leaf residues in treated perennial shrubs and trees treated with imidacloprid were initially low. Residue levels were below the limit of detection for several weeks after application, but increased to levels above 10 ppm over the next several months in some instances, illustrating that expression of residues in pollen and nectar may follow a curve dependent upon numerous variables. Regardless of the timing of application, it is important that the analysis phase of field studies include sampling of the most important bee-relevant matrices (i.e., pollen, nectar) and characterize the level of residues during plant bloom. Consideration may also need to be given to characterizing the persistence of residues over time, that is, accumulation from one year to the next (depending upon environmental fate properties).

7.11.4 Field Treatments for Honey Bee Colonies, Spiked Sucrose, and Spiked Pollen

For evaluating the distribution of a pesticide throughout a hive, sucrose, pollen, or protein (pollen substitute) supplements spiked with the proposed test compound (e.g., pesticide active ingredient) should be considered as a potential method of exposure in semi-field and field tests. Spiked pollen, protein (pollen substitute), or sucrose can also be utilized in laboratory and field tests to ensure and accurately quantify exposure to the hive.

When spiked sucrose solution is used as the route of exposure for three or more days, a protein supplement is recommended to ensure that effects observed are due to treatments and not insufficient nutrition. If exposure to the compound is expected to be through pollen collection and feeding, spiked protein can be fed to the test bees. An alternative is to collect and homogenize pollen from a pollen trap, spike the pollen samples with the compound being evaluated, and press the spiked pollen into empty combs. However, for some lipophilic compounds, pressing the pollen into a comb could end up extracting the compound if it partitions to the wax. An alternative would be to prepare pollen cake on which the bees can forage. Also, certain pollens should be avoided because they may contain contaminants such as flavonoids that are toxic to bees. In addition, the pollen used should be pesticide free. Finally, the protein content of some pollen, and differences in preference may reduce feeding. In some cases, researchers have used spiked protein supplements. One recommendation is to provide a 500 g protein supplement to the colony each week during a brood cycle (e.g., 21 days). Palatability or toxicity of the test compound may result in the need to alter the size of the supplement. A pollen trap may be used to significantly reduce the quantity of pollen that foraging bees bring into the hive (field studies), thus, encouraging consumption of the spiked protein supplement. A local sucrose feeder may also be used to reduce long distance foraging.

An advantage of using spiked protein supplements is that treated crops are not required and the field size where hives are placed is not relevant as long as there is adequate forage for the number of hives. In these studies, pollen traps can be used to reduce any extraneous pollen from entering the hive. Spiked protein supplements ensure that the hives are exposed to the test substance. Since the protein supplement is not specific to a particular crop, exposure is applicable to any plant where pollen is a food source.

As discussed earlier, appropriate steps should be taken to validate the proper handling of residue samples during collection, shipping, and processing. Validated results indicate that the field handling is appropriate and that the results from the field samples accurately represent actual field residues. See Chapter 8 for more discussion on considerations and conduct of field studies for measuring potential effects.

7.12 HEALTH OF HONEY BEE COLONIES CAN INFLUENCE EXPOSURE

In typically managed colonies, pests and pathogens are present in amounts not necessarily found in the simulated scenarios of laboratory-based or field studies. Honey bee pathogens such as *Nosema* (Fries et al. 2006; Chauzat et al. 2007) and various bee viruses (Chen et al. 2007, 2011; Ribière et al. 2007) are commonly present in managed honey bee colonies. When colonies are subjected to changes caused by pesticide exposure, the pathogen loads can change in honey bees (Alaux et al. 2010; Pettis et al. 2010), and in turn, influence biological and behavioral traits of honey bees. The behavior of diseased honey bees can be modified. For example, diseased honey bees may forage earlier in their life cycle (Ribière et al. 2008), or may be less vigorous foragers, leading to less overall foraging activity and consequently a lower pesticide exposure. Colonies used for testing should be healthy colonies, with minimal levels of pests and pathogens, as these can influence foraging behavior.

7.13 HIGHER TIER STUDIES WITH NON-*APIS* BEE SPECIES

If a screening-level risk assessment does not indicate a presumption of low risk to non-*Apis* bee species, exposure can be evaluated using higher tier studies. In many cases, exposure assessments for honey bee workers may address potential exposure for non-*Apis* bees. However, in some cases, non-*Apis* bees face unique exposure pathways not addressed by exposure assessments for honey bees (see Section 7.2) and consequently, exposure estimates for non-*Apis* bees should be pursued through higher tier studies. Higher tier studies may be pursued solely for exposure information but given their complexity and cost, they likely will be undertaken for information on both exposure and effects. A brief discussion regarding alfalfa leafcutter bees and mason bees provides an example.

7.13.1 Alfalfa Leafcutter Bees: Contamination of Nesting Materials

Alfalfa leafcutter bees (*M. rotundata*) and other species of *Megachile* and *Osmia* will collect leaf pieces from a variety of plants to either wrap or build partitions between their brood cells. Common examples of plants used by these non-*Apis* species include species such as rose (*Rosa* spp.), snow berry (*Symphoricarpos albus*), bindweed (*Convolvulus arvensis*), buckwheat (*F. esculentum*), honeysuckle (*Lonicera* spp.), wild grape (*Vitis vinifera*), or wild senna (*Senna hebecarpa*) (Mader et al. 2010). Alfalfa leafcutter bees deployed for alfalfa pollination also use materials collected from the fields in which they are pollinating or foraging. Whether the bees use the target crop or surrounding non-cropped area, there is a potential for exposure from direct application to the crop or drift to adjacent plants.

In the case of the alfalfa leafcutter bee used for alfalfa pollination, it is critical to understand the level of exposure from contaminated leaf pieces and, ultimately, the toxicity of this exposure. See also Chapter 8 on Laboratory Testing Approaches for a discussion on laboratory-based effects studies using treated foliage and see also Chapter 9 for a discussion on considerations with respect to effects information from either semi-field or field studies. One possible approach would be to use a modification of US EPA's guidelines for assessing the toxicity of pesticides on foliage, where alfalfa is sprayed and then brought into a laboratory at various post-application time points, and allowing bees to forage on the foliage. Another approach would be to use a semi-field or field study design as described in the section, Semi-Field Studies.

7.13.1.1 Semi-Field Studies

The following steps relate to assessing potential levels of exposure from contaminated mud, such as with mason bees (e.g., *Osmia cornifrons, Osmia cornuta, Osmia lignaria,* or *Osmia rufa*) that collect mud to build partitions between their brood cells.

1. Plant enclosed shelter (6 m by 2.5 m or larger) with *Phacelia* (*Phacelia tanacetifolia*), sweet clover (*Melilotus* spp.), or other favored forage plant. (Note: In this case, it is also possible to consider the use of artificial nectar or pollen feeder.)
2. Deploy incubated *Osmia* spp. cocoons as loose cells or natal tubes in the enclosure at least 15 days prior to pesticide application (see Bosch and Kemp 2001; Mader et al. 2010 for management advice).
   - Provided the bees have undergone appropriate diapause (generally 100–200 days at 1.7–4.4°C.), bees will begin emerging 5–10 days after initiating incubation at temperatures of at least 21°C. More rapid emergence can be stimulated by incubating cocoons at 29°C, until all bees have emerged.
   - Note that male emergence precedes female emergence, often by several days, and nesting typically will not begin until 1–2 days after mating (which usually occurs on the day of female emergence).
3. Provide a source of wet mud with high clay content in a 1 m wide shallow pan or tray. Water this tray on a daily basis from below in order not to wash pesticide from surface. Ensure that the moisture level is not excessive leading to drowning.
4. Use observation tunnel nests for the bees (i.e., boards with grooves routered into one side (8 mm for *O. cornuta*, 7.5 mm for *O. lignaria*, 6 mm for *O. cornifrons*), covered by a layer of clear acetate and sandwiched with a second piece of wood to create a dark tunnel that can be opened to allow for monitoring.
5. Open observation tunnel nest and note completed cells.

6. Temporarily close tunnel nests and apply the test material to the mud that is supplied for nest construction at the levels of interest.
7. Note the new cells created.
8. Open nests and remove the mud partitions that divide the brood cells in order to measure:
   - pesticide residue in pollen–nectar stores (pollen ball), and
   - pesticide residue in mud partitions.
9. Remove exposed cells at 15, 20, and 25+ days to assess the movement of the pesticide into bee bread, larval mortality, etc. Depending on the species, full development from egg hatching to adult emergence is completed between 60 and 125 days at 28–17°C. Higher temperatures will result in faster development, but should not exceed 28°C.

7.13.1.2 Field or Semi-Field Studies

1. Deploy leafcutter bees in closable or sealable shelters in an alfalfa field 10 days prior to pesticide application (see Chapter 8 for further discussion on proper incubation timing). Observation tunnel-nests for the bees can be constructed to facilitate monitoring by boring a 6.0 mm ($\frac{1}{4}$ inch) holes or grooves into one side of a wood plank, and covering the holes or grooves with clear acetate. The acetate on such nests should be covered with a removable opaque cover to increase nest attractiveness. The opaque cover can be removed temporarily in order to make notations on the acetate. See also Abbott et al. (2008).
2. During the active nesting period, close the shelter at night to prevent foraging in the green house, cage, or field until the following day. With the nest shelter closed, carefully enter it and note the constructed cells (pre-treatment) in the observation tunnels. With the shelter closed, pesticides can be applied to the field adjacent (at least 200 m radius) around the shelter.
3. After an appropriate time has elapsed (depending upon study goals and active ingredient being used), open the shelter to allow bees to forage, build, and provision the cells.
4. Note new cells created in the observation nests.
5. Newly constructed cells can be monitored for development: eggs will hatch in about 15 days at 15.6°C down to 1–2 days at 35°C. Prior to egg hatching, cells may also be dissected to separate leaf pieces from cell contents (bee bread and egg) to assess
   - pesticide residues in the pollen–nectar mixture (pollen ball), and
   - pesticide residues on leaf pieces.
6. At 15, 20, and 25+ days, cells can be sampled for the presence of pesticide residues in the pollen ball, monitored for larval mortality, and other parameters. Full development from egg hatching to adult emergence takes 35 days at 15.6°C, but only 11 days at 35°C.

7.13.2 Using Non-*Apis* Bees to Measure Pesticide Contamination of Pollen and Nectar

Using the techniques described here, pollen balls may be removed from the cells of solitary tunnel-nesting bees (e.g., *Osmia* spp. or *M. rotundata*) placed in shelters deployed in fields or orchards treated with pesticides, including systemic pesticides applied as drench or trunk injection. If sufficient forage is available, then these managed non-*Apis* solitary bees typically forage in the area immediately surrounding their nest (40–60 m), thereby helping to ensure that the study organism is coming in contact with the treated plants in well-designed field studies. These bees can also be used readily in semi-field studies as they forage readily in enclosures when provided with adequate forage and nesting material (Bohart and Pedersen 1963; Abel et al. 2003).

Female foragers of *Osmia* or *Megachile* spp. may also be netted in front of their nest shelters. If they are returning with pollen, it may be gently scraped or brushed from their abdomens or removed by holding the bee with entomological forceps and applying a vibrating tuning fork to the forceps. Note that, unlike honey bees, members of the family Megachilidae, which includes both *Osmia* and *Megachile*, carry pollen in long hairs (scopae) on the underside of their abdomens. This pollen is carried dry, unlike honey bees that carry wet pollen with nectar or honey in order to pack it onto their pollen baskets [corbiculae] (Vaissière and Vinson 1994). It is unknown if wetted pollen may interact with pesticides in the field differently compared to dry pollen.

With regard to nectar contamination, the crop portion of the alimentary track of non-*Apis* bees can be extracted just as easily as with honey bees. Clearly the amount of nectar that can be recovered will be a bit less in smaller species such as mason bees or leafcutter bees, but the procedure is the same as with honey bees. It may be advantageous to anesthetize the foragers prior to squeezing their abdomen gently so as to avoid being stung repeatedly at the same spot though the smaller non-*Apis* species are usually less prone to sting and less agile at doing so than honey bees (but this is not true with bumble bee workers).

Field techniques using non-*Apis* bees are presented in greater detail in Chapter 9 on semi-field and field approaches to testing pesticide risk to bees.

7.13.3 Non-*Apis* (Solitary Species) as an Exposure Surrogate for *Apis* Bees

In certain respects, non-*Apis* bees may serve as a useful surrogate for honey bees in exposure studies. Solitary bees, such as leafcutter (*Megachile* spp.) and mason (*Osmia* spp.) bees, typically forage over a much smaller area than honey bees. For example, solitary bees typically forage within a few hundred meters of a nest, rather than two miles (several kilometers) as is common with honey bees. Because of this smaller foraging area, it is possible that a field experiment may provide a more accurate picture of potential exposure, even chronic exposure. Where a honey bee colony will forage over potentially 500 hectares or more, if sufficient forage is present, solitary bees will visit flowers as close to the nests as possible and thus be exposed consistently to local field applications and residues.

7.14 SUMMARY AND RECOMMENDATIONS

Workshop participants agreed that the most significant route of exposure to bees from foliar-applied pesticides is from both contact and oral exposure (of foraging adults, hive adults, and larvae) to contaminated pollen, nectar, and processed food (e.g., bee bread, honey, and larval jelly). For systemic compounds (applied as a seed treatment, soil drench, or trunk injection), the most significant route of exposure is through oral ingestion of residues in pollen, nectar, and processed food (e.g., bee bread or larval jelly). Other potential routes of exposure include contaminated drinking water and hive material (e.g., contaminated comb wax) and inhalation. For non-*Apis* bee species, unique potential exposure routes include contaminated soil (for solitary ground-nesting species and tunnel-nesting species that use mud to build cell partitions), and contact with sprayed leaves and nesting material that may also be contaminated. Workshop participants agreed that when assessing the major routes of exposure, methods should be conservative enough to account for various potential exposure routes. Unique potential exposure routes, for systemic pesticides, include contaminated abraded dust from seed treatment scenarios, consumption of contaminated aphid honeydew, or possible consumption of contaminated guttation water.

7.14.1 Exposure Estimates

To estimate contact exposure for foliar-applied products, published insect data from direct application exposure studies with honey bees (Koch and Weißer 1997) can be used to estimate the PEDc of foraging honey bees. Using this data, a worst-case estimate of 1.79 μg/bee is predicted after an application of 1 kg/ha directly to foraging bees.

For non-*Apis* species, Workshop participants recommended using the data for leaf-dwelling and soil-dwelling arthropods from the data developed by Schabacker et al. (2005) to address exposure to leaf-dwelling and soil-nesting non-*Apis* bee species, respectively.

For predicting oral exposure to bees for products applied as spray solutions during crop bloom, there is a limited amount of public data available to make an exposure estimate based on predicted concentrations in pollen and nectar. There is, however, a larger set of proprietary data that may be available from semi-field studies conducted by pesticide registrants. Therefore, Workshop participants discussed the possibility and value of an industry coalition to compile pollen and nectar residue data from both published and proprietary studies to develop a nomogram that can be used to predict concentrations in pollen and nectar based on field application rates. Preferably, a nomogram such as this would contain both mean and 90th percentile predictions.

Pollen and nectar residue levels, reported as mg/kg, can be compared to results from oral exposure toxicity studies with bees if the results of the studies are based on concentrations in the diet, that is, LC50, or as a NOEC (also expressed as mg/kg bee diet). However, if the results from oral exposure toxicity studies are expressed as a median lethal dose (e.g., LD50 in μg/bee), then the predicted exposure dose (in μg/bee) can be calculated based on the concentrations in pollen and nectar, and reported as (adjusted per) consumption rates for different castes of honey bees.

For systemic compounds applied as seed treatment coating, soil applications, or trunk injections, the most significant routes of exposure for adult and larval bees will be through ingestion of pollen, nectar, and processed pollen (i.e., bee bread or larval jelly) and processed nectar (i.e., honey). Recognizing the limited field data available to develop exposure models, participants of the Workshop considered the proposal by the ICPBR for a default value of 1 mg/kg in pollen and nectar (Alix and Lewis 2010), as a potentially appropriate point estimate of exposure for a screening-level assessment for seed treatment and soil applications. Once again, if the results from oral exposure toxicity studies are expressed as a dose (e.g., μg/bee), then the predicted dose can be calculated based on the concentrations in pollen and nectar coupled with reported consumption rates from different castes of honey bees.

7.14.2 Higher Tier Studies to Refine Exposure Assessments

When a screening level assessment indicates potential risks, higher tier studies with applications to bee-attractive plant materials are an option to refine exposure estimates for a specific product. A Tier 2, (contact) toxicity study of residues on foliage with honey bees may be conducted. In this laboratory study a bee-attractive plant (e.g., alfalfa) is sprayed with the formulated product and the bioavailability and persistence of toxic residues are evaluated at various exposure time points after application. The results can be used to determine the length of time between application and when bees can be safely exposed to residues on leaves or flowers of a treated crop (i.e., RT).

7.14.3 Refining Oral Exposure of Honey Bees to Foliar-Applied Compounds

Tier 3 semi-field or tunnel tests are recommended to refine the oral exposure assessment for honey bee colonies to both systemic and nonsystemic products sprayed on foliage. As discussed in the Hazard–Field section, Workshop participants recommend that semi-field studies should use a bee-attractive crop such as *Phacelia*, oilseed rape (*Brassica napus*), mustard (*Sinapis hirta*), or buckwheat (family Polygonaceae). Use of these study/crop scenarios would provide a better opportunity to ensure exposure because the bees would only have the treated crop to forage on for a specified duration. Therefore, the results from a semi-field test would provide data for a realistic, worst-case prediction of exposure of limited duration resulting from labeled use conditions. In these studies, pollen, nectar, bee bread, honey, and if desired, larval jelly can be collected and analyzed for residue levels. Unlike honey bee larvae that consume mostly processed pollen and nectar in the form of brood food or larval jelly, many non-*Apis* bee larvae consume only raw pollen. As such, in studies using non-*Apis* bees, oral exposure measurements can be obtained directly via the pollen.

7.14.4 Refining Oral Exposure of Honey Bees to Soil-Applied and Seed Treatment Systemic Compounds

Once again, a semi-field study is recommended for assessing exposure of honey bee colonies to systemic pesticides delivered via seed dressings or through soil treatments. For studies with systemic compounds, the actual crop being assessed should be used, (or potential worst case when multiple crops are being considered) since there may be different rates of uptake, distribution, and metabolism of a compound in different plant species (i.e., between an attractive surrogate crop such as *Phacelia* and a commercial target crop such as melon). Residue analysis should be timed to coincide with the highest nectar or pollen residues expected in the treated crop based on application timing as well as peak residues during bloom. Residues of systemic pesticides in leaves of trees may be highest several months after soil application, indicating that individual characteristics of the treated crop should be considered in assessing the residues in pollen and nectar. Similar to semi-field studies conducted with foliar spray products, residues in pollen, nectar, bee bread, honey, and if desired, larval jelly can be collected and analyzed for residues. The measured residue levels can be used in a refined risk assessment.

7.14.5 Refining Exposure of Non-*Apis* Bees

If a screening-level risk assessment indicates potential risk, exposure as well as the effect of a compound to non-*Apis* bee species can be refined using field or semi-field study designs. For assessing exposure to pesticides in pollen and nectar, solitary nesting bees such as blue orchard bees (*O. lignaria*) or alfalfa leafcutter bees (*M. rotundata*), can be used. However, nectar and pollen residue data gained from honey bee trials can also be used to assess exposure for non-*Apis* bees. Similar to studies with honey bees, for foliar-applied pesticides, studies with non-*Apis* bees should be conducted using a bee-attractive crop such as *Phacelia* or sweet clover. Pollen and nectar can be collected directly from the foraging bees. Semi-field or field studies can also be conducted with *Megachile* to evaluate potential (dermal or oral) exposure via contaminated nesting material. For assessing exposure to systemic pesticides used as a seed treatment, or applied as a soil treatment or trunk injection, a field study design can be used with these non-*Apis* species to evaluate worst-case exposure because of the limited foraging range of these species. Potential exposure via soil can also be evaluated using these species.

REFERENCES

Abbott VA, Nadeau JL, Higo HA, Winston ML. 2008. Lethal and sublethal effects of imidacloprid on *Osmia lignaria* and clothianidin on *Megachile rotundata* (Hymenoptera: Megachilidae). *J. Econ. Entomol.* 101(3):784–796.

Abel CA, Wilson RL, Luhman RL. 2003. Pollinating efficacy of *Osmia cornifrons* and *Osmia lignaria* subsp. *lignaria* (Hymenoptera: Megachilidae) on three Brassicaceae species grown under field cages. *J. Entomol. Sci.* 38:545–552.

Alaux C, Brunet J, Dussaubat C, Mondet F, Tchamitchan S, Cousin M, Brillard J, Baldy A, Belzunces LP, Le Conte Y. 2010. Interactions between *Nosema* microspores and a neonicotinoid weaken honeybees (*Apis mellifera*). *Environ. Microbiol.* 12(3):774–782.

Alix A, Chauzat MP, Duchard S, Lewis G, Maus C, Miles MJ, Pilling E, Thompson HM, Willner K. 2009a. Guidance for the assessment of risks to bees from the use of plant protection products applied as seed coating and soil applications—conclusions of the ICPBR dedicated working group. *Julius-Kühn Arch.* 423:15–26.

Alix A, Chauzat MP, Duchard S, Lewis G, Maus C, Miles MJ, Pilling E, Thompson HM, Willner K. 2009b. Environmental risk assessment scheme for plant protection products—conclusions of the ICPBR dedicated working group. *Julius-Kühn Arch.* 423:27–33.

Alix A, Vergnet C, Mercier T. 2009c. Risks to bees from dusts emitted at sowing of coated seeds: concerns, risk assessment and risk management. *Julius-Kühn Arch.* 423:131–132.

Alix A, Lewis G. 2010. Guidance for the assessment of risks to bees form the use of plant protection products under the framework of Council Directive 91/414 and Regulation 1107/2009. *EPPO Bull.* 40:196–203.

Atkins EL, Kellum D, Atkins KW. 1981. *Reducing Pesticide Hazards to Honey Bees: Mortality Prediction Techniques and Integrated Management Strategies.* Leaflet 2883. Division of Agricultural Sciences, University of California, Berkeley, CA.

Beekman M, Ratnieks FLW. 2000. Long-range foraging by the honey-bee, *Apis mellifera* L. *Funct. Ecol.* 14:490–496.

Bireley R. 2008. Pesticide Evaluation Report Fish and Wildlife Review for Imidacloprid, No. 226060EA. Department of Pesticide Regulation, Sacramento, CA.

Bireley R. 2009. Pesticide Evaluation Report Fish and Wildlife Review for Imidacloprid, No. 233237EA. Department of Pesticide Regulation, Sacramento, CA.

Bohart GE, Pedersen MW. 1963. The alfalfa leaf-cutting bee *Megachile rotundata* for pollination of alfalfa in cages. *Crop Sci.* 3:183–184.

Bosch J, Kemp W. 2001. *How to Manage the Blue Orchard Bee as an Orchard Pollinator.* Sustainable Agriculture Network, Beltsville, MD, p. 88.

Brewer LW, Sullivan JP, Atkins JM, Kamiri LK, Mihaich EM. 1997. Measured pesticide residues on insects in relation standard EPA estimates. Presented at the 18th Annual Meeting of the Society of Environmental Toxicology and Chemistry, San Francisco, CA, November 16–20, 1997.

Cane J. 2008. Bees (Hymenoptera: Apoidea: Apiformes). In: Vincent HR, Ring TC (eds), *Encyclopedia of Entomology*, Volume 2. Springer Verlag, pp. 419–434.

Cane JH, Griswold T, Parker FD. 2007. Substrates and materials used for nesting by North American *Osmia* bees (Hymenoptera: Apiformes: Megachilidae). *Ann. Entomol. Soc. Am.* 100:350–358.

Chauzat MP, Faucon JP. 2007. Pesticide residues in beeswax samples collected from honey bee colonies (*Apis mellifera*) in France. *Pest Manag. Sci.* 63:1100–1106.

Chauzat MP, Higes M, Martín-Hernández, Aranzazu Meana R, Nicolas Cougoule N, Faucon, JP. 2007. Presence of *Nosema ceranae* in French honey bee colonies. *J. Apic. Res.* 46:127–128.

Chauzat MP, Martel AC, Cougoule N, Porta P, Lachaize J, Zeggane S, Aubert M, Carpentier P, Faucon JP. 2010. An assessment of honeybee colony matrices, *Apis mellifera* (Hymenoptera: Aphidae) to monitor pesticide presence in continental France. *Environ. Toxicol. Chem.* 30(1):103–111.

Chen YP, Pettis JS, Collins A, Feldlaufer M. 2007. Prevalence and transmission of honeybee viruses. *Appl. Environ. Microbiol.* 72:606–611.

Chen YP, Evans JD, Pettis JS. 2011. The presence of chronic bee paralysis virus infection in honey bees (*Apis mellifera* L.) in the USA. *J. Apic. Res.* 50(1):85–86.

Choudhary A, Sharma DC. 2008. Dynamics of pesticide residues in nectar and pollen of mustard (*Brassica juncea* (L.) Czern.) grown in Himachal Pradesh (India). *Environ. Monit. Assess.* 144:143–150.

Dinter A, Brugger KE, Frost NM, Woodward MD. 2009. Chlorantraniliprole (Rynaxypyr): a novel DupontTM insecticide with low toxicity and low risk for honey bees (*Apis mellifera*) and bumble bees (*Bombus terrestris*) providing excellent tools for uses in integrated pest management. *Julius-Kühn Arch.* 423:84–96.

Fischer DL, Bowers LM. 1997. Summary of field measurements of pesticide concentrations in invertebrate prey of birds. Presented at the 18th Annual Meeting of the Society of Environmental Toxicology and Chemistry, San Francisco, CA, November 16–20, 1997.

Fletcher JS, Nellessen JE, Pfleeger TG. 1994. Literature review and evaluation of the EPA food-chain (Kenaga) nomogram, an instrument for measuring pesticide residues on plants. *Environ. Toxicol. Chem.* 13:1383–1391.

Forster R. 2009. Bee poisoning caused by insecticidal seed treatment of maize in Germany in 2008. *Julius-Kühn Arch.* 423:126–131.

Free JB. 1977. *The Social Organization of Honey Bees.* North American Bee Books, Hebden Bridge.

Fries I, Martín R, Meana A, García-Palencia P, Higes M. 2006. Natural infections of *Nosema ceranae* in European honeybees. *J. Apic. Res.* 45:230–233.

Gibbs J, Sheffield CS. 2009. Rapid range expansion of the wool-carder bee, *Anthidium manicatum* (Linnaeus) (Hymenoptera: Megachilidae), in North America. *J. Kans. Entomol. Soc.* 82:21–29.

Girolami VM, Greatti M, Di Bernardo A, Tapparo A, Giorio C, Squartini A, Mazzon L, Mazaro M, Mori N. 2009. Translocation of neonicotinoid insecticides from coated seeds to seedling guttation drops: a novel way of intoxication for bees. *J. Econ. Entomol.* 102(5):1808–1815.

Goulson D, Stout JC. 2001. Homing ability of the bumblebee, *Bombus terrestris*. *Apidologie* 32:105–112.

Greenleaf SS, Williams NM, Winfree R, Kremen C. 2007. Bee foraging ranges and their relationship to body size. *Oecologia* 153:589–596.

Hart A, Thompson H. 2001. Estimating pesticide residues on invertebrates eaten by birds and mammals. Poster Presentation at the SETAC 22nd Annual Meeting, Baltimore, MD, November 11–15, 2001.

Hoeger F, Kenaga E. 1972. Pesticide residues on plants: correlation of representative data as a basis for their estimation of their magnitude in the environment. In: Korte F (ed.), *Environmental Quality and Safety: Chemistry, Toxicology and Technology*. George Thieme Publishers, Stuttgart, pp. 9–25.

Johansen C, Mayer D. 1990. *Pollinator Protection: A Bee and Pesticide Handbook*. Wicwas, Cheshire, CT.

Kim J, Williams N, Kremen C. 2006. Effects of cultivation and proximity to natural habitat on ground-nesting native bees in California sunflower fields. *J. Kans. Entomol. Soc.* 79:309–320.

Koch H, Weißer P. 1997. Exposure of honey bees during pesticide application under field conditions. *Apidologie* 28:439–447.

Konrad RN, Ferry A, Gatehouse, Babenreier D. 2008. Potential effects of oilseed rape expressing oryzacystatin-1 (OC-1) and of purified insecticidal proteins on larvae of the solitary bee *Osmia bicornis*. *PLOS ONE* 3(7):e2664. doi:10/1371/journal.pone.0002664

Mader E, Spivak M, Evans E. 2010. *Managing Alternative Pollinators*. SARE handbook 11, University of Maryland, College Park, MD, 170 p. http://www.sare.org/publications/pollinators/pollinators.pdf (accessed January 20, 2014).

Malone L, Burgess E, Stefanovic D, Gatehouse H. 2000. Effects of four protease inhibitors on the survival of worker bumblebees, *Bombus terrestris* L. *Apidologie* 31:25–38.

Martel AC, Lair C. 2011. Validation of a highly sensitive method for the determination of neonicotinoid insecticide residues in honeybees by liquid chromatography with electrospray tandem mass spectrometry. *Int. J. Environ. Anal. Chem.* 91:978–988.

Martel AC, Zeggane S, Aurière C, Drajnudel P, Faucon JP, Aubert M. 2007. Acaricide residues in honey and wax after treatment of honey bee colonies with Apivar® and Asuntol®. *Apidologie* 38:534–544.

Mason CE. 1986. Progression of knockdown and mortality of honey bees (Hymenoptera: Apidae) sprayed with insecticides mixed with Penncap-M. *Environ. Entomol.* 15:170–176.

Mayer D, Johansen C. 1990. *Pollinator Protection: A Bee and Pesticide Handbook*. Wicwas Press, Cheshire, CT, p. 161.

Michener CD. 2007. *The Bees of the World*, 2nd edn. John Hopkins University Press, Baltimore, MD, p. 913.

Mullin C, Frazier M, Frazier J, Ashcraft S, Simonds R, vanEngelsdorp D, Pettis J. 2010. High levels of miticides and agrochemicals in North American apiaries: implications for honey bee health. *PLOS ONE* 5(3):e9754.

Murphy CM, Breed MD. 2008. Nectar and resin robbing in stingless bees. *Am. Entomol.* 54:37–44.

OEPP/EPPO. 2001. OEPP/EPPO: EPPO Standards PP1/170(3). Test methods for evaluating the side effects of plant protection products on honeybees. *Bull. OEPP /EPPO Bull.* 31:323–330.

Omer A. 2008. Pesticide Evaluation Report Pest and Disease Protection Review for Imidacloprid. Department of Pesticide Regulation, Sacramento, CA.

Orantes Bermejo FL, Gomez Pajuelo A, Megias Megias M, Fernandez Pinar CT. 2010. Pesticide residues in beeswax and beebread samples collected from honey bee colonies in Spain. Possible implications for colony losses. *J. Apic. Res.* 48(1):246–250.

O'Toole C, Raw A. 1999. *Bees of the World*. Blandford, London, p. 192.

Papathakis M. 2008. Pesticide Evaluation Report Chemistry Review for Imidacloprid. Department of Pesticide Regulation, Sacramento, CA.

Pasquet RS, Peltier A, Hufford MB, Oudin E, Saulnier J, Paul L, Knudsen JT, Herren HS, Gepts P. 2008. Long distance pollen flow assessment through evaluation of pollinator foraging range suggests transgene escape distances. *Proc. Natl. Acad. Sci. U.S.A.* 105:13456–13461.

Pereboom JJM. 2000. The composition of larval food and the significance of exocrine secretions in the bumblebee *Bombus terrestris*. *Insectes Sociaux* 47(1):11–20.

Pettis JS, vanEnglesdorp D, Johnson J, Gively G. 2012. Pesticide exposure in honey bees results in increased levels of the gut pathogen *Nosema*. *Naturwissenschaften*. 99(2):153–158.

Pistorius J, Bischoff G, Heimbach U, Stähler M. 2009. Bee poisoning incidents in Germany in Spring 2008 caused by abrasion of active substance from treated seeds during sowing of maize. *Julius-Kühn Arch.* 423:118–126.

Potts SG, Willmer P. 1998. Compact housing in built-up areas: spatial patterning of nesting aggregations of a ground-nesting bee. *Ecol. Entomol.* 23:427–432.

Pouvreau A. 1984. Biologie et écologie des bourdons. In: Pesson P, Louveaux J (eds), *Pollinisation et Productions Végétales*. INRA, Paris, pp. 595–630.

Ribière M, Ball B, Aubert M. 2008. Natural history and geographical distribution of honey bee viruses. In: Aubert M, Ball B, Fries I, Moritz R, Milani N, Bernadinelli I (eds), *Virology and the Honey Bee*. European Commission, Bruxelles, pp. 15–84.

Ribière M, Lallemand P, Iscache AL, Schurr F, Celle O, Blanchard P, Oliver V, Faucon JP. 2007. Spread of infectious chronic bee paralysis virus by honeybee (*Apis mellifera* L.) feces. *Appl. Environ. Microbiol.* 73(23):7711–7716.

Romaniuk K, Spodniewska A, Kur B. 2003. Residues of chlorinated hydrocarbons in propolis from Warmia and Muzuria voivodship apiaries. *Medycyna Wet.* 11:1023–1026.

Rortais A, Arnold G, Halm MP, Touffet-Briens F. 2005. Modes of exposure of honeybees to systemic insecticides: estimated amounts of contaminated pollen and nectar consumed by different categories of bees. *Apidologie* 36:71–83.

Roubik DW. 1989. *Ecology and Natural History of Tropical Bees*. Cambridge University Press, New York, p. 514.

Sampson BJ, Knight PR, Cane JH, Spiers JM. 2007. Foraging behavior, pollinator effectiveness, and management potential of the new world squash bees *Peponapis pruinosa* and *Xenoglossa strenua* (Apidae: Eucerini). *HortScience* 42:459.

Sapir Y, Shmida A, Ne'eman G. 2005. Pollination of *Oncocyclus irises* (Iris: Iridaceae) by night-sheltering male bees. *Plant Biol.* 7:417–424.

Schabacker J, Barber I, Ebaling M, Edwards P, Riffel M, Welter K, Pascual J, Wolf C. 2005. Review on initial residue levels of pesticides in arthropods sampled in field studies. Report from the European Crop Protection Organization.

Schenke D, Joachimsmeier I, Pistorius J, Heimbach U. 2010. Pesticides in guttation droplets following seed treatment—preliminary results from greenhouse experiments. Presented at the 20th Annual Meeting of SETAC Europe, Seville, May 23–27, 2010.

Seiber JN, McChesney MM. 1987. Measurement and computer model simulation of the volatilization flux of molinate and methyl parathion from a flooded rice field. Final Report to the California Department of Food and Agriculture.

Seiber JN, McChesney MM, Majewski MS. 1991. Volatilization rate and downward contamination from application of dacthal herbicide to an onion field. Final Report to the California Department of Food and Agriculture.

Shuler RE, Roulston TH, Farris GE. 2005. Farming practices influence wild pollinator populations on squash and pumpkin. *J. Econ. Entomol.* 98:790–795.

Suchail S, Guez D, Belzunces LP. 1999. Toxicity of imidacloprid and its metabolites in *Apis mellifera*. In: Belzunces LP, Pélissier G, Lewis, GR (eds), *Hazards of Pesticides to Bees*, Vol. 98. INRA, Avignon, pp. 122–126.

Tapparo A, Marton D, Giorio C, Zanella A, Soldà L, Marzaor M, Vivan L, Girolami V. 2012. Assessment of the environmental exposure to honeybees to particulate matter containing neonicotinoid insecticides coming from corn coated seeds. *Environ. Sci. Technol.* 46:2592–2599.

Tew JE. 1997. *Protecting honey bees from pesticides*. Factsheet HYG-2161-97. The Ohio State University, Wooster, OH.

Thompson HM. 2001. Assessing the exposure and toxicity of pesticides to bumblebees (*Bombus sp.*). *Apidologie* 32:305–321.

Thompson HM. 2007. Assessment of the risk posed to honeybees by systemic pesticides. DEFRA Research Project PS2322.

Thompson HM, Hunt LV. 1999. Extrapolating from honeybees to bumblebees in pesticide risk assessment. *Ecotoxicology* 8:147–166.

Tremolada P, Bernardinelli I, Colombo M, Spreafico M, Vighi M. 2004. Coumaphos distribution in the hive ecosystem: case study for modeling applications. *Ecotoxicology* 13(6):589–601.

Vaissière BE, Merritt SJ, Keim DL. 1985. *Melissodes thelypodii* Cockerell (Hymenoptera: Anthophoridae), an effective pollinator of hybrid cotton on the Texas High Plains. In: Brown JM (ed.), *Proceedings of the Beltwide Cotton Production Research Conference*, National Cotton Council of America, Memphis, TN, pp. 398–399.

Vaissière BE, Vinson SB. 1994. Pollen morphology and its collection effectiveness by honey bees, *Apis mellifera* L. (Hymenoptera: Apidae), with special reference to upland cotton, *Gossypium hirsutum* L. (Malvaceae). *Grana* 33:128–138.

Van der Steen JJM, Bortolloti L, Chauzat MP. 2008. Can pesticide acute toxicity for bumblebees be derived from honeybee LD50 values. Hazards of pesticides to bees—10th International Symposium of the ICPBR Bee Protection Group, October 8–10, Bucharest, Romania.

Vicens N, Bosch J. 2000. Weather-dependent pollinator activity in an apple orchard, with special reference to *Osmia cornuta* and *Apis mellifera* (Hymenoptera: Megachilidae and Apidae). *Environ. Entomol.* 29:413–420.
Visscher PK, Seeley TD. 1982. Foraging strategy of honeybee colonies in a temperate deciduous forest. *Ecology* 63:1790–1801.
Waller G. 1969. Susceptibility of an alfalfa leafcutting bee to residues of insecticides on foliage. *J. Econ. Entomol.* 62:189–192.
Wallner K. 2009. Sprayed and seed dressed pesticides in pollen, nectar and honey of oil seed rape. *Julius-Kühn Arch.* 423:152–153.
Winfree R, Williams NM, Dushoff J, Kremen C. 2007. Native bees provide insurance against ongoing honey bee losses. *Ecol. Lett.* 10:1105–1113.
Winfree R, Williams NM, Gaines H, Ascher JS, Kremen C. 2008. Wild bee pollinators provide the majority of crop visitation across land-use gradients in New Jersey and Pennsylvania, USA. *J. Appl. Ecol.* 45(3):793–802.
Winston ML. 1987. *The Biology of the Honey Bee*. Harvard University Press. Cambridge, MA.

8 Assessing Effects Through Laboratory Toxicity Testing

J. Frazier, J. Pflugfleder, P. Aupinel, A. Decourtye, J. Ellis, C. Scott-Dupree, Z. Huang, H. Thompson, P. Bachman, A. Dinter, M. Vaughan, B.E. Vaissière, G. Maynard, M. Kasina, E. Johansen, C. Brittain, M. Coulson, and R.C.F. Nocelli

CONTENTS

8.1 INTRODUCTION

Toxicity testing in support of a risk assessment process for determining the potential impacts of chemicals to pollinator insects, and, more specifically, honey bees has typically involved both laboratory and field studies. Initially, tests are conducted that are intended to serve as a screen for whether a chemical represents a potential hazard. These tests are typically laboratory-based studies conducted on individual bees and are intended to provide conservative estimates of toxicity based on acute exposures of individual organisms under highly controlled environmental conditions. Based on the likelihood of exposure and the degree of sensitivity of the test species in the initial laboratory tests, higher-tiered tests may be required to understand whether the

Pesticide Risk Assessment for Pollinators, First Edition. Edited by David Fischer and Thomas Moriarty.

effects observed in laboratory studies conducted on individual insects extend to the colony/population level under environmentally relevant exposure conditions.

For reasons discussed earlier, testing to determine the potential effects of chemicals on non-target organisms has typically relied on the use of surrogate test species. Selection of a surrogate species must consider the availability of the species and its ability to thrive under laboratory testing conditions. As such, the husbandry and environmental needs of the test species must be documented so that tests can be readily conducted and reproduced or replicated. Ideally, the test species should be a relatively sensitive indicator of toxicity; however, it is generally recognized that the test species is unlikely to be the most sensitive of all species it is intended to represent. Although the European honey bee (*Apis mellifera*) has been used extensively in testing chemicals for potential effects, it is recognized that its biology is different from non-*Apis* bees (e.g., solitary bees) and other pollinating insects and that these differences may translate into significant differences in how the organism may be exposed and affected. The extent to which data from any surrogate test species are considered biased can only be elucidated through equally rigorous studies using other species. Currently, data for non-*Apis* bee species are limited; however, differences in the sensitivity of *Apis* and non-*Apis* bees may not be as pronounced as differences in potential exposure between honey bees and non-*Apis* bees. As an example, solitary ground-nesting bees of similar sensitivity to honey bees may be more vulnerable to exposure to soil treatments compared to honey bees.

The intent of toxicity tests is to provide measurement endpoints that can be used to assess the adverse effects from exposure to a particular stressor, for example, pesticides. Endpoints measured at the individual level are intended to provide insight on effects that are likely to impact entire populations or communities. In doing so, measurement endpoints drawn from laboratory-based tests should be readily linked to assessment endpoints (i.e., impaired survival, growth, or reproduction) that, in turn, are linked to protection goals. These assessment endpoints relate directly to maintenance of insect pollinators at the population or community level.

To ensure greater consistency in toxicity testing across chemicals, regulatory authorities have established guidelines that outline study design elements that should be considered, as well as the nature of data to be collected. To conserve resources (i.e., focusing resources where they are most needed), and limit the number of animals required for testing, regulatory authorities have approached ecological risk assessment in a tiered manner. Laboratory-based studies (Tier 1), which can be conservative, relatively rapid and economical, are the first tier in evaluating chemicals for their potential (toxic) effects. Tier 1 tests provide an understanding of acute lethality and potential sublethal effects. This information should guide the decision of the assessor whether additional testing is needed. If, based on the outcome of Tier 1 laboratory-based studies, more refined studies are required, then their design should be informed by the Tier 1 study. A higher-tier study, such as a semi-field study, should be designed to answer questions identified in the lower-tier study(ies), which are limited. As such, a linkage should begin to be drawn between different tiers, that is, as moving from studies that look at the individual to studies that begin to look at the colony, and ultimately look at the colony in an environmentally realistic setting.

Considerable testing has been conducted with honey bees under relatively standardized conditions resulting in a sizeable database on the acute contact toxicity of a wide range of chemicals. This toxicity data generated through relatively standardized testing enables risk assessors to compare the relative toxicity of chemicals to bees across chemical classes with highly divergent modes of action. Workshop participants believed that since Tier 1 laboratory studies often serve as the basis on which further testing is or is not required, these studies are relied upon to be accurate, informative, and efficient. Further, studies must be designed and harmonized to provide the highest quality data with the least amount of variability. This chapter provides an overview of existing toxicity tests and their strengths and weaknesses, and discusses proposed modifications to existing studies, or additional studies that could address limitations in the current battery of studies.

8.2 OVERVIEW OF LABORATORY TESTING REQUIREMENTS AMONG SEVERAL COUNTRIES

8.2.1 Overview of Honey Bee Laboratory Testing in the European Union

Regulatory agencies in different world regions have developed varied approaches and requirements for hazard test results used in ecological risk assessment to evaluate the potential hazard of pesticides to honey bees. The requirements for regulatory testing on honey bees in the European Union (EU) can be found in Annex II and III of EU Directive 91/414 (EU Directive 91/414), and additional regulatory guidance has also been provided. (SANCO 2002; OECD 2007; EPPO 2010, 2011). A new EU Regulation (EC 1107/2009), intended to replace EU Directive 91/414, was published in October 2009, but the data requirements and risk assessment criteria to support this new directive have not been established.

European testing has always followed a sequential testing scheme, that is, starting with laboratory-based testing and then moving on to higher-tier studies if warranted. Where there is only one route of exposure (e.g., oral exposure in case of soil application of systemic products), the acute testing can be restricted to that route (i.e., contact or oral). Since oral exposure can be a relevant route of exposure for systemic products applied as a seed dressing, the acute oral toxicity of such a substances has to be determined. However, in recent years, information and incidents have indicated that the contaminated dust associated with planting pesticide treated seed is an exposure route that should be considered. (Alix et al. 2009; Forster 2009; Pistorius et al. 2009). In such a case, potential routes of exposure would include oral and contact and, therefore, effects testing would be required to account for both routes of exposure. Acute tests with the formulated product, that is, active ingredients (a.i.) plus inerts, are required if the product contains more than one active substance, or if the toxicity of a new formulation cannot be reliably predicted to be either the same or lower than a tested formulation (EU 91/414).

In the EU, regulatory authorities may require a bee brood feeding test to assess potential hazard of a pesticide on honey bee larvae. Currently, this testing must be carried out when the active substance may act as an insect growth regulator, or when available data indicate that there are effects on development at immature stages. Larval testing may be carried out according to the method described by Oomen et al. (1992) in which colonies are fed pesticide concentrations in sugar syrup. Dose levels used in this test should reflect maximum levels (of active ingredient) expected in the applied product.

If results of either the adult or larval tests indicate that a presumption of minimal risk cannot be made, then further testing such as a semi-field or field testing is triggered in order to determine whether any toxicity is observed under realistic exposure conditions. OECD guidance document No 75 (OECD 2007) and EPPO 170 (EPPO 2011) provide recommendations on testing honey bee brood under semi-field and field conditions.

8.2.2 Overview of Honey Bee Laboratory Testing for Regulatory Purposes in North America

Similar to the EU, North America (United States Environmental Protection Agency (USEPA), and Canada's Pest Management Regulatory Agency (PMRA)) employs laboratory-based tests as a first step for evaluating the potential toxicity of chemicals to insect pollinators. The USEPA's data requirements for insect pollinator testing are defined in the US Code of Federal Regulations 40 (CFR 40; 2012). Similar to the European process, the North American process also follows a tiered approach.

Tier 1 consists of an acute contact toxicity test with young adult honey bees (USEPA 2012a). Until recently, the USEPA has typically required just the acute contact toxicity test; however, in efforts to better harmonize with its counterparts in Canada and Europe, and in recognition that exposure occurs through ingestion of pesticide residues as well as through contact, the United States has begun to require oral toxicity tests consistent with OECD guidelines (OECD 1998b). Higher tier studies may be required if the results of the acute toxicity tests indicate that the LD50 <11 μg a.i./bee toxicity, and/or if other lines of information, such as data in the open literature and incident data indicate that additional information is needed.

Currently, higher tier tests include laboratory-based toxicity of residues on foliage test (USEPA 2012b) and field-based pollinator study (USEPA 2012c). The toxicity of residues on foliage test is based on the work of Johansen et al. (1977) and Lagier et al. (1974) and is intended to provide data on the residual toxicity of a compound to honey bees. In this study, the test substance is applied to a sample of crop material (alfalfa is preferred) at the typical label rate and placed with caged test bees which forage on the treated plant material. Mortality and adverse effects are recorded after 2, 8, and 24 hours of exposure to the treated foliage. If the mortality of bees exposed to 24-hour-old residues is greater than 25%, sampling is continued at 24-hour intervals until mortality of bees exposed to treated foliage is not significantly greater than the controls.

Beyond the toxicity test of residues on foliage, if any of the following conditions are met, EPA may require a pollinator field study (USEPA 1996):

- Data from other sources (e.g., open literature, beekill incidents) indicate potential adverse effects on colonies, especially effects other than acute mortality (reproductive, behavioral, etc.).
- Data from toxicity of residue on foliage studies indicate extended residual toxicity.
- Data derived from studies with terrestrial arthropods other than bees indicate potential chronic, reproductive, or behavioral effects.

Field pollinator testing is intended to examine the potential effects of a chemical on the whole honey bee colony, and the nature of these studies is discussed in Chapter 9. USEPA testing requirements stipulate the use of technical grade active ingredient (purity >95%) in acute contact toxicity tests, while higher-tier tests are typically conducted using the formulated product.

8.3 UNCERTAINTIES IN CURRENT TESTING PARADIGMS

Laboratory-based acute toxicity testing of honey bees in the United States has not formally included studies examining the potential effects of pesticides on honey bee larvae (brood). In addition, while test guidelines stipulate that sublethal effects must be reported in acute tests, the typical endpoint reported from these tests is the median lethal dose (LD50) and rarely is a median effect concentration (EC50) based on sublethal effects reported. Given that the current US test guidelines are designed to yield regression-based endpoints, that is, LD_X values, endpoints such as no-observed-adverse-effect concentrations (NOAEC) and lowest-observed-effect concentrations (LOAEC) which require hypothesis testing are not likely attainable since treatments are not sufficiently replicated.

Also, as noted earlier, under the US testing process, the honey bee is used as a surrogate for other pollinator insects and for terrestrial invertebrates. In the EU, however, specific test guidelines are available for examining the effects of pesticides on non-target arthropods and beneficial insects based on the ESCORT 2 guidance (Candolfi et al. 2000) independent of the studies examining toxicity to honey bees. Uncertainties regarding the use of honey bees as surrogates for other non-*Apis* bees were identified at the Workshop. These uncertainties are centered on the fact that the life history and social biology of honey bees are significantly different from those of other bees and arthropods. At this time, there are insufficient data to determine whether or not honey bees serve as reasonable surrogates for other non-*Apis* bees or insect pollinators in general (i.e., whether laboratory studies conducted with *A. mellifera* provide endpoints sufficiently protective of the range non-*Apis* bees or other insect pollinator insects and/or terrestrial invertebrates). However, it was noted by Workshop participants that since laboratory studies are intended to examine the intrinsic toxicity of a chemical to a particular test organism, differences in the biology of the test organism relative to those species for which it is intended to serve as a surrogate may not be critical. Table 8.1 provides a comparison of the acute laboratory toxicity tests (OECD 1998a, OECD 1998b, and USEPA 2012a) currently required by regulatory authorities in the EU and United States.

8.4 LIMITATIONS AND SUGGESTED IMPROVEMENTS FOR TIER 1 TESTING

8.4.1 ADULT APIS *MELLIFERA* WORKER ACUTE TOXICITY

Exposure of honey bees can be from direct overspray while the bees are foraging, by contact with contaminated surfaces of the plant, or by intake of contaminated pollen and nectar. The hazard posed by short-term exposures can be assessed using acute toxicity tests. As discussed in the preceding section, acute honey bee testing under laboratory conditions has been conducted for some time according to several different test guidelines and published methods (EPPO 2010a; SETAC 1995; Stute 1991; USEPA 2012a). Workshop participants considered the OECD test guidelines (OECD 1998a, 1998b) to be the most detailed of those available for assessing the acute toxicity of pesticides to honey bees for the reasons presented below.

Acute honey bee tests performed according to OECD guidelines (OECD 1998a, 1998b), can be designed as limit tests or as dose–response studies (with a minimum of five doses and a minimum of 3 replicates of 10 bees at each dose). The bees are held under controlled temperature and humidity conditions while mortality and behavior is monitored for a minimum of 48 hours (this is extended if effects are prolonged). The reported data include the LD50 (with 95% confidence limits), at 24 and 48 hours and, if relevant 72 and 96 hours time points (in μg test substance per bee), the slope of dose–response curves, and any other observed abnormal bee responses. Both tests include a control (treated with the same concentration of solvent as in the treated doses) and a toxic standard (e.g., dimethoate) with defined acceptance criteria.

The acute contact test (OECD 1998b) involves direct application of the test substance (active ingredient or formulation), usually as a 1 μL drop, diluted in an organic solvent or water as required, applied directly to the dorsal thorax of the bee. Among the advantages of the acute contact test guidelines are:

- replication (at least three replicates);
- no in-hive treatments for 4 weeks prior to use in a study are permitted;
- higher number of test organisms is specified (30 bees);
- prescriptive environmental conditions;
- stringent control mortality is specified (less than or equal to 10% mortality);
- a toxic standard is required and validity criteria are stated; and,
- test duration is prolonged in case of delayed effects.

There is only one internationally accepted oral acute toxicity test guideline (OECD 1998a). The test is similar in design to the acute contact toxicity test described above, but consists of group feeding. Caged replicate bees are fed a known volume of treated sucrose solution over a maximum period of 6 hours and then untreated sucrose is supplied *ad libitum*. Group feeding can be used to administer the dose of test substance because honey bees exhibit trophallaxis, i.e., the transfer of food among colony members. The applicability and repeatability of this is demonstrated by the toxic reference chemical (e.g., dimethoate), which is stable within a testing facility. Some pesticides, such as pyrethroids, are repellent and the total dose may not be consumed, so careful monitoring of the intake of the test substance per bee is required.

TABLE 8.1
Comparison of Acute Contact Test Guidelines (OECD 1998b and USEPA 2012a) and Acute Oral Test Guideline (OECD 1998a)

	OECD 1998b (Acute Contact)	EPA OPPTS 850.3020 (Acute Contact)	OECD 213 (Acute Oral)
Status and background	Adopted September 21, 1998 Based on EPPO GL 170 (1992) and improvements considered made by ICPBR (1993) Other GLs considered: SETAC (1995), Stute (BBA) (1991), EPA OPPTS 850.3020 (2012a)	Public draft April, 1996 Based on OPP 141-1 (1982)	Adopted September 21, 1998 Based on EPPO GL 170 (1992) and improvements considered made by ICPBR (1993) Other GLs considered: SETAC (1995), Stute (BBA) (1991), EPA OPPTS 850.3020 (1995)
Test species and test organisms	Young, healthy, adult worker bees (*Apis mellifera*), same race, similar age and feeding stage, from queen-right colony, known history Bees collected from frames without brood are suitable Bees should not have been treated chemically for at least 4 weeks	Young test bees, 1–7 days old (*A. mellifera*), may be obtained directly from hives or from frames kept in an incubator, from same source	Young, healthy, adult worker bees (*A. mellifera*), same race, similar age and feeding stage, from queen-right colony, known history Bees collected from frames without brood are suitable Bees should not have been treated chemically for at least 4 weeks
Test cages	Clean and well-ventilated, made of any appropriate material, for example, stainless steel, wire mesh, plastic, disposable wooden cages Groups of 10 bees	Test chambers may be constructed of metal, plastic, wire mesh, or cardboard, or a combination of these materials Groups of at least 25 bees	Clean and well-ventilated, made of any appropriate material, for example, stainless steel, wire mesh, plastic, disposable wooden cages Groups of 10 bees
Handling, feeding, preparation	Food–*ad libitum*—as sucrose solution (50% w/v),for example, via glass feeders Bees may be anaesthetized with carbon dioxide (CO_2) or nitrogen (N_2) for application. Amount should be minimal Moribund bees should be rejected before testing	A 50% sugar/water solution should be provided *ad libitum* (purified or distilled water should be used) Bees may be anaesthetized with carbon dioxide (CO_2) or nitrogen (N_2) for application	Food–*ad libitum*—as sucrose solution (50% w/v), for example, via glass feeders Feeding system should allow recording of food intake (e.g., glass tubes 50 mm long, 10 mm wide, and narrow end) Bees may be starved for up to 2 hours before test initiation Moribund bees should be rejected before testing

TABLE 8.1
(Continued)

	OECD 214 (Acute Contact)	EPA OPPTS 850.3020 (Acute Contact)	OECD 213 (Acute Oral)
Solvents	Test substance applied as solution in a carrier, that is, organic solvent—acetone preferred—or a water solution with a (commercial) wetting agent Two separate control groups, that is, water and solvent /dispersant	A solvent is generally used to administer the test substance. The solvent of choice is acetone (or other volatile organic solvents) Two concurrent control groups, that is, water and solvent (or carrier) control	Test substance applied as 50% sucrose solution in a carrier, that is, organic solvent (e.g., acetone), emulsifiers or dispersants at low concentration up to max 1% should not be exceeded Two separate control groups, that is, water and solvent/dispersant
Test and control groups	Normally five doses in geometric series with a factor ≤ 2.2 covering the range of LD50 for definitive test (ranger-finder proposed) Minimum of three replicates with 10 bees for each dose rate and control (Minimum of 30 bees for each dose) Max. ≤ 10% control mortality at test end	A minimum of five dosage levels spaced geometrically. Recommended spacing for each dosage level to be at least 60% of the next higher level. Three or more dosages should result between 0 to 100% mortality Minimum of 25 bees for each dosage Max. ≤ 20% control mortality during the test	Normally five doses in geometric series with a factor ≤2.2 covering the range of LD50 for definitive test (ranger-finder proposed) Minimum of 3 replicates with 10 bees for each dose rate and control (Minimum of 30 bees for each dose) Max. ≤ 10% control mortality at test end
Limit test	100 μg a.i./bee in order to demonstrate that the LD50 is greater than this value	25 μg a.i./bee in order to demonstrate that the LD50 is greater than this value	100 μg a.i./bee in order to demonstrate that the LD50 is greater than this value
Toxic standard	At least 3 dose rates with 3 × 10 bees to demonstrate, for example, the toxic standard, dimethoate, is within the reported contact LD50 of 0.10–0.30 μg a.i./bee (Gough et al. 1994). Other toxic standards are acceptable	A concurrent positive control is not required A lab standard is recommended; also when there is a significant change in source of bees	At least three dose rates with 3 × 10 bees to demonstrate, for example, the toxic standard, dimethoate, is within the reported contact LD50 of 0.10–0.35 μg a.i./bee (Gough et al. 1994). Other toxic standards are acceptable

(*Continued*)

TABLE 8.1
(Continued)

	OECD 214 (Acute Contact)	EPA OPPTS 850.3020 (Acute Contact)	OECD 213 (Acute Oral)
Exposure	1 μL per bee applied on dorsal side of thorax (higher volumes, if justified) via micro-applicator Temperature: 25 ± 2°C Relative humidity: 50–70% Test duration: 48 hours (If mortality increases by >10% between 24 hours and 48 hours, the duration is prolonged to maximally 96 hours provided that the control does not exceed 10%)	5 μL per bee should not exceeded Temperature: 25–35°C Relative humidity: 50–80% Test duration: 48 hours	100–200 μL per 10 bees of 50% sucrose solution in water (or higher) provided for 3–4 (max. 6) hours Amount consumed is measured Temperature: 25 ± 2°C Relative humidity: 50–70% Test duration: 48 hours (If mortality increases by >10% between 24 hours and 48 hours, the duration is prolonged to maximally 96 hours provided that the control does not exceed 10%)
Observations	Mortality at 4 hours, 24 hours, 48 hours, and potentially at 72 hours and 96 hours Abnormal behavioral effects during the test period should be recorded	Mortality at 4 hours, 24 hours, 48 hours All signs of intoxication and other abnormal behavior (e.g., ataxia, lethargy, hypersensitivity) during the test period should be recorded	Mortality at 4 hours, 24 hours, 48 hours, and potentially at 72 hours and 96 hours Amount of diet consumed per group should be measured to determine palatability of diet Abnormal behavioral effects during the test period should be recorded
Data reporting	Range-finding data LD50 plus 95% confidence limits, that is, at 24 hours, 48 hours and, if relevant 72 hours and 96 hours (in μg test substance per bee) and slope of curves Mortality statistics (e.g., probit analysis, moving-average, binominal probability) Other biological effects and any abnormal bee responses Deviations from test guideline	Range-finding data LD50 plus 95% confidence limits, that is, at 24 hours, 48 hours, and slope of curves, goodness-of-fit test results Mortality statistics (e.g., probit analysis, moving-average, binominal probability) Signs of intoxication and other abnormal behavior Deviations from test guideline	Range-finding data LD50 plus 95% confidence limits, that is, at 24 hours, 48 hours, and if relevant 72 hours and 96 hours (in μg test substance per bee) and slope of curves Mortality statistics (e.g., probit analysis, moving-average, binominal probability) Other biological effects and any abnormal bee responses Deviations from test guideline

Participants of the Workshop discussed the limited number of cases which would compel specific deviations from the OECD acute test guidelines, such as when working with the Africanized bee. However, changes in study design can affect outcomes and reliability of the resulting data. Before data generated from modified study designs can be used reliably in risk assessment, the methodology and the resulting data should undergo a separate validation exercise (e.g., determination of appropriate toxic reference and control data).

8.5 ADULT ORAL CHRONIC TOXICITY—*APIS* BEES

Undertaking an adult oral chronic toxicity study is a refinement step in the proposed risk assessment scheme. Currently, there is no standardized guideline for chronic toxicity testing with bees, but method proposals and study design elements from acute toxicity tests which may be applicable to longer-term studies can be found in a number of publications, (Schmuck 2004; Suchail et al. 2001; Moncharmont et al. 2003; Aliouane et al. 2009; USEPA 2012a). While a detailed list of design elements in a chronic toxicity test can be found in Appendix 1, Workshop participants also identified the factors below as considerations:

- There is no standardized duration for the study considering that the longevity of honey bees differs between summer and winter. However, if the study aims at representing the typical exposure period of a forager on plants, then a 10-day period will cover most of the cases. Indeed, these bees will have already reached 14 days of age prior to being recruited as foragers, that is, the last activity of female worker bees. For summer bees, with their shorter life span and greater likelihood of being in the immediate vicinity of a treated crop, it is unlikely that their lifespan would last any longer than 10 days on the treated crop. Should the treated crops not be in their immediate vicinity, then it is likely that exposure will take place over a more limited period as the number of possible foraging trips per day declines as the distance increases. It is currently recommended that the study be performed over a10-day duration to ensure the most likely constant exposure period as well as high control survival (longer study durations may result in reduced control survival that can limit the ability of the study to detect treatment effects).
- To achieve a 10-day study duration, a mixed pollen (protein source) and sucrose (carbohydrate source) diet may be required.
- Some pesticides may induce reduced food intake due to repellency (e.g., pyrethroids) and the longevity of the bees may be affected by the reduced food intake due to repellency rather than reflecting a toxic effect of the pesticide. Therefore, food intake has to be assessed in parallel with mortality on a daily basis. The pattern of exposure may affect the observed toxicity for example, a single dose per day versus continuous exposure. Continuous exposure could mean: 1) dosed diet *ad libitum* or, 2) a fixed amount of dosed diet daily (e.g., 2 hours plus untreated diet during the rest of the time). Research is still underway to determine which approach is most appropriate.

8.6 HONEY BEE BROOD TESTS IN THE LABORATORY

The *in vitro* honey bee brood test provides quantitative oral or contact toxicity data on larvae for active ingredients or formulated products. These data should be used in an appropriate brood risk assessment scheme. *In vitro* larvae tests have been developed by Rembold and Lackner (1981) and used for the assessment of pesticides by Wittmann (1981). Some years later, Aupinel et al. (2005) improved this method in several aspects. Participants of the Workshop discussed brood tests, specifically the study design by Aupinel et al. (2005), and weighed further design considerations and improvements. A detailed list of suggested modifications to the Aupinel et al. study design can be found in Appendix 2.

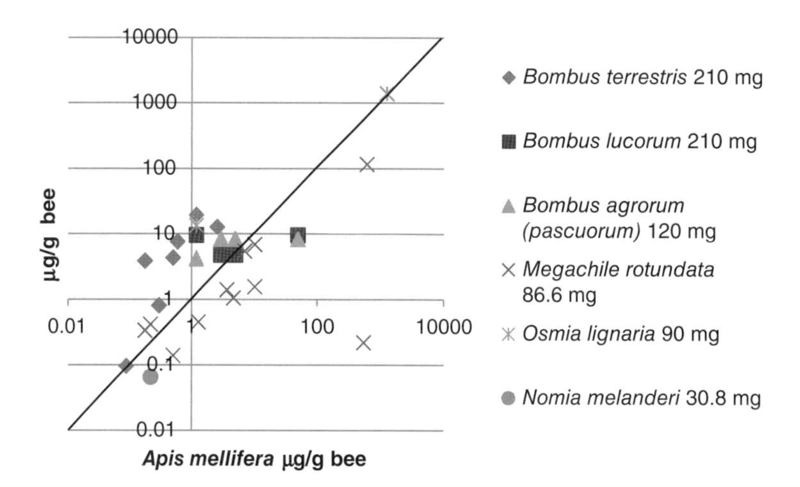

FIGURE 8.1 Comparison of the contact toxicity (LD50) of 21 pesticides to adults of *Apis mellifera,* three species of the social bee Bombus and three species of solitary bees (Osmia, Megachilidae, and Nomia). Points below the diagonal line indicate greater sensitivity than *Apis mellifera*, while points above the diagonal line represent lower sensitivity than *Apis mellifera* (Johansen et al. 1983). (For a color version, see the color plate section.)

8.7 ADULT TOXICITY TESTING WITH NON-*APIS* BEES

As discussed previously, there is always an uncertainty regarding the extent to which a surrogate test species, such as the honey bee, is a sensitive indictor of the many other species it represents. Data currently available suggest that adult non-*Apis* bees are similar in pesticide sensitivity to *A. mellifera* when bodyweight is taken into account. This conclusion is based on the analysis of a data set composed mainly of test results for pesticides of older chemistries, so some caution may be in order when considering compounds of new chemical classes. Figure 8.1 shows the relative toxicity (contact LD50 normalized to 1 g body weight) of 21 pesticides to bumble bees and solitary bees in comparison to the honey bee. Figure 8.2 depicts the decline in

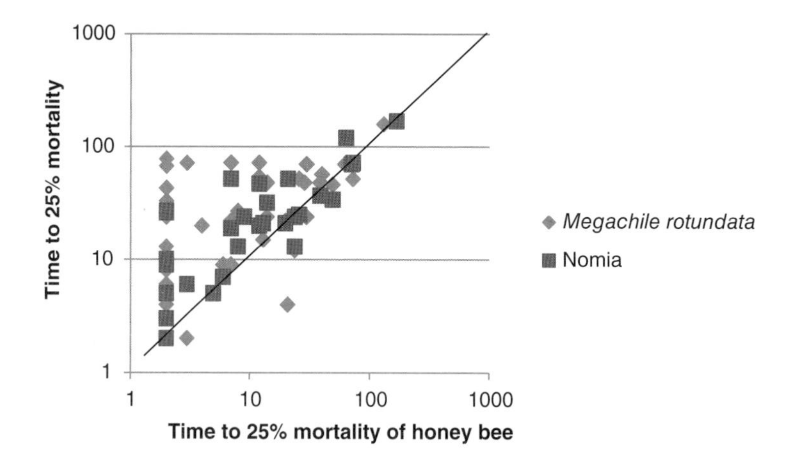

FIGURE 8.2 Comparison of the toxicity of pesticides to adults of *Apis mellifera* with the solitary bees *Megachile rotundata* and *Nomia melanderi* based on time for sprayed residues to decline to a concentration causing 25% or less mortality. Points below the diagonal line indicate greater sensitivity than *Apis mellifera*, while points above the diagonal line represent lower sensitivity than *A. mellifera* (Johansen et al. 1983). (For a color version, see the color plate section.)

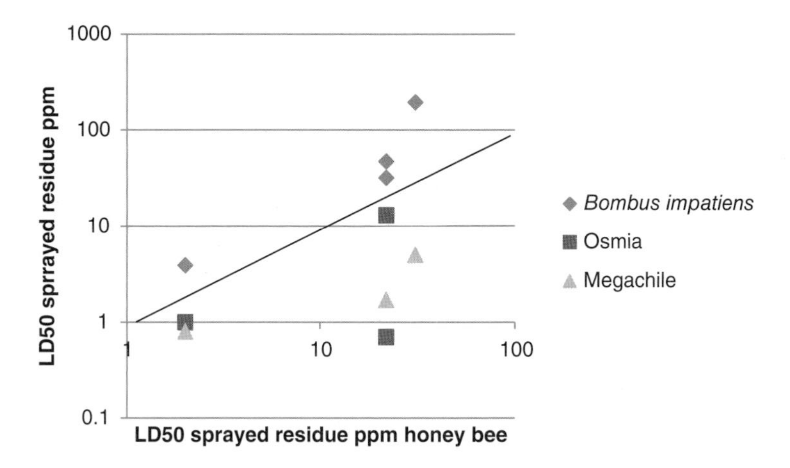

FIGURE 8.3 Comparison of the toxicity (LD50) of sprayed residues of clothianidin, imidacloprid, lambda-cyhalothrin and spinosad to adults of *Apis mellifera, Megachile rotundata*, and *Osmia lignaria*. Points below the diagonal line indicate greater sensitivity than *A. mellifera*, while points above the diagonal line represent lower sensitivity than *A. mellifera* (Johansen et al. 1983). (For a color version, see the color plate section.)

toxicity of residues on foliage for honey bee adults compared to the solitary alfalfa leafcutter bee (*Megachile rotundata*) and the alkali bee (*Nomia melanderi*). Figure 8.3 depicts the median lethal doses of sprayed residues of four pesticides (clothianidin, imidacloprid, lambda cyhalothrin, and spinosad) to *A. mellifera, M. rotundata*, and *Osmia lignaria*. The data suggest that the toxicity of these pesticides falls within an order of magnitude of the values for *A. mellifera*. This indicates that an assessment factor of 10 may be adequate to account for interspecies differences in sensitivity when acute toxicity values for honey bees are used in risk assessments.

As part of the problem formulation for an ecological risk assessment, risk assessors and risk managers can consider whether testing should include non-*Apis* species, such as when evidence or information suggests that the honey bee is not likely to be a reasonable surrogate for a crop, landscape, or region owing primarily to concerns regarding marked differences in potential exposure rather than in toxicity per se, that is, susceptibility rather than sensitivity. When selecting species to be used in the laboratory, it is important to consider their availability, ease of handling, and survival under controlled laboratory conditions. Therefore, it is recommended that both relevance (to a risk assessment and attendant protection goals), sensitivity and susceptibility are considered when determining whether to employ non-*Apis* species in an assessment.

Owing to differences in potential exposure, non-*Apis* bees may provide a means of examining the potential effects of these differences in the susceptibility of a species. For example, honey bees are capable of foraging over long distances and may have a wide range of forage available to them. However, non-*Apis* bees, for example, orchard mason bees (*O. lignaria*), are limited in the area in which they forage and may be confined to a particular treated area where the likelihood of exposure is increased.

8.7.1 Non-*Apis* Bee Testing Methods

As discussed earlier, toxicity tests intended to support regulatory decisions typically involve highly standardized testing protocols and rely on test species that are readily available and lend themselves to testing under laboratory conditions. The test species must be available in large enough numbers and have well-defined husbandry conditions to support replicate testing and thrive under specified test conditions used to examine particular routes of exposure. As with honey bees, the endpoints measured in toxicity tests with non-*Apis* bees have frequently focused on lethality; measures of sublethal effects on non-*Apis* bees would require similar linkages to assessment endpoints as those identified for honey bees. The development of these linkages may be more challenging though, as sublethal effects on individual solitary bees may have a considerably different impact at the population level than similar effects to social bees that form large colonies where the colony may have sufficient redundancy to buffer it from such effects.

The social non-*Apis* bee species most readily manipulated in the laboratory are the genera Bombini and the Meliponini (stingless bees). Some *Bombus* species are also readily available as they are used in commercial pollination of greenhouse crops. Several laboratory studies with non-*Apis* species have been published which reflect a range of methods (Table 8.2). As mentioned earlier, the ability of one non-*Apis* bee species to act as a surrogate for others involves the ready availability, and ability for that species to tolerate testing conditions. This then would indicate that the husbandry needs of that organism are well understood.

8.7.2 Non-*Apis* Larval Testing

Although toxicity testing with some species of adult non-*Apis* bees have been reported with some frequency, published laboratory studies conducted with non-*Apis* larvae are more limited, these are listed below (Table 8.3).

8.8 SUBLETHAL EFFECTS AND TEST DEVELOPMENTS

Sublethal effects are defined as reactions to an exposure not causing death. As discussed, while not specifically designed for such, current acute tests include the recording and measuring of sublethal effects. The laboratory-based (10-day) chronic study, however, is designed (i.e., longer exposure duration) with the intent of providing more specific information on sublethal effects. Beyond these, experimental research published in the open literature has gone further into investigating sublethal effects of pesticides to bees. This research has revealed insights on physiology and behavior (Desneux et al. 2007). Most experimental research regarding the behavioral effects of pesticides on bees has occurred over the last 10 years. While these test methods, and results are interesting, further work is needed not only to standardize test methods but also to be able to understand the impact of a sublethal effect in the context of the whole colony. A sublethal effect at the individual level is only relevant to protection goals when it can be linked to a resulting effect at the colony level. This section discusses some of the methods that have been developed to measure the potential sublethal effects of pesticides on honey bees.

8.8.1 Proboscis Extension Response in Laboratory

When a bee lands on a flower, it extends its proboscis as a reflex stimulated by nectar. This reflex leads to the uptake of nectar and induces the memorization of the floral odors diffusing concomitantly. Thus, the memorization of odors plays a prominent role in flower recognition during subsequent forage trips by the same individual (Menzel et al. 1993). Under laboratory conditions, learning and memory can be analyzed using a bioassay based on the olfactory conditioning of the proboscis extension response (PER) on restrained individuals.

The PER assay is based on the temporal paired association of a conditioned stimulus (CS) and an unconditioned stimulus (US). During conditioning, the PER is elicited by contacting the gustatory receptors of the antennae with a sucrose solution (US) while an odor (CS) is simultaneously released. The proboscis extension is immediately rewarded (Reward R) by the uptake of the sucrose solution. Bees can develop

TABLE 8.2
Published Laboratory Tests with Non-*Apis* Bees and Associated Methodologies

Species	Oral	Contact	Reference
Megachile rotundata *Osmia lignaria*	Individually housed adult bees with access to plastic ampoule containing pesticide inserted at base of periwinkle flower 87–90% success rate		Ladurner et al. 2003, 2005
M. rotundata	Group feeding of 10 newly emerged bees on 1 mL	1. Direct application— held at 25°C for 20 minutes to reduce activity, 1 μL applied to dorsal thorax 2. Filter paper soaked in pesticide and dried	Huntzinger et al. 2008
Bombus impatiens, *M. rotundata*, *O. lignaria*		Contact with treated filter paper	Scott-Dupree et al. 2009
M. rotundata (4–5-day-old adults); *Nomia melanderi* (2–3 weeks old)		Direct application to mesoscutum	Mayer et al. 1998
O. lignaria	Individually fed using flower (cherry) method For delayed activity fed on fresh sucrose	Cooled to 4°C before dosing, 1 μL applied to thorax	Ladurner et al. 2005
N. melanderi, *M. rotundata*	Placed into tubes inserted in caps of glass vials with individual bees, group-housed after dosing	Direct application to dorsal thorax	Johansen et al. 1983
M. rotundata		1 μL applied to thorax of males and females	Tasei et al. 1988
Bombus terrestris	Individually dosed and then group-housed	1 μL applied to ventral thorax	Thompson 2001

the PER as a conditioned response (CR) to the odor alone after even a single pairing of the odor with a sucrose reward.

The PER assay with restrained workers has been used to investigate the behavioral effects of a number of pesticides (Decourtye et al. 2002; Weick and Thorn 2002; Abramson et al. 2004; Decourtye et al. 2004). An acute exposure to a test compound can be applied before, during, or after the PER conditioning, and long-term

TABLE 8.3
Larval Test Methods for Non-*Apis* Bee Species

Species	Test Elements	Measurement Endpoints	Reference
Osmia lignaria	Eggs raised on treated pollen in 24-well culture plates; cocoons overwintered and emerged 29°C	Timing and completion of larval development; mortality; emergence, sex and weight	Abbott et al. 2008; Tesoriero et al. 2003; Peach et al. 1995
Megachile rotundata	Eggs collected from leaf tunnels, separated into 96-well plates and dosed pollen; cocoons overwintered and emerged	Timing and completion of larval development; mortality; emergence, sex and weight	Abbott et al. 2008
Osmia cornuta	Eggs placed on provisions in gelatin capsules, 1 μL applied to surface of provisions	Mortality	Tesoriero et al. 2003
M. rotundata	Leaf envelope opened and provision dosed	Weight of emerged adults	Peach et al. 1995
Nomia melanderi, *M. rotundata*	Eggs and young larvae directly dosed	Completion of cocoons	Johansen et al. 1983
M. rotundata	Male immature stages, dosed pollen provision	Number developing, cocoon completion	Tasei et al. 1988
Bombus terrestris	Larvae kept 10/egg cup with three adults 28°C, and 50% relative humidity, tested 1-, 4- and 6-day old larvae, fed treated pollen dough or sucrose 24 hours,	Mortality	Gretenkord and Drescher 1996

scenarios may be explored with this method for compounds that are expressed in the pollen and nectar. The PER assay has been used to investigate how a chemical treatment can interfere with medium-term (Decourtye et al. 2004) or long-term olfactory memory (El Hassani et al. 2008). PER tests have recorded reduced learning performances for bees after 11 days of treatment with insecticides administered orally (Decourtye et al. 2003) and topically (Aliouane et al. 2009).

PER assays can provide useful information that can be related to the memory and olfactory discrimination abilities of free-flying foragers. However, there is uncertainty regarding the extent to which the PER assay reflects what would occur under more typical settings (e.g., the bees are not restrained, or the exposure is not constant). PER testing that results in statistically significant effects on olfactory learning should be followed up with additional testing, for example, semi-field testing using intact colonies and tests such as those described in Chapter 9.

8.8.2 Artificial Flowers in Semi-Field Cage

Olfactory processing can be investigated using free-flying foragers visiting artificial flower feeders. The use of artificial flower feeders simulates a natural foraging situation more closely than does the laboratory tests on restrained worker bees using the conditioned PER procedure.

In artificial flower experiments, a nucleus colony (about 4000 workers and a fertile queen) is placed in an outdoor flight cage. Each artificial flower feeder is a plastic Petri dish containing glass balls (allowing landing of foragers on the feeding sites) and filled with a sucrose solution that is or is not treated with the test chemical. To limit the influence of visual or spatial cues, the artificial feeder is rotated slowly (e.g., rpm), and an odorant (e.g., pure linalool) is allowed to diffuse. The device is placed in front of the hive entrance. The conditioning (pairing odor and sucrose reward) is conducted for 2 hours on the first day. Testing is then carried out on the following days. For each observation event, the number of forager visits on either the scented sites or the unscented artificial flowers, is recorded. (For a more detailed list of design elements for the artificial flower experiment, please see Appendix 3.)

The comparison of responses of honey bees before and after exposure to the test chemical on the same colony is a potential limitation. Moreover, there are many unknown points, such as reliability, and sensitivity to large panel of pesticides with various modes of action. Another uncertainty is the actual exposure to individual bees, as bees are not restricted in the length of time they feed at the artificial flowers. Therefore, it is very difficult to characterize the concentration–response relationship.

8.8.3 Visual Learning Performance in a Maze

Orientation performance of bees in a complex maze relies on associative learning between a visual mark and a reward of sugar solution. In a visual learning performance maze, bees fly through a sequence of boxes to reach a feeder containing a reward of sugar solution. The path through the maze spans a number of boxes, including decision boxes (i.e, a box with three holes, each in a different wall, where the bee enters through one hole and is then expected to choose between the two other holes), and non-decision boxes (i.e., a box with two holes, each in a different wall, where the bee entered through one hole and is then expected to leave through the other hole) (Figure 8.4).

During conditioning, bees are collectively trained to associate a mark (designating the correct hole/path) with food. To that end, an identical mark is fixed in front of the correct hole/path as well as the sucrose solution feeder outside the maze for 1 hour. After conditioning, the capacity of an individual bee to negotiate a path through the maze is tested. An observer notes the number of correct and incorrect decisions, and then number of turns back. Finally, the bees are captured and placed in rearing cages equipped with a water supply and sugar syrup. Oral delivery of the treatment chemical is via the sucrose solution (50% w/w) available to the bees. After consumption of the treated sugar solution, and a starvation period, the bees are released at the test maze entrance. The effect of the treatment solution on performance is then compared with that of an untreated sucrose solution.

Menzel et al. (1974) demonstrated that honey bees in flight can associate a visual mark to a reward and this associative learning is used by bees to negotiate a path in a complex maze (Zhang et al. 1996). After treatment with a sublethal dose of a chemical, the ability of bees to perform the task can be impaired compared to untreated control bees (Decourtye et al. 2009). The maze test relies on the visual learning of foragers in relation to navigation. However, while the maze test has demonstrated neurotoxic effects with pesticides, there are insufficient data at this time to determine whether the test will provide useful information for chemicals with other modes of action. Additionally, bee navigation in the field relies upon several guidance mechanisms, (e.g., position of sun, magnetism, etc.), whereas in the maze test, performance is based on the use of a limited number of pertinent cues. Additional experiments are needed to establish whether effects on

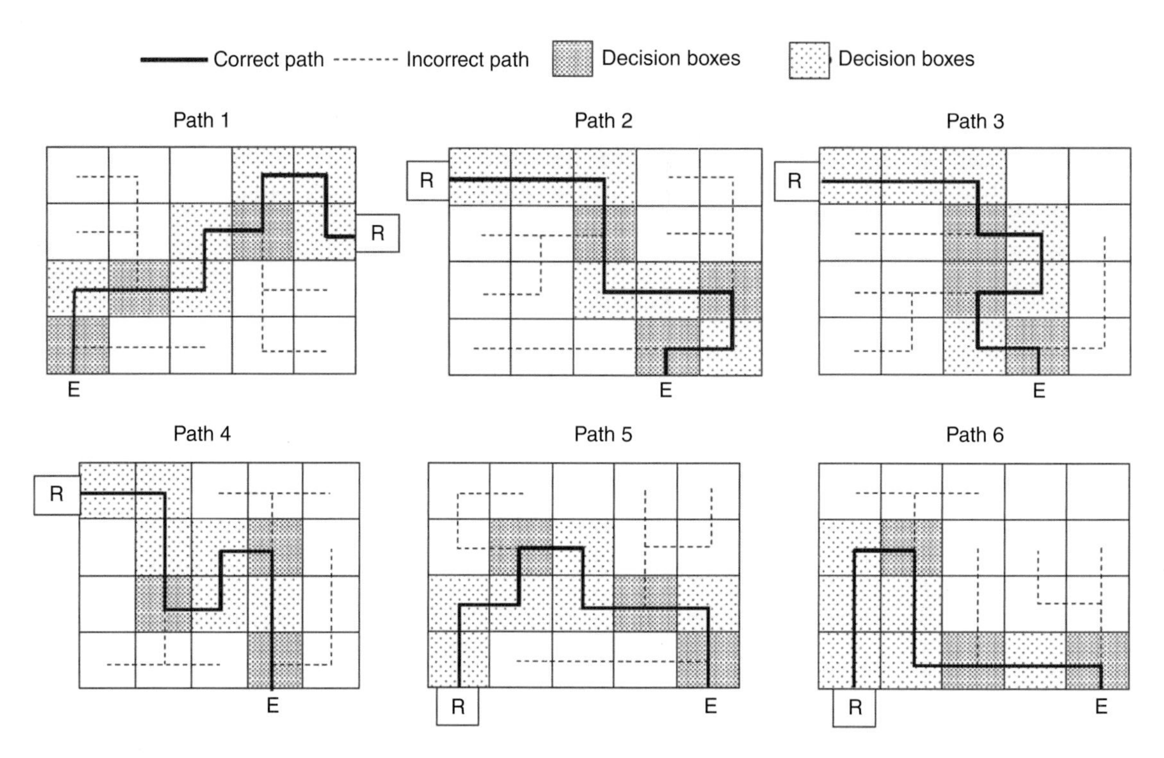

FIGURE 8.4 Maze paths used before, during, and after treatment. Path 1 was used for the conditioning procedure and other paths were used for the retrieval tests. Each path started with the entrance (E), contained three decision boxes, six no-decision boxes, and finished with the reward box (R).

maze performance reflect what may actually occur when foragers are exposed to pesticides in the field and are then confronted with complex environmental cues. (For a more detailed discussion of Visual Learning Test, please see Appendix 4.)

8.8.4 Radio-Frequency-Identification-Tagged Bees to Measure Foraging Behavior

Experimental test situations have been designed in relation to feeding behavior and social communication (Schricker and Stephen 1970; Cox and Wilson 1984; Bortolotti et al. 2003; Yang et al. 2008). Initial experiments that looked at field-level navigation were limited by the number of individual bees that could be simultaneously monitored (using bees marked with paint or colored number tags). To address this limitation, automated tracking and identification systems have been developed using radio frequency (RF) transponder technology. The use of transponders has the potential to revolutionize the study of insect life-history traits, especially in behavioral ecotoxicology.

Different transponder devices have been employed on honey bees: harmonic radar (Riley and Smith 2002) and radio frequency identification (RFID) (Streit et al. 2003). Currently, the RFID tags seem to be the technology offering the most advantages. Advantages of RFID include:

- the large number of individual insects that can be tracked;
- the number of detections which can be monitored rapidly and simultaneously (milliseconds);
- limited transpondence interference from matrices such as propolis, glue, plastic, or wood;

- absence of the need for time-consuming visual observations; and,
- reduced disruption to bee behavior given the small size of the RFID tags compared to what is needed for harmonic radar tracking.

Using this test technology, the experimental colony is maintained in an outdoor tunnel. A feeder placed away from a hive can deliver sucrose solution. A tag-equipped bee passing underneath the reader is identified by the reader and is sent to a database with real-time recording. By passing underneath the reader, both at the hive and at the feeder, the foraging bee is monitored twice, thus determining the direction of target and the travel time between the two recording points. The reader software records the identification code and the exact time of the detection in a database for later analysis of spatial and temporal information. Such analyses may include time spent within the hive, the time spent at the feeder, the time spent between the feeder and the hive, the number of entries into and exits from the hive, and the number of entries into and exits from the feeder.

RFID devices allow the study of both the behavioral traits and the lifespan of bees, especially under biotic and/or abiotic stress. However, the large quantity of data obtained with this technique requires an interface for analyzing the data and providing the life-history traits of individual bees. Under semi-field conditions, RFID microchips have provided detectable effects due to exposure to an insecticide (Decourtye et al. 2011). (For a more detailed discussion of the RFID experimental test design, please see Appendix 5.)

8.9 CONCLUSIONS

Although laboratory toxicity tests are currently available for evaluating the potential effects of chemicals on bees, there is no single consistent approach used by different regulatory authorities and, therefore, the design and scope of these tests vary. For the purposes of screening-level risk assessments, many regulatory authorities rely on acute tests using young adult honey bees to evaluate toxicity through contact and oral exposure routes. While guidelines are becoming available that include acute toxicity tests with honey bee larvae, there is also a need to expand these laboratory test methods to examine the effects of chemicals from subacute and chronic exposure durations. Laboratory-based studies will likely continue to focus on individual test organisms; and, although laboratory-based toxicity testing has historically focused on mortality, tests are evolving to provide insight on sublethal effects such as impaired behavior. As the range of measurement endpoints continues to expand, there is a need to provide both qualitative and quantitative linkages between measurement endpoints and assessment endpoints on which regulatory authorities typically base decisions. Efforts are also underway to expand the range of test species to address concerns that *A. mellifera* may not be an adequate surrogate for non-*Apis* bees with considerably different life cycles.

REFERENCES

Abbott VA, Nadeau JL, Higo HA, Winston ML. 2008. Lethal and sublethal effects of imidacloprid on *Osmia lignaria* and clothianidin on *Megachile rotundata* (Hymenoptera: *Megachilidae*). *J. Econ. Entomol.* 101(3):784–796.

Abramson CI, Squire J, Sheridan A, Mulder PG. 2004. The effect of insecticides considered harmless to honey bees (*Apis mellifera*): proboscis conditioning studies by using the insect growth regulators tebufenozide and diflubenzureon. *Environ. Entomol.* 33:378–388.

Aliouane Y, El Hassani AK, Gary V, Armengaud C, Lambin M, Gauthier M. 2009. Subchronic exposure of honeybees to sublethal doses of pesticides: Effects on behavior. *Environ. Toxicol. Chem.* 28(1):113–122.

Alix A, Vergnet C, Mercier T. 2009. Risks to bees from dusts emitted at sowing of coated seeds: concerns, risk assessment and risk management. *Julius-Kühn Arch.* 423:131–132.

Aupinel P, Fortini D, Dufour H, Tasei JN, Michaud B, Odoux JF, Pham-Delègue MH. 2005. Improvement of artificial feeding in a standard in vitro method for rearing *Apis mellifera* larvae. *Bull. Insect.* 58:107–111.

Bortolotti L, Montanari R, Marcelino J, Medrzycki P, Maini S, Porrini C. 2003. Effects of sublethal imidacloprid doses on the homing rate and foraging activity of honey bees. *Bull. Insectol.* 56:63–67.

Candolfi M, Barrett K, Campbell P, Forster R, Grandy N, Huet M-C, Lewis G, Oomen P, Schmuck R, Vogt H. 2000. Guidance document on regulatory testing procedures for pesticides with non-target arthropods. ESCORT Workgroup, Wageningen, The Netherlands. Society of Environmental Toxicology And Chemistry—Europe (SETAC).

[CFR 40] Code of Federal Regulations 40. 2012. Protection of the Environment. Part 158 (Data Requirements for Pesticides. Subpart G (Ecological Effects) § 158.630 (Terrestrial and aquatic nontarget organism data requirements table. http://ecfr.gpoaccess.gov/cgi/t/text/textidx?c=ecfr&sid=e2fa3dd8d45333c0c4427f3d556c30f9&tpl=/ecfrbrowse/Title40/40cfr158_main_02.tpl.

Cox RL, Wilson WT. 1984. Effects of permethrin on the behavior of individually tagged honey bees, *Apis mellifera* L. (Hymenoptera: *Apidae*). *Environ. Entomol.* 13:375–378.

Decourtye A, Armengaud C, Renou M, Devillers J, Cluzeau S, Gauthier M, Pham-Delegue MH. 2004. Imidacloprid impairs memory and brain metabolism in the honey bee (*Apis mellifera* L.). *Pestic. Biochem. Physiol.* 78(2):83–92.

Decourtye A, Devillers J, Aupinel P, Brun F, Bagnis C, Fourrier J, Gauthier M. 2011. Honeybee tracking with microchips: a new methodology to measure the effects of pesticides. *Ecotoxicology* 20:429–437.

Decourtye A, Lacassie E, Pham-Delegue MH. 2003. Learning performances of honeybees (*Apis mellifera* L) are differentially affected by imidacloprid according to the season. *Pest Manag. Sci.* 59:269–278.

Decourtye A, Lefort S, Devillers J, Gauthier M, Aupinel P, Tisseur M. 2009. Sublethal effects of fipronil on the ability of honeybees (*Apis mellifera* L.) to orientate in a complex maze. In: Oomen PA, Thompson HM (eds), *Hazards of Pesticides to Bees*. Arno Brynda GmbH, Berlin, pp. 75–83.

Decourtye A, Pham-Delegue MH, Kaiser L, Devillers J. 2002. Behavioral methods to assess the effect of pesticides on honey bees. *Apidologie* 33:425–432.

Desneux N, Decourtye A, Delpuech JM. 2007. The sublethal effects of pesticides on beneficial arthropods. *Annu. Rev. Entomol.* 52:81–106.

[EC 1007/2009] EC No 1107/2009: http://eur-lex.europa.eu/LexUriServ/LexUriServ.do?uri=OJ:L:2009:309:0001:0050:EN:PDF (accessed January 21, 2014).

El Hassani AK, Dacher M, Gary V, Lambin M, Gauthier M, Armengaud C. 2008. Effects of sublethal doses of acetamiprid and thiamethoxam on the behavior of the honeybee (*Apis mellifera*). *Arch. Environ. Contam. Toxicol.* 54:653–661.

EPPO. 2010a. PP 1/170 (4): efficacy evaluation of plant protection products: side-effects on honey bees. *OEPP /EPPO Bull.* 40:313–319.

EPPO. 2010b. PP 3/10 (3): Chapter 10: environmental risk assessment scheme for plant protection products. Risk assessment to honey bees. *Bull. OEPP /EPPO Bull.* 40:1–9.

EPPO. 2011. PP1/170: EPPO standards—Test methods for evaluating the side effects of plant protection products on honeybees. *Bull. OEPP /EPPO Bull.* 31:323–330.

[EU 91/414] EU Directive 91/414: http://www.uksup.sk/download/oso/20030409_smernica_rady_91_414_eec.pdf (accessed January 21, 2014).

Forster R. 2009. Bee poisoning caused by insecticidal seed treatment of maize in Germany in 2008, pp. 126–130. In: Oomen PA, Thompson HM (eds), *Hazards of Pesticides to Bees*—10th International Symposium of the ICP-BR Bee Protection Group. Bucharest (Romania), October 8–9, 2008. Julius Kühn Archiv 423.

Gough HJ, McIndoe EC, Lewis GB. 1994. The use of dimethoate as a reference compound in laboratory acute toxicity tests on honey bees (*Apis mellifera* L.) 1981–1992. *J. Apic. Res.* 33(2):119–125.

Gretenkord C, Drescher W. 1996. Laboratory and cage test methods for the evaluation of insect growth regulators (Insegar, Dimilin) on the brood of *Bombus terrestris* L. Proceedings of the 6th ICP-BR Symposium, on Hazazrds of Pesticides to Bees, Braunschweig, Germany.

Huntzinger CI, James RR, Bosch J, Kemp WP. 2008. Fungicide tests on adult alfalfa leafcutter bees (Hymenoptera: *Megachilidae*). *J. Econ. Entomol.* 101(4):1088–1094.

Johansen C, Mayor D. 1977. *Bee Research Investigations*. Dept. of Entomology, Washington State University, unpublished, p. 22.

Johansen CA, Mayer DF, Eves JD, Kious CW. 1983. Pesticides and bees. *Environ. Entomol.* 12(5):1513–1518.

Ladurner E, Bosch J, Kemp WP, Maini S. 2005. Assessing delayed and acute toxicity of five formulated fungicides to *Osmia lignaria* Say and *Apis mellifera*. *Apidologie* 36:449–460.

Ladurner E, Bosch J, Maini S, Kemp WP. 2003. A method to feed individual bees (*Hymenoptera: Apiformes*) known amounts of pesticides. *Apidologie* 34:597–602.

Lagier RF, Johansen CA, Kleinschmidt MG, Butler LI, McDonough LM, Jackson DS. 1974. *Adjuvants Decrease Insecticide Hazard to Honey Bees*. College of Agriculture Research Center, Washington State University Bulletin 801, p. 7.

Mayer DF, Kovacs G, Lunden JD. 1998. Field and laboratory tests on the effects of cyhalothrin on adults of *Apis mellifera, Megachile rotundata* and *Nomia melanderi*. *J. Apic. Res.* 37(1):33–37.

Menzel R, Erber J, Masuhr T. 1974. Learning and memory in the honeybee. In: Barton-Browne L (ed),*Experimental Analysis of Insect Behaviour.* Springer, Berlin, pp. 195–217.

Menzel R, Greggers U, Hammer M. 1993. Functional organization of appetitive learning and memory in a generalist pollinator, the honey bee. In: Papaj DR, Lewis AC (eds), *Insect Learning: ecological and evolutionary perspectives*. Chapman & Hall, New York, pp. 79–125.

Moncharmont FD, Decourtye A, Hanfier CH, Pons O, Pham-Delegue M. 2003. Statistical analysis of honeybee survival after chronid exposure to insecticides. *Environ. Toxicol. Chem.* 22(12):3088–3094.

OECD. Test No. 213: Honeybees, Acute Oral Toxicity Test, OECD Guidelines for the Testing of Chemicals. Section 2. OECD Publishing; 1998a. doi: 10.1787/9789264070165-en

OECD. Test No. 214: Honeybees, Acute Toxicity Test, OECD Guidelines for the Testing of Chemicals. Section 2. OECD Publishing; 1998b. doi: 10.1787/9789264070189-en

OECD. 2007. Series on Testing and Assessment, Number 75. Guidance document on the honey bee (*Apis mellifera* L.) brood test under semi-field conditions. ENV/JM/MONO(2007)22.

Oomen PA, De Ruijter A, Van Der Steen J. 1992. Method for honeybee brood feeding tests with insect growth-regulating insecticides. *Bull. OEPP /EPPO Bull.* 22:613–616.

Peach ML, Alson DG, Tepedino VJ. 1995. Sublethal effects of carbaryl bran bait on nesting performance, parental investment and offspring size and sex ratio of the alfalfa leafcutting bee (Hymenoptera: Megachilidae). *Environ. Entomol.* 24(1):34–39.

Pistorius J, Bischoff G, Heimbach U, Stähler M. 2009. Bee poisoning incidents in Germany in spring 2008 caused by abrasion of active substance from treated seeds during sowing of maize, pp. 118–126. In: Oomen PA, Thompson HM (eds), *Hazards of Pesticides to Bees*—10th International Symposium of the ICP-BR Bee Protection Group, Bucharest (Romania), October 8–9, 2008. Julius Kühn Archiv 423.

Rembold H, Lackner B. 1981. Rearing of honeybee larvae in vitro: effect of yeast extract on queen differentiation. *J. Apic. Res.* 20:165–171.

Riley JR, Smith AD. 2002. Design considerations for a harmonic radar to investigate the flight of insects at low altitude. *Comput. Electron. Agric.* 35:151–169.

[SANCO 2002] SANCO/10329/rev 2 final, 2002: http://ec.europa.eu/food/plant/protection/evaluation/guidance/wrkdoc09_en.pdf (accessed January 21, 2014).

Schmuck R. 2004. Effects of a chronic dietary exposure of the honeybee *Apis mellifera* (Hymenoptera: *Apidae*) to Imidacloprid. *Arch. Environ. Contam. Toxicol.* 47:471–478.

Schricker B, Stephen WP. 1970. The effects of sublethal doses of parathion on honeybee behavior. Oral administration and the communication dance. *J. Apicult. Res.* 9:141–153.

Scott-Dupree CD, Conrol L, Harris CR. 2009. Impact of currently used or potentially useful insecticides for canola agroecosystems on *Bombus impatiens*, (Hymenoptera: Apidae), *Megachile rotundata* (Hymenoptera: Megachilidae) and *Osmia lignaria* (Hymenoptera: Megachilidae). *J. Econ. Entomol.* 102(1):177–182.

SETAC. 1995. *Procedures for Assessing the Environmental Fate and Ecotoxicity of Pesticides*. Dr. Mark R. Lynch (ed), SETAC-Europe, Belgium.

Streit S, Bock F, Pirk CWW, Tautz J. 2003. Automatic life-long monitoring of individual insect behaviour now possible. *Zoology* 106:169–171.

Stute K. 1991. Auswirkungen von Pflanzenschutzmitteln auf die Honigbiene. Richtlinien für die Prüfung von Pflanzenschutzmitteln im Zulassungsverfahren, Teil VI, 23–1, Biologische Bundesanstalt für Land- und Forstwirtschaft (BBA), Braunschweig, Germany.

Suchail S, Guez D, Belzunces LP. 2001. Discrepancy between acute and chronic toxicity induced by imidacloprid and its metabolites in *Apis mellifera*. *Environ. Toxicol. Chem.* 20:2482–2486.

Tasei JN, Carre S, Moscatelli B, Grondeau C. 1988. Recherche de la D.L. 50 de la deltamethrine (Decis) chez *Megachile rotundata* F. Abeille pollinisatrice de la lucerne (*Medicago sativa* L.) et des effets de doses infralethales sure les adules et les larves. *Apidologie* 19(3):291–306.

Tesoriero D, Maccagnani B, Santi F, Celli G. 2003. Toxicity of three pesticides on larval instars of *Osmia cornuta*: preliminary results. *Bull. Insectol.* 56(1):169–171.

Thompson HM. 2001. Assessing the exposure and toxicity of pesticides to bumblebees (*Bombus* sp.). *Apidologie* 32:305–321.

USEPA. 1996. Ecological Effects Test Guidelines. OPPTS 850.3040: Field Testing for Pollinators. EPA 712-C-96-150. http://www.epa.gov/ocspp/pubs/frs/publications/OPPTS_Harmonized/850_Ecolgical_Effects_Test_Guidelines/Drafts/850-3040.pdf.

USEPA. 2012a. Ecological Effects Test Guidelines. OCSPP 850.3020: Honey Bee Acute Contact Toxicity. EPA 712-C-019. http://www.regulations.gov/#!documentDetail;D=EPA-HQ-OPPT-2009-0154-0016 (accessed January 21, 2014).

USEPA. 2012b. Ecological Effects Test Guidelines. OCSPP 850.3030: Honey Bee Toxicity of Residues on Foliage. EPA 712-C-018. http://www.regulations.gov/#!documentDetail;D=EPA-HQ-OPPT-2009-0154-0017 (accessed January 21, 2014).

USEPA. 2012c. Ecological Effects Test Guidelines. OCSPP 850.3040: Field Testing for Pollinators. EPA 712-C-017. http://www.regulations.gov/#!documentDetail;D=EPA-HQ-OPPT-2009-0154-0018 (accessed January 21, 2014).

Weick J, Thorn RS. 2002. Effects of acute sublethal exposure to coumaphos or diazinon on acquisition and discrimination of odor stimuli in the honey bee (Hymenoptera; *Apidae*). *J. Econ. Entomol.* 9:227–236.

Wittmann D, Engels W. 1981. Development of test procedures for insecticide-induced brood damage in honeybees. *Mitt. Dtsch. Ges. Allg. Angew. Entomol.* 3:187–190.

Yang EC, Chuang YC, Chen YL, Chang LH. 2008. Abnormal foraging behavior induced by sublethal dosage of imidacloprid in the honey bee (Hymenoptera: *Apidae*). *J. Econ. Entomol.* 101(6):1743–1748.

Zhang SW, Bartsch K, Srinivasan MV. 1996. Maze learning by honeybee. *Neurobiol. Learn. Mem.* 66:267–282.

9 Assessing Effects Through Semi-Field and Field Toxicity Testing

J. Pettis, I. Tornier, M. Clook, K. Wallner, B. Vaissiere, T. Stadler, W. Hou, G. Maynard, R. Becker, M. Coulson, P. Jourdan, M. Vaughan, R.C.F. Nocelli, C. Scott-Dupree, E. Johansen, C. Brittain, A. Dinter, and M. Kasina

CONTENTS

Pesticide Risk Assessment for Pollinators, First Edition. Edited by David Fischer and Thomas Moriarty.

9.1 INTRODUCTION

Semi-field and field studies may be conducted for regulatory purposes if lower tier assessments trigger further evaluation of a chemical's potential to cause adverse effects. For example, a regulatory trigger value may have been breached in the lower tier assessment that in turn means that a protection goal may not be met based on the findings at that level. One way to ensure that a protection goal is met is to modify the use of the subject compound such that it may no longer pose an unacceptable risk to the honey bees *Apis mellifera*[1] or non-*Apis* bees.[2] However, modifying or restricting the use of a compound may be undesirable or unnecessary if further information is obtained from either a semi-field or field study that demonstrate otherwise. Such a study or studies should provide greater insight into whether adverse effects to *Apis* or non-*Apis* bees are likely to occur under real-world field use of the pesticide in question. As such, the objective of the regulatory study(ies) may be to try to indicate, both quantitatively and qualitatively, what the possible effects may be under more environmentally realistic or relevant conditions. Such studies should be predicated on a well-developed problem formulation that builds on lower tier studies as well as the associated risk assessment.

As part of problem formulation, there should be a clear idea of the regulatory concern, identification of protection goals, assessment endpoints and related measurement endpoints on which to base judgments. For the purpose of developing guidance relative to higher tier tests, the participants of the Workshop assumed the protection goals stated at the outset of the conference, which include:

1. Protection of managed pollination in agricultural or horticultural-based crops (i.e., *Apis* and non-*Apis* species)
2. Protection of honey production and other hive-products; and
3. Protection of biodiversity (primarily non-*Apis* bees).

This chapter provides an overview of what to consider when planning or assessing either a semi-field or field study. As regards the honey bee, much use has been made of European Mediterranean Plan Protection Organization (EPPO) 170 (EPPO 2010) and Organization for Economic Cooperation and Development (OECD) 75 (OECD 2007). Participants during the Society of Environmental Toxicology and Chemistry (SETAC) 2011 Workshop used their collective practical and regulatory experience to provide further information on how a study should be conducted. Therefore, the following is seen as a development of both EPPO 170 and OECD 75 based on the experience of the experts present at the Workshop. If the risk assessor indicates the need for either a semi-field or field study, then it is recommended that this chapter along with the information provided in EPPO 170 and OECD 75 be consulted. The information in these references may also be consulted when such studies are being evaluated for regulatory purposes.

[1] We consider *Apis mellifera*, to be composed of approximately 17 subspecies that originated in Europe.

[2] Non-*Apis* bees are highly varied in terms of social and solitary lives, the duration of their activity in the field, the amount of pollen and nectar they store, and where they nest. For details, see Chapter 3 and Chapter 8.

9.2 DEFINITION OF SEMI-FIELD AND FIELD STUDIES

Elements in the design of semi-field and field studies encompass the study's objectives, the test organism, a study site, methods, endpoints, sample design, quality assurance and quality control standards, and the statistical analysis of the data. In discussing the elements of a semi-field study, the participants of the Workshop defined a semi-field study as the following.

A semi-field study is designed to measure exposure or effects and is performed on a crop that is grown outdoors in an enclosed test system with controlled or confined exposure. The crop is subject to good agricultural practices (i.e., grower standard practices), and therefore, there will or could be weeds present but the predominant plant, and thus the source of nectar or pollen, will be the crop. The test system could nevertheless be designed to reflect a desired exposure system and specific foraging environments, for example, a mixture of crop and weeds, or flowering margins. The details of the test design (such as application parameters or measurement endpoints) will depend upon the regulatory questions being asked. However, semi-field studies generally attempt to maximize exposure by confining bees to a particular source of treated nectar or pollen.

For species (both non-*Apis* and *Apis*) that are used to pollinate plants grown in greenhouses, it may also be necessary to carry out a higher tier study. A semi-field study will be enclosed with controlled or confined exposure but will be of reduced size compared to a commercial glasshouse. The size of the test environment is related to the species being studied, and the questions or issues being investigated.

A semi-field study, therefore, provides for a potentially worst-case exposure scenario.

A field study is designed to measure exposure or effects and is performed on a crop that is grown outdoors with no enclosure. The crop is established and maintained following good agricultural practices. While the bees are free-flying and able to seek out alternative food sources, alternative sources of pollen and nectar should be minimized. The study design elements (e.g., selection of crop, duration of the study, or environmental conditions) will depend upon the questions being asked. A field study for a greenhouse situation should be conducted in a commercial greenhouse.

9.3 DESIGN OF A SEMI-FIELD STUDY

When deciding whether a semi-field study is appropriate, it is necessary to consider various strengths and weaknesses of this type of study to ascertain whether it is the most appropriate way to refine the understanding of the potential risks from the use of a compound (Table 9.1 and Table 9.2).

9.3.1 WHEN WOULD A SEMI-FIELD STUDY BE APPROPRIATE?

Consistent with the tiered approach to toxicity testing and risk assessment, semi-field studies may be triggered when lower tier assessments (relying on laboratory results) indicate potential risks that are inconsistent with protection goals. In such cases, higher tier tests may provide information that reduces the uncertainty about risk, allowing for a more informed decision. Outlined below are scenarios when a semi-field study may be appropriate; and when a semi-field study may not be an appropriate option.

- If, as a result of the initial laboratory assessment, acute mortality or sub-lethal effects are considered to be the main concern, then a semi-field study may be appropriate.
- If repellency or an impact on foraging activity is predicted, either on the basis of efficacy data (e.g., a compound is known to act via an antifeedant effect) or from any observations, laboratory or any other relevant studies, then a semi-field study may give the risk assessor useful information on the potential short-term effects the compound may exert on foraging behavior. Because the confined

TABLE 9.1
Strengths and Weaknesses of Semi-Field Tests with *Apis Mellifera*

Strengths

Exposure is known since the bees are enclosed and there is usually a toxic reference treatment group. (The toxic reference treatment (using chemical of known toxicity to bees) is used to confirm that the bees are exposed to the treatment and to calibrate the ability of study to detect treatment effects known to be associated with the reference chemical.)

Provides realistic exposure both inside and outside the hive, that is, to both materials available at the target crop as well as concentrations in the hive.

The test system can also be designed to determine the residual toxicity. Weathering of the applied material and natural exposure of honey bees are inherent in the design.

Irrigation of the crop (via drip irrigation to avoid wash-off) is possible, hence potentially reducing the likelihood of the study being adversely affected by drought.

In contrast to laboratory studies, semi-field studies present a more realistic scenario of interaction between the bees and the environment.

Due to their smaller size and shorter duration, semi-field studies are less affected by fluctuations in ecological variables.

Potential for sublethal effects can be observed more easily than in either laboratory or field studies.

Brood can be considered in specifically designed semi-field studies (see OECD 75).

Semi-field tests are relatively quick and easy to perform.

Semi-field environments are smaller scale in operation than field studies, making it feasible to test greater numbers of replicates, which in turn should allow for more robust statistical designs.

As the bees are enclosed and have no alternative foraging environment, the exposure is potentially a "worst-case" scenario.

Certain exposure scenarios that are difficult to study under real field conditions, for example, aphid honeydew, can be studied under semi-field conditions.

Weaknesses

Experience with performing these studies has shown that it is difficult to keep colonies in an enclosed structure for long periods and, as a result there is a limited amount of time that a colony of *A. mellifera* can survive in the enclosure. The correct stage of crop bloom is critical to the study and, as a result it is only appropriate to assess the effect of short-term exposure including potential effects on brood (OECD 2007). Where exposure is either repeated over a sustained period (e.g., where there are repeated applications of the pesticide), or where exposure is continuous (e.g., from the use of systemic seed treatment), semi-field studies may be of limited usefulness in determining long-term effects.

Semi-field studies tend to use colonies with only 3000–5000 bees (EPPO 2010), which is smaller than a full size (managed) colony. Due to the current state of knowledge, it is not possible to determine whether an observed effect in a semi-field study would reliably indicate either an effect in a standard full size colony or no effect under field conditions. Hence, extrapolation of adverse effects to a full size unenclosed colony under more realistic field conditions may not be possible.

Due to the small size of the colony, it is not as easy to assess pollen and nectar storage and hive weight development; therefore, it is difficult to assess potential effects on honey production (i.e., a potential protection goal identified at the Workshop) when adverse effects are observed on other parameters.

Because the size of the colonies used in semi-field studies prohibit their ability to successfully overwinter, these studies may not provide information on overwintering success.

Due to the nature of the enclosed test design, not all crop scenarios are possible to test, (e.g., size of plants, area required, and nutritional value of crop to bees).

There is potentially limited foraging area; therefore, care is needed to ensure that sufficient nutrition (i.e., enclosed crop area) is available.

There is a possible stress on bees due to enclosed nature of the study, that is, bees have a desire to escape, consequently reducing their foraging activity on the crop. However, balance of tent size or crop field size and colony size should ensure foraging and exposure (EPPO 2010).

TABLE 9.2
Strengths and Weaknesses of Semi-Field Tests with Non-*Apis* Bee Species

Strengths
Individual colonies, or aggregations of individual solitary bees (such as *Meloponini* or *Bombus*) can be used and thus the pesticide effects are readily interpreted. Increased replications are possible and readily performed so statistical analysis may be easier.
Product use on a wide range of crops, including those that are not readily pollinated by honey bees (e.g., eggplant), can be assessed.
Some social non-*Apis* bees amenable to these tests, such as *Meloponini* (stingless bees) and *Bombus,* are easier to handle than *Apis* as they are reluctant to sting. Additionally, many of the solitary non-*Apis* bees, although capable, are reluctant to sting. Solitary bee species amenable to semi-field studies (e.g., *Osmia* and *Megachile* species) will not sting.
The area of the enclosure of a semi-field study can support full colonies of non-*Apis* species (*Bombus* or *Meloponini*) or a collection of independent individuals (solitary bees), hence an extended study can be done. These bees have a complete life cycle in 3–6 weeks (solitary bees) or one season (*Bombus*) in temperate climates.
Individual solitary bees typically provision nests over a 3–6-week period, thus allowing for a complete (or at least almost complete) lifecycle study for solitary bees if the forage crop flowers for more than 3 weeks.
It is possible to do larval exposure tests with solitary bees because pollen or nectar is brought straight to a cell and an egg is laid on the nest. This behavior leads to a potentially conservative assessment since the progeny has direct exposure, dermal and oral, with food resources that potentially contain the test pesticide.
Non-*Apis* bees can be used and maintained efficiently in small enclosures.
Non-*Apis* bees will forage under less optimal conditions in terms of temperature, relative humidity, and wind. This is especially true for *Osmia* and *Bombus* spp., which are quite hardy.
In solitary species such as those in Megachilidae, the larvae are in direct contact with nectar and pollen, and so there is the possibility of contact and oral exposure. This is not the case with *Apis* larvae that require a special larval test to ensure adequate expose larvae to a given pesticide and route of exposure.
Weaknesses
Resource supplements may be needed for crops that do not provide both pollen and nectar, which may reduce bee activity.
In temperate areas, the annual life cycle of solitary bees limits the window in which adult or larval testing may be conducted.
There is significant uncertainty as to how representative the current commercially available non-*Apis* bees are for other non-*Apis* species. For all non-*Apis* bees, there is enormous variation in the use of resources, behaviour, habitat requirements, life cycles, etc.

nature of the semi-field study can have a potentially confounding effect on the behavior of honey bees (e.g., stress), effects observed in a semi-field study may not be indicative of effects that would occur under field situations. However, if a potentially significant adverse effect on foraging behavior is observed, then there could be long-term effects and it may be necessary to extend the semi-field study or conduct a full-field study.

- A semi-field test may be used to validate or test a safe re-entry time for bees. Based on information gathered from a foliar residue toxicity study (see Chapter 8), a semi-field test can be used to provide additional information on the residual toxicity of a compound under more environmentally relevant conditions. For example, a semi-field study may provide information on the test compound residues, (i.e., when residues are dry and therefore "safe" for bees). This information can be applied to risk mitigation.
- If a pesticide is systemic and intended to be used as either a seed treatment, solid formulation (e.g., granule or pellet), or soil treatment, then a semi-field study can provide detailed information regarding exposure levels both in the target crop and in the hive associated with the specific application

parameters. Care is required in selecting a study site to ensure that environmental conditions (e.g., soil conditions (moisture, pH), duration from soil treatment to drilling, and flowering) are appropriately representative of the proposed use. The study can also provide an indication of the likelihood of initial mortality and initial behavioral effects following exposure. Since confinement may affect bee behavior per se, it is necessary to compare effects seen in treatment groups with those observed in the control. If there is a possibility of long-term effects resulting from this type of exposure, then it may be possible to modify this study appropriately or alternatively it may be preferable to conduct a field study. It is also important to target the exposure position of the study (i.e., the portion when the colonies are confined under an enclosure) with the time when the test plant is flowering and the highest expected residues are present in pollen and nectar.

- If the compound is an insect growth regulator, or exhibits insect growth regulatory characteristics, then a test according to Oomen et al. (1992) or a semi-field study over a 28-day period (OECD 2007) can provide information on the potential effects on growth or development.

One of the advantages of a semi-field study, in comparison to a field study, is that it allows for the inclusion of a toxic standard (i.e., one replicate is run with a test material that is known to elicit adverse effects to the test organism). However, since there are occasions where it is not possible to use a toxic reference chemical (e.g., systemic seed treatments[3]), the absence of a toxic reference does not greatly compromise the utility of the test. When testing seed treatment application scenarios, the residues on treated seed should be determined as well as the residues in pollen and nectar; exposure to the bees is assumed as the test system is closed and exposure cannot be avoided.

- Semi-field studies are also useful studies for non-*Apis* species such as *Megachile rotundata* as they may provide information on alternative routes of exposure, that is, leaves which are used for nest building, in addition to conventional routes of exposure such as nectar and pollen.
- It is possible to determine colony effects in a semi-field study over an extended period (e.g., for 3 months or longer) with species such as stingless bee and bumble bee colonies. For example, a bumble bee colony may be housed in a box with two connected chambers (one chamber for the colony's nest and one chamber from which the colony may be fed (Kearns and Thompson 2001). The nest box may be opened and the colony allowed to forage outside in a semi-field enclosure. After this exposure period, the nest may be closed and the colony fed in the nest box's feeding chamber for a month or two to look at delayed lethal or sublethal effects on reproduction and colony growth. After a couple of months, bumble bee colonies will switch from raising workers to raising drones and queens. Similarly, one can expose foragers from a stingless bee colony for several days in a semi-field enclosure and then close up the nest box. The colonies in this case can then be fed by placing food (sugar water and vitamins) at regular intervals into the nest box. Stingless bees have perennial colonies (much like honey bees) and may be fed *in situ* for many months.

As noted above, semi-field studies address mortality from short-term exposure as well as short-term behavioral effects. However, there is a concern whether they are able to address, (i) long-term effects from either short-term or sub-lethal exposure or, (ii) long-term effects from long-term or continual (i.e., via hive products) exposure or long-term chronic exposure.

[3] The lack of a toxic standard for a systemic seed treatment or solid formulation is due to the lack of a compound that causes known effects.

9.4 OUTLINE OF A SEMI-FIELD STUDY FOR *APIS* AND NON-*APIS* BEES

9.4.1 Design of a Semi-Field Study for *Apis* Bees

The following section is based largely on EPPO 170 and OECD 75 and should be seen as an extension of both the guidance documents, and considered along with the details of these guidances. In developing the elements of this chapter, the Workshop participants relied upon their experience as well as information included in EPPO 170 and OECD 75. The aim of the following section is to highlight further issues to be considered when planning and carrying out a semi-field study as well as issues that should be considered when evaluating a semi-field study for risk assessment purposes.

It is important that the aims of any semi-field study are clearly determined prior to the conduct of these studies. Clear problem formulation is required to ensure that the study is appropriately designed and focused to address the regulatory questions being asked. All semi-field studies should be designed to address specific concerns highlighted at lower tiers. EPPO 170 and OECD 75 are relatively flexible guidance documents and consequently allow studies to be designed to address specific issues. The considerations of the participants of the workshop, and of this chapter, do not remove or reduce that flexibility of the referenced guidance documents, rather they highlight areas or elements that are thought to be important considerations for incorporation into a semi-field study.

9.4.1.1 Size of Semi-Field Study

The minimum size of a semi-field study enclosure according to EPPO 170 is 40 m^2. Recommendations in this section are based on professional experience and are considered appropriate in terms of practicality of conducting the study and for determining effects of mortality and behavior. However, this area is only appropriate in terms of certain field crops (e.g., *Phacelia*, oilseed rape or canola, mustard). For other crops (e.g., melons, apples), the area (40 m^2) may need to be amended due to issues such as the number, density, and attractiveness of flowers, availability of nectar and pollen, or the size of the plants. The area of the test enclosure may also need to be amended depending upon the size of the colonies being used.

When studying bee brood, an increased (enclosed) crop area (>60 m^2) may be preferable to ensure the colony has access to adequate floral resources. However, the precise area depends on colony size, crop, and duration of confinement; 40 m^2 (OECD 2007) may be acceptable for a small colony that is confined for no more than 10 days.

9.4.1.2 Crop

The standard crops (i.e., oilseed rape or canola, mustard, and *Phacelia*) are easy to cultivate and manage but more importantly are highly attractive to honey bees. *Phacelia* has an open flower that it is highly attractive. The openness of its flower will mean that bee-relevant parts of the flower will be fully exposed to the spray application; hence, honey bees foraging after the spray application will be exposed to residues. Oilseed rape and mustard are both highly attractive to honey bees so a high level of exposure can be ensured. Results from studies carried out on these crops can be extrapolated to other crops, provided that the application parameters in terms of application rate, timing of applications, and number of applications used on the surrogate crops are comparable (ideally identical) to that of the subject product. If effects are observed on these standard crops then it may be possible to further refine the assessment by using the target crop species.

When considering systemic soil or seed treatments, it is preferable to use the actual crop. A crop other than the target crop needs to be justified on the basis of exposure (e.g., it may be appropriate to select a crop that is attractive and has high residues in nectar and pollen as a "model" crop rather than the actual crop of concern).

9.4.1.3 Size of Colony

Each tunnel, cage, or tent should include one, healthy queenright (i.e., a fertile, laying queen) colony per cage. Precise size of the colony used will depend upon the study design; EPPO 170 recommends a size of 3000–5000 bees.

It is important to have sufficient nutritional resources within an enclosure to ensure that the bees do not starve. Generally, feeding will not be necessary; however, if there is concern regarding the attractiveness of a specific crop or situation, then supplemental feeding may be needed. For example, if testing maize, then additional food will be required as maize produces no nectar.

9.4.1.4 Test Treatment

Sprays Only Test treatments and water (negative) controls are required; ideally a positive control (reference toxicant) is also required. It is customary to test only the proposed field rate. If, however, a model crop is used, for example, *Phacelia*, then it may be appropriate to have more than one treatment rate. This may enable the data to be extrapolated to other crops and other application rates. Additional tunnels or cages could be used to address different application rates as well as effects from treating at different times of the day. However, at a minimum, a study at the maximum proposed rate should be carried out.

A positive (reference toxicant) control: (i) provides an indication of the sensitivity of the test system; (ii) demonstrates exposure; and (iii) indicates the magnitude of response to a known toxin. However, positive controls kill bees unnecessarily and can add to the cost and complexity of study design; therefore, their use should be considered carefully. Positive control compounds are useful if it is unclear if any dose of the tested pesticide will have effects. If a positive control is used, it is necessary to select a compound whose toxicity profile is known and consistent with that under consideration, for example, for assessment of a potential acutely toxic compound, then there is a need to use a similar compound. Historically, dimethoate has been used as a reference chemical when studying acutely toxic compounds on adult forage bees. If insect growth regulatory effects are expected, then a known insect growth regulator with similar effects should be used. When a positive control is used, there should always be clear effects. There should not be sustained mortality at high levels in the water control. There should be an appropriate number of replicates for the treatment groups to provide sufficient power to discriminate treatment effects with a level of precision.

For Systemic Solid Formulation, Seed Treatments, or Soil Treatments While there is a need to have both the treatment and a water (negative) control, currently it is not possible at this time to identify a suitable positive (reference toxicant) control for most systemic solid formulation, seed treatments, or soil treatments.

9.4.1.5 Pre-Application

Sprays Only Healthy colonies should be used and transferred to the test site a minimum of 2–3 days prior to treatment. This is due to mortality that inevitably occurs when a colony is moved and subsequently confined. If the hive is moved during the day, the hive will tend to acclimate more quickly. There should be a measurement of mortality over the acclimation period; the greater number of measurements of mortality will provide greater confidence that the effects after treatment are attributable to the treatment rather than due to the hive acclimation. It is likely that there will be variability between colonies and every effort should be made to ensure that they are as consistent as possible. This can be partly achieved by moving the colonies at the same time. Attempts should be made to ensure the colonies are as similar as possible, in terms of the number of bees, at the start of the study. Excessive variation at the start of the study will make the study difficult to interpret and hence potentially limit its usefulness.

Further work is required to determine the range of background levels of mortality once the colony(ies) is situated at the test location in order to establish acceptable levels or ranges of mortality. These background levels could be used to help interpret whether the level of mortality observed in the treatment is treatment-related or not, providing an indication as to the overall reliability of the study.

With spray treatments, the colony is placed in the semi-field setting when the crop is just about, or at flowering. The effects of the pesticide to honey bees foraging that crop are then determined. With systemic chemistries, there is a potential for exposure to occur over a longer period of time; therefore, the honey bees should be present during the whole flowering period of the plant. Acclimation as outlined above is, therefore, not possible as exposure of the bees to the pesticide will occur as soon as they are introduced into the treatment area. However, a consideration of mortality due to moving the colony is still required. One potential way around this problem is to compare the mortality that occurs with the untreated crop to that with the treated crop. Nevertheless, the significance should be determined statistically. Semi-field studies may be most effective for determining acute effects related to systemic chemistries. If sublethal effects are predicted, then a modified semi-field test designed to ascertain any long-term effects, or simply a full-field test may be more appropriate.

9.4.1.6 Post-Treatment Assessments

Assessments of mortality via the placement of dead bee traps, sheets, or tarps at the front of the hive and within the enclosure should ideally be carried out daily but at least on days 0, 1, 2, 4 and 7 post-treatment. This frequency is not appropriate for in-hive assessments as the disturbance could cause significant effects.

Sublethal Behavioral Tests There is a need to standardize and refine the number and type of tests or observations that can be made to document potential behavioral changes due to sublethal pesticide exposure. In these tests, it is typical to report abnormal behavior in foraging or other behaviors that may be exhibited during the test; but, definitive and meaningful quantifiable measures may often be lacking. Rather than making general observations on bee behavior, it is proposed that more detailed measurements can be made in addition to the general observations used to date. Of these, perhaps the most obvious is in measuring foraging activity.

When measuring foraging activity, the number of returning foragers should be counted before treatment and at regular intervals post treatment. The number of returning foragers with pollen loads should constitute a separate count from those returning without pollen (nectar and water foragers). Observations should last for 1–3 minutes. The observation periods should be equally divided across all test groups so that measurements are taken at approximately the same time with the controls as with treatments.

A second observation that could be quantitatively measured in a semi-field test is the average flower handling time. This measure is made by recording the time taken for the bee to work a flower (i.e., to remove pollen or nectar). The observer simply records the total flower handling time for bees collecting pollen and nectar. If the flower type is such that distinct pollen and nectar foraging is possible, then these forager types should be kept separate. The exact number of required measurements should be determined or justified statistically. The time of day at which measurements are taken should be randomized between plots to avoid time of day and or weather bias. As with previous studies of this type, general observations of any unusual bee behavior should be noted and quantified if possible (e.g., 30 bees were seen twitching and exhibiting excessive grooming on the landing board during the 1–3-minute foraging counts). In addition, it may be possible to determine foraging behavior in front of the hive.

Due to the confined nature of semi-field studies, consensus of the Workshop participants was that an adverse effect on behavior compared to the control should be interpreted with caution and should trigger additional consideration. The relevance of an effect, or lack thereof, in a semi-field study may not be assumed to be directly translated to the field scale. Interpretation of effects, or lack thereof, must be done with care. Additional information could be obtained to aid interpretation of any effects seen. This information could come from a variety of sources; however, the Workshop participants considered that field studies were the most appropriate source to validate any effects or lack of effects that are considered significant.

Depending upon the regulatory question being asked, it may be necessary to determine residues in fresh pollen, stored pollen, nectar, honey, and wax. The types of samples to be collected depend on the study and the questions to be answered. Residues in foraging honey bees may also be ascertained and this information could be used in interpreting potential incidents.

9.4.1.7 Results

Traditionally when determining if a study is acceptable, there is consideration of whether it has met various quality criteria, such as adequate controls or chain of custody. In addition, there should be consideration as to how the study compares to the above guidance. The use of a toxic reference chemical can help meet the need for quality assurance measures; however, it is not essential for the reasons stated above.

Based upon the study objectives, key outputs from a standard semi-field study could be the following.

- Mortality in the crop: use of sheets or tarps in the crop.
- Mortality at the hive: use of dead bee traps or sheets in front of the hives.
- Foraging activity and other behavior.
- Measures of exposure: residues in pollen, nectar, pollen pellets, and dead bees.
- Pollination deficit: it may be possible to determine if there is a difference in the degree of pollination success (e.g., via fruit set) of the treated versus untreated crop.
- Assessment of the brood, including an estimate of adults, the area containing cells, larvae, and capped cells (if this is a key area then methods outlined in OECD 75 should be followed).

9.4.2 Design of a Semi-field Study for Non-*Apis* bees

At present, there is no equivalent EPPO 170 or OECD 75 guideline for using non-*Apis* bees in semi-field or field studies. As a result, the Workshop participants suggest that if there is a regulatory question regarding a pesticide that requires the inclusion of a non-*Apis* species as a result of triggers activated by laboratory effects bioassays, the study design should be developed on a case-by-case basis with consideration of the specific endpoints described for semi-field honey bee studies and the overall regulatory question. Care should be taken when evaluating and interpreting results from these studies until protocols are sufficiently vetted through ring-testing.

When selecting non-*Apis* species to be used for semi-field studies, attention needs to be paid to their availability, ease of handling, and survival under experimental conditions. Therefore, it is recommended that the species used are those that are either commercially available or can be readily reared under laboratory conditions.

9.4.2.1 Semi-Field Studies—Solitary Bees

Three solitary non-social bee species are recommended for use in semi-field studies in temperate zones: *Osmia lignaria*, *Osmia Bicornis,* and *M. rotundata* (Johansen et al. 1984; Tasei et al. 1988; Konrad et al. 2008; Ladurner et al. 2008). *M. rotundata* will be used as the descriptive species in this section.

M. rotundata, the alfalfa leafcutting bee, is a non-social Eurasian bee species that is widely managed as a pollinator of alfalfa for seed production in the United States and Canada, and is occasionally deployed for the pollination of other specialty crops (e.g., canola, carrot—for seed, blueberries). Dormant alfalfa leafcutting prepupae are sold as loose cells in 4 L increments (approximately 10 000 individual cells).

Due to standard field production cycles, dormant loose cells are usually only available from late fall through early winter. Cells should be maintained at 1.7°C–4.4°C and 50% relative humidity (RH) until natural emergence during early summer in most of the northern hemisphere. Bees maintained in cold storage beyond this point begin to deplete stored energy reserves and may fail to emerge upon incubation (210 total days is the general upper limit for diapause before viability declines significantly). Cells should be stored in open or ventilated containers and tumbled periodically to reduce the growth of molds. Bees can be incubated to adulthood with as few as 150 days of cold storage diapause. Careful control of temperature (i.e., 29°C) and humidity (70% RH) will cause most of the incubated bees to emerge from their cocoons at approximately the same time (50% emergence in 23 days and complete emergence after 32 days).

Few release rates (density rates) exist for crops with the exception of alfalfa, where 74 000 to 100 000 bees per hectare are recommended, and canola and blueberries, where 50 000 bees per hectare (Mader et al. 2010a) are recommended. Release rates will vary based on the size of enclosure and crop to be utilized in the semi-field study but could be as few as 200–500 solitary bees per tunnel site of 40 m^2.

Site selection for the study should use the same criteria as those for semi-field *Apis* studies. Once an enclosure is ready, a wooden nest shelter containing enough styrofoam nesting boards to accommodate all the *M. rotundata* to be released for the study should be placed in the test enclosure (2–3 nest tunnels per bee), facing the morning sun, 3–4 days in advance of the initiation of the study (i.e., before the pesticide is to be sprayed in the semi-field enclosure). Bees ready to emerge or already emerged should be placed in front of the nest shelter and left to orient to the nest. Bees should not require supplemental feed as long as there is sufficient crop in bloom. These bees do not require a water source so long as enough flowers or a nectar feeder is available. However, if mason bees (*O. lignaria*) are used, a drip bucket and excavated damp mud pit are needed inside a test enclosure (i.e., tunnel) cage. The mud pit should be excavated so the bees can access the soil profile layer with the best clay-water content. Nectar is not sufficient for wetting mud.

Key Outputs

- Mortality in the crop: same as for *Apis*.
- Mortality in the hive or nest shelter: use of a tarp placed on the ground in front of the nest shelter may allow some assessment of *M. rotundata* mortality. However, solitary bees may die within the nest material, making mortality assessment more difficult. Assessment schedule should be the same as those for *A. mellifera*.
- Foraging activity: same as for *Apis*.
- Reproductive success (colony health): once it is known that the released female *M. rotundata* have successfully mated and started to provision cells (i.e., individual cells or eggs are present or tunnels are sealed), assessments on increasing brood nest (e.g., brood development) can begin. Nest boxes can be monitored on the first day once cell provisioning has commenced and continued on a weekly or bi-weekly basis. Count and mark completed tunnels. Observation nests (grooved boards with clear acetate or glass covering the grooves) can be used to observe nest, cell, and brood development without disturbing the bees. At 15.6°C (60°F), eggs of *M. rotundata* take 15 days to hatch and then an additional 35 days are required for the larvae to reach the prepupal stage. At 35°C (95°F), it takes 2–3 days for the eggs to hatch and 11 days for the larvae to reach the prepupal stage (Mader et al. 2010a). Therefore, if flowering of the study crop ends prior to either 14 days at 35°C or 50 days at 15.6°C, then the nest box needs to be removed from the study site and placed in a growth chamber that simulates the average temperatures experienced by the bees while they were in the enclosure. Once the prepupal stage has been reached, a segment of the styrofoam nest needs to be dismantled, and cells per tunnel counted and weighed, and then dissected to determine the number of cells with prepupae and those that are provisioned but with no larvae present. If there are no larvae present (i.e., these cells are called "pollen balls"), it indicates that larvae have died in the first or second larval instar, which may be related to the exposure to extreme temperatures (cold and hot) during that stage in development (Mader et al. 2010a). The remaining styrofoam nest sections can be dismantled, cells counted and then placed in storage at 2°C–5°C (35°F–40°F) at 50% RH until the following

spring. At that time, the diapause can be broken and the number of emerged adults can be counted and compared to the total number of cells. This allows for determination of mortality in progeny (sublethal effects).

9.4.2.2 Semi-Field Studies—Social Non-*Apis* Bees

Bombus spp. will be used as the descriptive species in this section.

Bumble bee colonies are readily available from commercial sources.[4] A colony consisting of 50–300 workers and a queen can efficiently pollinate 1000–3000 m^2 (Morandin et al. 2001) of tomatoes, yet should also perform as well as a honey bee nucleus hive in a smaller enclosure (40–60 m^2). The 40–60 m^2 foraging area, and considerations for supplying alternative forage (e.g., nectar or pollen) are relevant considerations for bumble bees, as they are for honey bees. In addition, feeding bumble bees colonies can be done in a much more controlled way than *Apis*. When *Bombus* are commercially reared, they are fed in the nest, and the same could be done for colonies used in a semi-field test. Colonies should be provided with identical amounts of supplemental pollen or nectar, helping to minimize differences between treatments. Also, when changing food stores, the pollen or nectar that was not consumed can be removed and weighed in order to determine how much the colony consumed. A colony population of at least 100 workers and a queen should be used for semi-field studies, and exposure duration should be 10 days followed by supplemental feeding. If the colony is movable, then it may be appropriate to move it to a non-agricultural and pesticide-free landscape to continue development outside the tunnel, rather than keep them inside the tunnel with artificial food.

When extracting bees for sampling, or for mark and release, it is necessary to distinguish the queen (usually the largest bee) from the workers. Harm to the queen is likely to result in defensive behavior on the part of the workers and a rapid reduction in colony life span. Similarly, it may be desirable to distinguish between male bees and female workers. In general, male bumble bees have larger eyes, longer antennae, no pollen baskets (corbiculae), and depending on the species, may have a notable patch of yellow hair on the front of their face.

One to two *Bombus* spp. colonies of similar age and with approximately 300 workers per colony should be moved to the semi-field study enclosure with entrances closed in the morning. Each colony should be placed on a concrete block with the entrance facing the morning sun. This should be done 2–3 days prior to the initiation of the study.

Key Outputs

- Mortality in the crop: same as for *Apis*.
- Mortality at the hive: same as for *Apis*. A small tarp can be placed under the colony extending outward from the entrance so that any dead adults or drone larvae discarded by the colony can be counted over time. The tarp should be cleaned of all discarded adults and drone larvae after each assessment. Endpoints such as discarded dead adults and drone larvae are indicators of colony condition.
- Foraging activity: same as for *Apis*.
- Reproductive success (colony health). Before placing colonies in the semi-field enclosure, a (close-up) photograph should be taken of the brood nest and food stores through the plastic inner cover at night when most of the bees are back in the nest. The photograph should be labeled with date and time and assessed for presence of brood in all phases of development by marking the cells with a marker on the photograph.

[4] Worldwide, different bumble bee or alternative social non-*Apis* species are commercially reared for pollination purposes and, therefore, in most regions will not require import procedures (Mader et al. 2010b).

9.4.2.3 Semi-Field Studies—Stingless Species

The stingless bees (Meliponini) consist of approximately 24 genera of bees with around 400 species (the number is not clear as many species still remain to be described). They are important social bees in the subtropics and tropics (Nogueira-Neto 1997). Meliponini occur mainly in Neotropical America, Australia, Indonesia, Malaysia, India, and Africa (Proni 2000). These bees are and have been important cultural components of many communities in the tropics and they are managed for their pollination services and honey production.

Stingless bees have varied nesting sites, from aerial parts of trees to underground. They differ from *Apis* spp. in that their combs or cells are arranged horizontally and are mass provisioned by the nurse bees with nectar, hypopharyngeal gland secretions, and pollen before the queen lays the egg after which the cell is closed. Full development of the adult takes place within these cells without any further input by the nurse bees; hence each cell is representative of the conditions that existed during the construction and provisioning of the cells. A newly emerged bee destroys its cell immediately. Honey and pollen stocks are usually stored at the periphery of the nest with the brood in the middle of the colony. However, the arrangement of the brood and storage pots varies between species, and for many species these details remain unknown. It is believed that the adult workers have a similar life span to that of *A. mellifera*, that is, they live 30–40 days.

Meliponini range in length from 1.8 to 13.8 mm (Michener 2007) and, because of this, the choice of the species is important for risk assessment tests. For example, in the past few years, *Melipona scutellaris* has been tested in greenhouses on tomato plants; and in tropical areas, some species such as *Trigona carbonaria* live or are managed in semi-domesticated situations. See Table 3.1 for a list of species and references for non-*Apis* species that have been employed in laboratory or field tests.

Individual bees or the inner colony are easily accessed for testing. Individual bees can be chilled for several minutes in a freezer to slow down their movement for ease of handling (the entire hive box should not be chilled). Heard (1999) and others have developed various hive box systems that can be used to manage these bees.

As regard to size of semi-field study, it is proposed that the approach used for the honey bee be adopted for the stingless non-*Apis* species.

Key Outputs Details are similar to *Bombus* above.

9.4.3 Interpretation of Effects in Semi-Field Studies

As stated at the outset of this chapter, the interpretation of effects (i.e., a statistically or biologically significant difference from the control) is linked to the protection goals and, in particular, whether the results indicate that protection goals are likely to be met or not.

If the protection goal is pollination activity or function, then a semi-field study with measurements of foraging activity is capable of determining whether pollination activity is related to treatment. If there is an adverse effect on foraging activity in the semi-field study, then further information is required to determine whether the effects are realized at the field level. It was the view of the Workshop participants that this would be best addressed via a field study. Alternatively, consideration of risk mitigation may be elements of consideration in determining how to proceed.

If the protection goal is honey production, then the results from a semi-field study can be interpreted as follows:

- If effects are clearly *not seen* on any parameters then it can be inferred that there will be no impact on honey production at the field scale when fullsized colonies are exposed. This assumes that long-term effects from short-term exposure are not an issue.

- If effects *are seen* or observed, for example, mortality or reduction in foraging or behavioral effects, then it may not immediately be assumed that honey production will be adversely impacted at the full-field scale. Since the semi-field test is potentially a worst case exposure scenario, the assessor needs to determine whether similar or any effects would be realized at the full-field level and hence whether honey production could be impacted.

If the protection goal is maintenance of biodiversity in terms of the ecosystem service of pollination by other non-*Apis* bees, then no negative impact on populations is the protection goal. Semi-field studies showing statistically significant effects that are expected to result in high levels of mortality should be considered for more refined field studies.[5]

9.4.4 Assessment of the Variability and Uncertainty in an *Apis* Semi-field Study

As with any experimental testing, there are sources of variability and uncertainties associated with the studies. Confining organisms to a restricted study environment can confound efforts aimed at reflecting more environmentally realistic conditions. Sources of variability and uncertainty should be considered in evaluating the conclusions from such studies (Table 9.3). To the extent that researchers can recognize and limit these potential confounding effects, the data generated from semi-field studies will likely improve, as well as their utility in regulatory decision making.

9.5 DESIGN OF A FIELD STUDY

9.5.1 When would a Field Study be Appropriate?

Field trials may be carried out if an acceptable risk is not estimated by either lower tier tests or the proposed risk mitigation is undesirable. Questions to be answered from a field test should be based on the results of lower tier studies, whether laboratory or semi-field. For example, if behavioral effects are observed in a semi-field study, it may be desirable to see if these are observed under more realistic field conditions. It may also be more appropriate to conduct a field study where a semi-field study is not considered to be appropriate (i.e., it is not necessary always to follow the tiered approach). For example, it may be relevant when there is the likelihood of long-term effects following short-term exposure. As with any test involving animals, the need for and intent of the study should be clearly articulated. This is particularly true for field pollinator studies given the number of variables that must be managed, and the considerable resources they require both on the part of the regulated community to conduct the study as well as the regulatory authority tasked with reviewing the study.

9.6 OUTLINE OF A FIELD STUDY FOR *APIS* AND NON-*APIS* SPECIES

9.6.1 Design of a Field Study for *Apis mellifera*

Field trials can be used to address a range of exposure scenarios and effects. The results can be used by the risk assessor to determine whether significant uncertainties have been sufficiently addressed and if the protection goals may be met. However, there are various strengths and weaknesses of field studies that need to

[5] In determining whether the protection goal of maintaining biodiversity has been met, it is necessary to determine whether it is possible to extrapolate from studies on one non-*Apis* species and conclude whether the pollination services (and any other services) that are supplied by non-*Apis* bees have been adversely affected. Further work is required to develop an appropriate risk assessment scheme around those goals and hence address issues such as the potential to extrapolate from one non-*Apis* species to others.

TABLE 9.3
Variability and Uncertainty in Semi-Field Studies with *Apis Mellifera*

Parameter	Discussion of Uncertainty
Enclosed population of bees	Under natural conditions, bees are free flying; enclosing them introduces a stressor that could lead to uncertainty in interpreting the results from a semi-field study. Enclosing bees in a semi-field setting causes two main issues, which may raise uncertainty when interpreting the results—(i) effects on behavior and (ii) availability of food and therefore, on foraging activity. Food availability and foraging issues can be addressed through design considerations to ensure sufficient food is available. This can be achieved by balancing the size of the colony with the size of the enclosed crop. Details regarding possible colony size and area of crop combinations are discussed above. Providing a study designed to ensure that ample food is readily available and that there are comparable controls should account for this potential confounding variable. Enclosing the bees in a semi-field setting could translate into behavioral effects, which could reduce exposure. For example, some bees will try to forage outside and as a result remain on the tent or cage wall rather than in the treated crop. It is not known what proportion of bees will exhibit this behavior. If the compound does not exhibit repellency effects on bees, it is thought that the same proportion of bees will potentially exhibit this characteristic in the controls as in the test groups. As there will be a proportion of bees that will not be exposed then this could potentially *underestimate* the risk. However, it is also not known what proportions of bees in the field are not exposed to the pesticide, that is, the proportion that will forage elsewhere. Provided that the population size is measured as a parameter, significant differences in comparison to controls indicate whether it is treatment related or not. It is considered that on the one hand, exposure is confined and controlled; however, there will be a proportion of bees that try to forage elsewhere. Overall, participants of the Workshop believe that this parameter is likely to over-estimate potential risk, that is, it will be worst case.
Size of colony	The colony of bees that is used in semi-field studies is small compared with those used in the field; and the way that a small colony reacts is different than the way fullsize colonies react. Extrapolating effects related to mortality and sub-lethal behavior from a small colony to a standard colony is uncertain and should be approached with caution. Due to this uncertainty, if any effects are noted then further studies should be considered.
Measure of mortality	Due to the confined nature of the study, it is likely that a semi-field study will yield a relatively accurate assessment of mortality. This is in contrast to the field, where detecting an accurate level of mortality within the crop is more difficult.
Density of bees in the treated crop	It is likely that the density of bees will be higher in a semi-field study compared to the field study. Due to the potential higher density of bees in a semi-field study compared to the field situation where alternative sources of food will be available, it is considered that bees are likely to have a higher level of exposure in a semi-field study, and therefore a semi-field test potentially over-estimates any effect.

TABLE 9.3
(Continued)

Parameter	Discussion of Uncertainty
Representativeness of the study site, agricultural practices, and conditions	It is unlikely that there will be a study to represent every crop and geographical and agricultural combination being considered in the specific regulatory context. Hence, there will be uncertainty regarding the representativeness of the selected study site in comparison with possible combinations under regulatory consideration. Ideally the study site, in terms of weather, flower availability, and forage, should be designed to ensure that the bees are exposed. Uncertainty regarding the representativeness of the crop will be reduced if a surrogate is chosen that ensures that bees are suitably exposed. Addressing uncertainty based on agricultural and geographical variability is more problematic.
Residues in pollen and nectar	For pollen and nectar residue sampled from the plants, there is no reason to believe that these should vary any more or less than what would occur under field conditions, with the exception of no or limited exposure to rain (wash-off), wind, or dew. Typically, semi-field studies have some latitude to make applications during periods of good weather. If poor weather is anticipated, then applications may be delayed several days provided the colonies are not already in the enclosure. However, semi-field studies are intended to reflect real world conditions, and if it rains, then such studies can still provide useful information. Typically, residue studies are conducted on the treated plants and in pollen or nectar to ensure that some level of exposure is achieved and the results are expressed relative to these residues.
Collected nectar, pollen pellets, bee bread, and dead bees	Regarding nectar, there may be a high turnover rate (foragers) in a semi-field study and, therefore, there may be difficulties in extrapolating this information to the field situation. Pollen and associated residues should be representative of what is likely to occur in the field and, therefore, the uncertainty associated with this parameter is low. Bee bread is difficult to collect in a semi-field study and the study has to be managed to ensure that this occurs. There is, therefore, some uncertainty regarding this parameter compared to what would happen in the field. Uncertainty exists if the study is extrapolated to other crops; for example, if one crop produces pollen and nectar whereas another species produces only pollen.
Assessment of the brood	This is possible only via OECD 75 and associated procedures.
Overall	Due to the confined nature of the semi field study, there is high confidence that exposure will occur compared to a full-field study. It is also likely that any adverse behavioral effects will be seen. Therefore, if either increased mortality compared to the control or behavioral effects are not observed then it is considered highly likely that these will not occur in the field. Uncertainty exists regarding the potential effects on brood development; however, it is considered that this will lead to potential overestimation of the risk. Due to the duration of the exposure in the semi-field study, determination of long-term effects requires special consideration.

TABLE 9.4
Strengths and Weaknesses of Field Studies for Both *Apis* and Non-*Apis* Bee Species

Strengths
Provides a realistic exposure scenario of bees foraging on a crop, provided test plot size is sufficient.
The realistic exposure scenario is likely to allow realistic behavior of the bees.
Can be designed to be consistent with good agricultural practice and grower standard practice.
Can be designed and used to assess long-term exposure and effects.
Ecologically (field level effects) and biologically (standard size colonies) more relevant than lower tier studies.
Measurement of certain protection goals can only be, or are more accurately, determined in field studies (e.g., pollination deficit or honey production) assuming that lower tier studies are insufficient to this end.
Weaknesses
Difficulty in finding appropriate sites, that is, there are practical issues in finding a site that is sufficiently isolated from other potentially attractive crops or pesticide treatments.
Because field studies are open, controlling nutritional sources may be difficult as bees may not forage exclusively within the treated field.
Expensive to establish treatment area of a size suitable for indicating "worst case" exposure. Field studies are logistically complex and are expensive since so many factors must be accounted for.
Potential difficulty related to background levels of pesticides in the foraging area.
Difficult to use toxic standard which in turn potentially raises concerns regarding sensitivity of the test system.
Potential high level of variability including weather, mortality away from the hive, replication, and interpretation of results.

be considered before they are used in risk assessments intended for use in a regulatory context. In Table 9.4, the strengths and weaknesses of the field study are listed. Qualities of the field study, with respect to either *Apis* or non-*Apis* bee species, are relatively generic and so are listed together in one table.

9.6.1.1 Study Design Considerations

For all types of application (i.e., spray, systemic solid formulation, seed treatments, or soil treatments applications).

The study should use colonies with a minimum of 10 000–15 000 foraging bees. Colonies should consist of 10–12 frames and include 5–6 brood frames. If colonies are of a different size then they should be evenly distributed between treatments. According to EPPO 170, an area of 2500–10 000 m^2 (0.25–1 ha) is recommended with a larger area proposed if the crop is not particularly attractive (e.g., 0.25 ha for *Phacelia* and 1 ha for mustard and oilseed rape). EPPO 170 also recommends that there should be a minimum of four colonies per field. It may be appropriate or necessary depending upon the regulatory question being asked, to consider the use of larger field sizes as this may provide a greater degree of realism when compared to the eventual use of the product. If larger fields are used, then more colonies may be required, depending upon the attractiveness of the crop. It is important to determine, from scientific literature, the proper colony loading rates based on the crop and size of field. In determining the size of individual fields, consideration must be given to the total number of treatments (i.e., the treated crop) and replicates per treatment (i.e., colonies per treated field).

While it is potentially desirable to use a positive control in a semi-field study, it is discouraged in a full-field study. This recommendation is based on extensive discussion among the International Commission for Plant-Bee Relationships (ICPBR) and EPPO. A negative control, however, is always required.

Participants of the Workshop agree that bees generally tend to forage on sources close to the colony, but that some bees will forage further afield and these individuals could bring additional residues into the colony.

Consequently, in order to ensure adequate isolation from other sources of pollen and nectar, the site should be located at least 2–3 km from alternative cultivated agricultural sources of pollen and nectar, including pollen and nectar from orchard trees. As regards confirming exposure, the following measurements should be considered.

- Bees/m^2—at least 5 bees per m^2 on *Phacelia* spp. or 2–3 bees per m^2 on oilseed rape and mustard (EPPO 2010). These are potentially only relevant for these crops and EU conditions and should be used with caution in other regions. It should also be noted that these densities are related to the number of colonies and size of treated area.
- Pollen identification—it is recommended to have additional colonies with pollen traps fitted. Identification of pollen can be difficult and sometimes identification only is possible to family level.

If appropriate, there should be an assessment of the degree of flowering, that is, the proportion of the crop actually in flower at any one time, for example, BBCH 60 onward for oilseed rape (see Meier 2001 for further details). This is particularly relevant for crops such as melons. Under certain conditions, it may be possible to manage the crop to prolong flowering so that a longer exposure period could result.

For systemic compounds, it is not possible to identify a suitable positive (reference) standard. In addition, similar to considerations with systemic compounds under a semi-field design, exposure will occur over a longer time. Therefore, the honey bees should be present during the whole flowering period of the plant. Acclimation to the pesticide will occur as soon as they are introduced into the treatment area. However, a consideration of mortality due to moving the colony is still required. One potential way around this is to compare the mortality that occurs on the untreated crop to that in the treated crop.

9.6.1.2 Pre-Application

For all application types, pre-application considerations are similar to that for semi-field studies. Refer to these sections above.

9.6.1.3 Post-Treatment Assessments

All types of application (i.e., spray, systemic solid formulation, seed treatments, or soil treatments applications).

Depending upon the regulatory question being asked, it may be necessary to assess behavioral effects in the field. Mortality, however, should always be determined. While this may be done via the use of dead bee traps, these may not always be appropriate, in which case sheets or tarps outside the hive should be used.

A key issue with field studies is ensuring that sufficient exposure occurs. If possible, studies should be designed to minimize alternative forage. However, it is inevitable that there will be some alternative sources present. In order to determine whether exposure has occurred, there is a need to monitor the activity of bees within the treated crop. This can be done in several ways.

- Measuring forage activity.[6]
- Measuring flight activity: aided through the use of marked bees.
- Identifying pollen from outside the colony.
- Measuring residues in pollen and nectar in bees and inside the colony.[7]

[6] See previous discussion on measuring foraging activity, and see similar discussion under the semi-field section.

[7] (Closely related to this point is whether the exposure measured in this study will be representative of the wide-scale use of the pesticide.)

9.6.1.4 Results

The following measurement endpoints and outputs are possible from a field study.

- Colony strength: ascertained through measurements of foraging activity, flight activity, and number of dead bees.
- Weight of the hive.
- Pollen, honey, and nectar stores: ascertained through measurement of percent comb coverage.
- Mortality at the hive: ascertained through measurements with dead bee traps or collecting sheets.
- Mortality of drones and pupae: ascertained through visual inspection of frames.
- Mortality in the crop: ascertained through collection sheets in the treatment site.
- Presence of the same queen.
- Foraging activity in the crop: measured in the test crop, or at the hive entrance where it can be recorded automatically.
- Returning foraging bees: can be counted automatically at the hive entrance.
- Behavioral abnormalities.
- Measurement of residues in pollen or nectar, or via pollen pellets, as well as in wax, bee bread, and dead bees: measurements of exposure inform assessment of risk.
- Assessment of the brood: see OECD 75; this measurement may also include an estimate of the number of adults, the area containing cells, eggs, larvae, or the capped cells.
- Disease or pest levels.

9.6.2 Long-term Risk to Honey Bees from Short-term Exposure

If potential overwinter effects are identified during the problem formulation step, then it is proposed that the field study can be modified in order to examine measurement endpoints that will address this uncertainty. (Generally, field studies are more appropriate to assess the impact of overwintering than extended semi-field studies.)

If a field study is to be conducted to determine whether the use of a product has any adverse effects on overwintering survival, then it is proposed that in addition to the considerations discussed above, the following points are also considered.

Following the exposure phase, the colonies (treatment and controls) should be relocated to an area that has limited, or no agricultural crops but an abundance of natural vegetation. This is necessary to ensure that exposure to additional pesticides does not occur.

At the end of the winter period, it is proposed that the following assessment endpoints should be determined, (the exact endpoints however, will depend upon the issues highlighted in problem formulation).

- Condition of the colonies or colony strength,
- Brood development,
- Brood assessment, including:
  - Number, density, or pattern of brood
  - Presence of healthy egg-laying queen
  - Estimate of pollen and nectar storage areas
  - Estimate of areas containing eggs, larvae, and capped cells
- Analysis for disease, (e.g., *Nosema apis*, *Varroa destructor*, American foulbrood, bee viruses)
- Weight of the colonies,
- Residue samples from the hive (e.g., pollen, wax, honey, bees).

9.6.3 Interpretation of Effects

As for semi-field studies, the interpretation of effects is linked to the protection goals. It should be noted that while a full-field test is the highest tier of testing it is important that final determination of potential risk and whether the use of the compound is consistent with protection goals should be based on the entire body of evidence across all tiers.

If the protection goal is pollination activity or pollination function, then the full-field study is capable of determining whether this is achieved via use of measurements on foraging (which can include foraging for nectar and pollen), behavior and, mortality. If no effect is observed on any of these parameters then the protection goal will be met. If effects are seen on any of these parameters, it may be unlikely that the protection goal will be met. Risk mitigation measures may enable the protection goal to be met; it is, however, essential to ensure that an assessment of the appropriateness and practicality of the risk mitigation measures can be made, and that the protection goal is met. (It should be noted that none of the measurement endpoints directly measure pollination activity *per se*, but are surrogate measures and indicative of pollination activity. That is, in using foraging activity it is assumed that a decrease in foraging activity will result in a decrease in pollination; e.g., decreased fruit set.)

If the protection goal is honey production by the colony, then this study can provide useful information. For example, if there are clearly no effects (either biologically or statistically) then it can be inferred that there will be no impact on honey production. If significant effects are observed over the course of the study, then it may be appropriate to explore risk mitigation measures to determine whether the protection goal of honey production can be met.

9.6.4 Design of a Field Study for Non-*Apis* Bees

Given the lack of investigation into a field level test for non-*Apis* species, it is assumed that all non-*Apis* bee testing will be in conjunction with field studies that are designed primarily for *Apis* bees.

Outlined below are draft protocols that could form the basis of field studies conducted to address specific regulatory questions.

9.6.4.1 Field Studies—Solitary Bees

M. rotundata will be used as the descriptive species in this section. It is also important to note that *M. rotundata* and *Osmia* spp. have a very restricted foraging range (approximately 300 m) compared to that of *A. mellifera* (2–3 km); therefore, it is much easier to ensure that their foraging will be restricted to the crop at the study sites. Preparation of *M. rotundata* for these studies should be undertaken using the same maintenance and handling protocols described for *M. rotundata* in the semi-field study.

Key outputs include mortality (in the crop and at the hive or nest) foraging activity, and reproductive success (as a measure of colony health). Assessment of these endpoints is similar to that for *Apis* tests.

9.6.4.2 Field Studies—Social Non-*Apis* Species

Bombus spp. will be used as the descriptive species in this section. It is also important to note that *Bombus* spp. have a much more restricted foraging range (400–750 m) (Knight et al. 2005) than *A. mellifera* (2–3 km) so it is much easier to assure that their foraging will be restricted to the crop at the study sites. Preparation of *Bombus* spp. for these studies should be undertaken using the same maintenance and handling protocols described for this species group in the semi-field study.

Key Outputs Key outputs include mortality (in the crop and at the hive) foraging activity, and reproductive success (as a measure of colony health). Assessment of these endpoints is similar to that for *Apis* tests.

9.6.4.3 Field Studies—Stingless Species

Stingless bees (Meliponini) have a social life similar to the honey bees albeit in much smaller colonies. There is an increasing body of literature (Heard 1999; Amano 2004) showing the value of stingless bees in pollination of crops in tropical and temperate countries. The stingless bees are native to tropical and subtropical areas, and more than 400 species have been recorded from these regions. The ease of handling these species (small colony sizes, and hesitance to sting) makes them ideal candidates for pollination in greenhouse conditions. However, in terms of their use for pesticide tests, there is very little information and thus the information below should be taken as a guide with allowances for improvement. It is expected that this guidance document will create interests among the practitioners to develop and validate methods and create a forum for revisions in the future, if required.

Hives Hives for stingless bees are box-shaped (commercial units) but smaller compared to those of honey bees. They do not have frames but rather are hollow, containing the whole colony component. Opening the hive, therefore, should be done gently to avoid damaging or destroying the nest structure. (Honey and pollen are stored in pots made of beeswax. The pots are typically arranged around a central set of horizontal brood combs.) When the young worker bees emerge from their cells, they tend to remain inside the hive, performing different jobs. As workers age, they become guards or foragers. Unlike the larvae of honey bees, meliponine larvae are not fed directly. The pollen and nectar are placed in a cell, an egg is laid, and the cell is sealed until the adult bee emerges after pupation (i.e., mass provisioning). At any one time, hives can contain 300–80 000 workers, depending on the species.

Stingless bee colonies can be purchased from beekeepers that specialize in stingless bee production and management. Stingless bees that are currently commercially available in tropical countries include, but are not limited to: *Melipona beecheii; Melipona quadrifasciata; Trigona carbonaria; Tetragonula fuscobalteata; Scaptotrigona bipunctata; Tetragonisca angustula; Meliponula ferruginea; Hypotrigona gribodoi;* and, *Meliponula bocandei.* See Chapter 3 for details on which species are appropriate for specific countries.

Care should be taken to acquire strong colonies with sufficient workers, each with about 10 000 healthy foragers; however, this will depend upon the species used. Up to eight colonies per hectare may be used. Stingless bee hives can be placed at strategic positions similar to operating with honey bees (e.g., either in the middle or edge of the field); and hives should be sheltered with a wooden cover placed on top of the hive to avoid direct rainfall on the hive.

Stingless bees have a wide foraging range, foraging up to 2.1 km (Kuhn-Neto et al. 2009), but on average they restrict their activity to within 1 km of the colony. The isolation distance from other forage sources recommended for honey bees (2–3 km) can thus be used.

The number of individuals per hive and per unit area recommended for honey bees can also be applied to the stingless bees. However, noting that there have been no field tests of this kind done for stingless bees, there is a research requirement to validate the protocol.

Treatment Application, Sampling, Data Analysis, and Interpretation Same as for *Apis*

Key Outputs The endpoints for the stingless bees in the field tests are similar to the honey bees and include:

- Colony strength
- Hive weight
- Pollen, honey, and nectar stores

- Mortality at the hive (via the use of dead bee traps or collecting sheets)
- Mortality of drones and pupae
- Mortality in the crop
- Presence of the same queen
- Foraging activity in the crop
- Returning foraging bees
- Behavior
- Residues in pollen, nectar, pollen pellets, wax, bee bread, and dead bees (i.e., measures of exposure)
- Assessment of the brood (including an estimate of adults, the area containing cells, eggs, larvae and capped cells)

9.6.5 Assessment of the Uncertainty in a Field Study

Unlike lower tier studies with insect pollinators, environmental conditions are far less easy to control in full-field studies. Additionally, although sources of variability and uncertainty may exist, there may be fewer options available for researchers to address these issues under full-field conditions. While many of the options available for semi-field studies may apply to full-field studies, the logistics of stratifying designs and increasing the number of replicates become logistically difficult to implement in light of the uncertainties pertaining to *Apis* and non-*Apis* bees in the field (Table 9.5).

9.7 ROLE OF MONITORING AND INCIDENT REPORTING

Some countries have incident monitoring schemes aimed at providing information that can inform regulatory decisions. These schemes provide some feedback on the quality and accuracy of regulatory decisions and, therefore, by association elements of that decision such as measurement endpoints, assessment endpoints, up through protection goals. In addition, some regulatory authorities require monitoring of bee colonies as a condition of registration where there is uncertainty whether the risk, or the risk mitigation meets the protection goals.

Monitoring schemes, for example, the U K Wildlife Incident Investigation Scheme (WIIS), rely on incidents being reported to a central organization. This scheme has provided much information on incidents associated with correct use, accidental incorrect or misuse, as well as abuse of pesticide products. These data, along with usage data, have been useful to determine the appropriateness of various regulatory restrictions as well as to provide information on the appropriateness of the regulatory trigger values (see Aldridge and Hart 1993, and Mineau et al. 2008). In North America (under the United States Environmental Protection Agency (EPA) system), pesticide registrants are required to report (adverse) incidents or adverse impacts from the use of their compound or products when they become aware of them. Other stakeholders may also report incidents to the EPA.

These schemes do, however, have limitations in that they are rely on the public to both find an incident and to report it. This can potentially lead to under-reporting if the beekeepers fear retribution, or the citizen is unaware of the process of reporting. The conditions of commercial agriculture verses that of native wildlife bias reporting toward *A. mellifera*. Consequently, incidents involving non-*Apis* bee species may be under-recorded. Nonetheless, monitoring schemes are a useful tool to the regulator to better understand the use and effects of pesticide compounds. Cost-effective reporting schemes need to be developed that provide incentives to applicators to help increase reporting of experiences from the field. This is critical for improving risk assessment and mitigation.

TABLE 9.5
Variability and Uncertainty in Field Studies with *Apis* and Non-*Apis* Bee Species

Parameter	Discussion of Uncertainty
Exposure	Uncertainty of exposure should be minimized by proper location of the site in relation to other foraging sites, ensuring that the target crop is maximally attractive to bees. Determination of exposure can be made through measurements (as discussed above for *Apis* species). As with *Apis* tests, it is essential that there is information on the degree of exposure in determining the usefulness of the study.
Location of sites	The location should be relevant for the crop and environmental conditions (climatic, botanical, and edaphic) both when and where the study is conducted. The likely reality is that tests cannot be conducted for all crop, formulation, or geographic combinations and so there may be uncertainty when extrapolating the results. The uncertainty could over- or under-estimate the risk depending upon the actual study in question and the uses or situations to which it is being extrapolated.
Difference between the treatment areas and the controls	It is possible that the control and the treatment areas may differ both in terms of climate and edaphic conditions. Any differences in the testing environment (i.e., vegetative surroundings, climatic, or soil conditions) should be minimized.
Extrapolation between different varieties and subspecies of bee	Only one bee species or subspecies will be tested in one study. Uncertainty will exist when extrapolating inter-species, but may also exist when extrapolating intra-species. For example, while there is information indicating that effects on *Apis mellifera mellifera* and *Apis mellifera carnica* are minimal, that is, they are of relatively similar sensitivities, the differences in sensitivity between *Apis mellifera scutellata* and subspecies of European honey bee are unknown, and *Apis mellifera scutellata* may be more or less sensitive than the European honey bee.
Mortality away from the hive	Measurement of mortality away from the hive will be difficult and, therefore, there will be much uncertainty in this parameter. It would not be reasonable to expect that any measurement endpoint can be thoroughly documented and in most cases, the best the study can do is detecting the relative differences between control and treated colonies. Dead bee traps are likely prone to the same biases in control and treated fields. It might be argued that predatory or scavenger insects may be reduced in treated fields relative to untreated fields and that there is a lower likelihood that dead bees may be removed from traps whereas in control fields greater scavenging may occur, making it appear as though mortality was lower in the untreated field. This underscores the need to calibrate dead bee traps to determine the efficiency of recovery. This parameter will potentially underestimate any level of mortality. However, other measurements, for example, colony health (strength and weight), will provide an indirect measure of mortality (i.e., if much mortality occurs away from the colony then it is likely that the overall hive health or colony development etc. will be adversely affected.)
Overall	A field study is an assessment of the potential effects on the colonies under more realistic climatic, botanical, and growing conditions. There are uncertainties regarding the degree to which bees are exposed, although the resulting exposure is likely to represent more normal conditions than those in a semi-field studies. There are uncertainties regarding the sensitivity of the bees tested as well as extrapolating the study to other sites, situations, and crops; however, these should be assessed on a case-by-case basis.

9.8 SUMMARY

Semi-field and field studies for *A. mellifera* are key components of the risk assessment process. They permit a further, more realistic and representative assessment of the potential risk and impacts from the use of pesticides. Due to the fact that these studies are higher tier, there are no standard guidelines available as there are for lower tier studies (e.g., the OECD acute oral toxicity 213). Each should be designed to address the concerns highlighted in the risk assessment. Currently in pesticide registration, semi-field and field studies tend to be conducted according to EPPO 170 or OECD 75. This chapter has built on the information in these guidance documents and provides further information and improvements regarding the conduct of semi-field and field studies. Information is also provided regarding how these studies can be interpreted and hence linked in a qualitative manner to protection goals. Areas of improvement are also included, with the main one being the use of statistics. While statistics are recommended in both EPPO 170 and OECD 75, no particular methodologies are proposed. This issue is further developed through other efforts connected with this Workshop.

Information is provided in this chapter on the design of semi-field and field studies for non-*Apis* bees. Due to the state of knowledge, these are not as well developed and are not currently incorporated into regulatory risk assessment. However, information in this chapter provides a useful starting point regarding what species can be tested under semi-field and field conditions and the design of such studies.

REFERENCES

Aldridge CA, Hart ADM. 1993. Validation of the EPPO/CoE risk assessment scheme for honeybees. In: Proceedings of the Fifth International Symposium on the Hazards of Pesticides to Bees, October 26–28, 1993 Plant Protection Service, Wageningen, pp. 37–41.

Amano K. 2004. Attempts to introduce stingless bees for the pollination of crops under glasshouse conditions in Japan, Food & Fertilizer Technology Center. http://www.fftc.agnet.org/library/article/tb167.html (accessed January 22, 2011).

[EPPO 170] EPPO. 2010. Efficacy evaluation of plant protection products: side-effects on honey bees. PP 1/170 (4). *OEPP/EPPO Bull.* 40:313–319. EU Directive 91/414. http://www.uksup.sk/download/oso/20030409_smernica_rady_91_414_eec.pdf

Heard TA. 1999. The role of stingless bees in crop pollination. *Annu. Rev. Entomol.* 44:183–206.

Johansen CA, Rincker CM, George DA, Mayer DF, Kious CW. 1984. Effects of aldicarb and its biologically active metabolites on bees. *Environ. Entomol.* 13:1386–1398.

Kearns CA, Thompson JD. 2001. *The Natural History of Bumblebees. A Sourcebook for Investigations.* University Press of Colorado, Boulder, CO.

Knight ME, Martin AP, Bishop S, Osborne J, Hale R, Sanderson R, Goulson D. 2005. An interspecific comparison of foraging range and nest density of four bumblebee (*Bombus*) species. *Mol. Ecol.* 14(6):1811–1820.

Konrad R, Ferry N, Gatehouse AM, Babendreier D. 2008. Potential effects of oilseed rape expressing oryzacystatin-1 (OC-1) and of purified insecticidal proteins on larvae of the solitary bee *Osmia bicornis*. *Plos One*. 3(7):e2664. doi:10.1371/journal.pone.0002664

Kuhn-Neto B, Contrera FAL, Castro MS, Nieh JC. 2009. Long distance foraging and recruitment by a stingless bee, *Melipona mandacaia*. *Apidologie*. 40:472–480.

Ladurner E, Bosch J, Kemp WP, Maini S. 2008. Foraging and nesting behavior of *Osmia lignaria* (Hymenoptera: Megachilidae) in the presence of fungicides: cage studies. *J. Econ. Entomol.* 101(3):647–653.

Mader E, Spivak M, Evans E. 2010a. Chapter 7: The alfalfa leafcutter bee. In: *Managing Alternative Pollinators—A Handbook for Beekeepers, Growers, and Conservationists*. USDA Sustainable Agriculture, Research and Education (SARE) Program, pp. 75–93.

Mader E, Spivak M, Evans E. 2010b. Chapter 5: Bumble bees. In: *Managing Alternative Pollinators—A Handbook for Beekeepers, Growers, and Conservationists*. USDA Sustainable Agriculture, Research and Education (SARE) Program, pp. 43–52.

Meier U. 2001. Growth Stages of Mono- and Dicotuledous Plants. BBCH Monograph. Federal Biological Research Centre for Agriculture and Forestry. http://lpub.jki.bund.de/index.php/BBCH/article/viewFile/470/420

Michener CD. 2007. *The Bees of the World*, 2nd edn. John Hopkins University Press, Baltimore, MD, 992 pp.

Mineau P, Harding KM, Whiteside M, Fletcher MH, Garthwaite D, Knopper LD. 2008. Using reports of bee mortality in the field to calibrate laboratory-derived pesticide risk indices. *Environ. Entomol.* 37(2):546–554.

Morandin LA, Laverty TM, Kevan PG. 2001. Bumble bee (Hymenoptera: Apidae) activity and pollination levels in commercial tomato glasshouses. *J. Econ. Entomol.* 94:462–467.

Nogueira-Neto P. 1997. *Vida e criação de abelhas indígenas sem ferrão*, 1st edn. Nogueirapis, São Paulo.

[OECD 75] OECD. 2007. Series on Testing and Assessment, Number 75. Guidance Document on the Honey Bee (*Apis mellifera*) Brood Test Under Semi-field Conditions. ENV/JM/Mono(2007)22.

Oomen PA, DeRuijter A, Van der Steen J. 1992. Method for honey bee brood feeding tests with insect growth-regulating insecticides. *EPPO Bull.* 22:613–616.

Proni EA. 2000. Biodiversity of indigenous stingless bees (Hymenoptera: Apidae: Meliponinae) in the Tibagi River Basin, Parana State, Brazil. *Arquivos de Ciencias Veterinarias e Zoologia da UNIPAR*. 3(2):145–150.

Tasei JN, Carre S, Moscatelli CB, Grondeau C. 1988. Recherche de la D L 50 de la deltamethrine (Decis) chez *Megachile rotundata* F. Abeille pollinistatrice de la luzerne (*Medicago sativa* L.) et des effets de doses infralethales sur les adultes et les larves. *Apidologie.* 19(3):291–306.

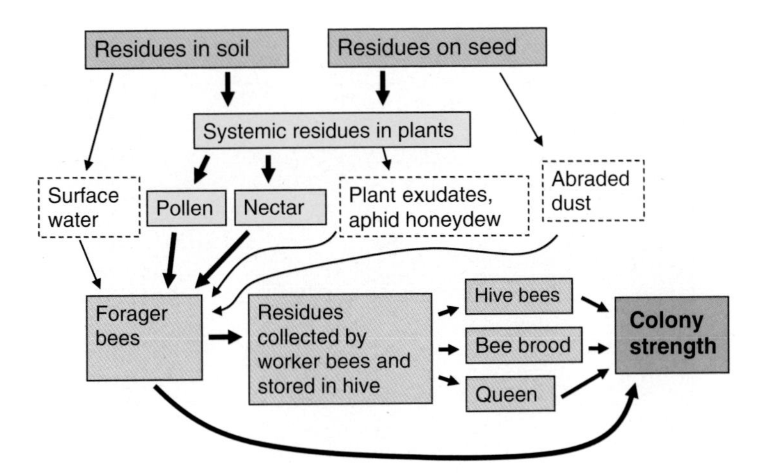

Plate 6.2 Depiction of stressor source, potential routes of exposure, receptors and attribute changes for a systemic pesticide applied to the soil or as a seed dressing.

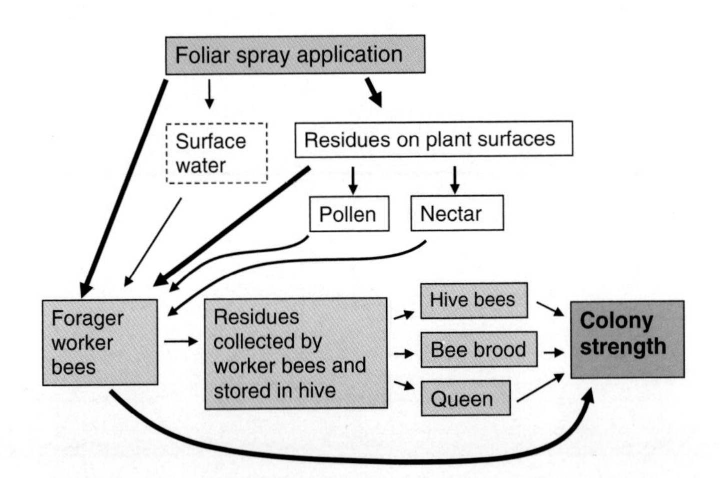

Plate 6.3 Depiction of stressor source, potential routes of exposure, receptors and attribute changes for a nonsystemic pesticide applied as a foliar spray.

Pesticide Risk Assessment for Pollinators, First Edition. Edited by David Fischer and Thomas Moriarty.

Plate 7.2 Leafcutter bee on blanket flower, photo by Mace Vaughan (Xerces Society for Invertebrate Conservation).

Plate 7.3 Micropipetting nectar samples, photo by Mike Beevers.

Plate 7.4 Hand collecting pollen by removing flower anthers, photo by Mike Beevers.

Plate 7.5 Honey bee semi-field study with *Phacelia*, photo provided by BASF SE.

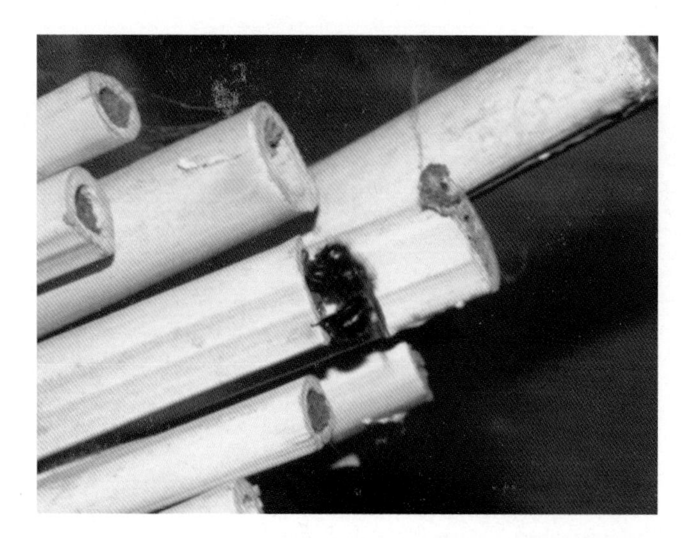

Plate 7.6 Mason bee, photo by Mace Vaughan (Xerces Society for Invertebrate Conservation).

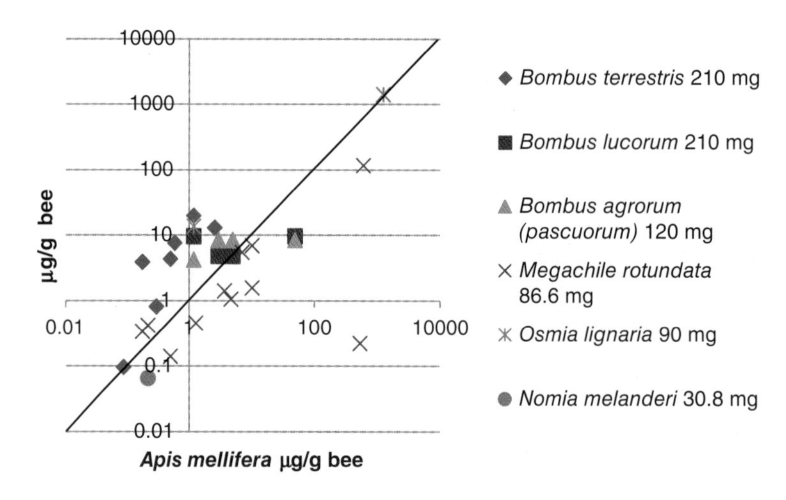

Plate 8.1 Comparison of the contact toxicity (LD50) of 21 pesticides to adults of *Apis mellifera*, three species of the social bee Bombus and three species of solitary bees (Osmia, Megachilidae, and Nomia). Points below the diagonal line indicate greater sensitivity than *Apis mellifera*, while points above the diagonal line represent lower sensitivity than *Apis mellifera* (Johansen et al. 1983).

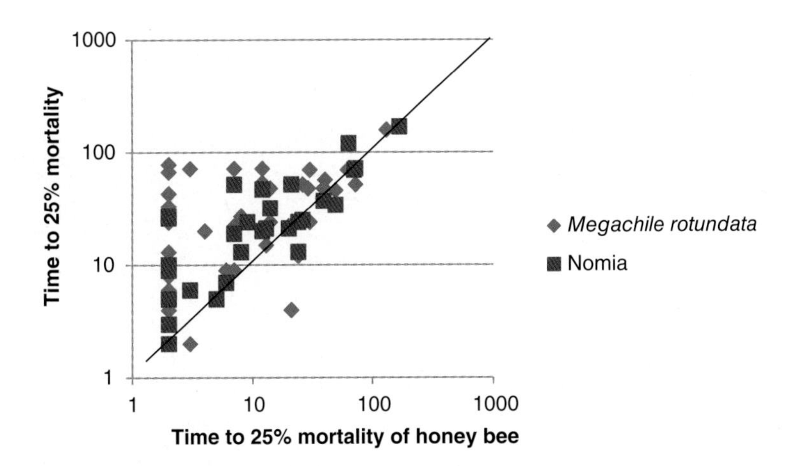

Plate 8.2 Comparison of the toxicity of pesticides to adults of *Apis mellifera* with the solitary bees *Megachile rotundata* and *Nomia melanderi* based on time for sprayed residues to decline to a concentration causing 25% or less mortality. Points below the diagonal line indicate greater sensitivity than *Apis mellifera*, while points above the diagonal line represent lower sensitivity than *A. mellifera* (Johansen et al. 1983).

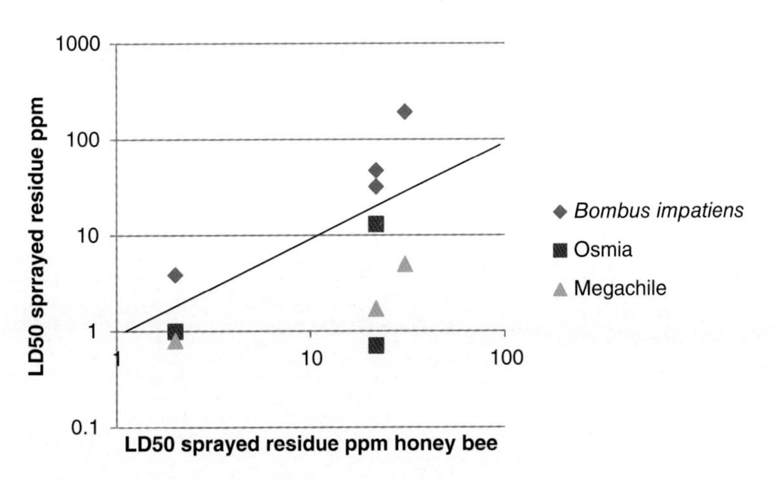

Plate 8.3 Comparison of the toxicity (LD50) of sprayed residues of clothianidin, imidacloprid, lambda-cyhalothrin and spinosad to adults of *Apis mellifera, Megachile rotundata*, and *Osmia lignaria*. Points below the diagonal line indicate greater sensitivity than *A. mellifera*, while points above the diagonal line represent lower sensitivity than *A. mellifera* (Johansen et al. 1983).

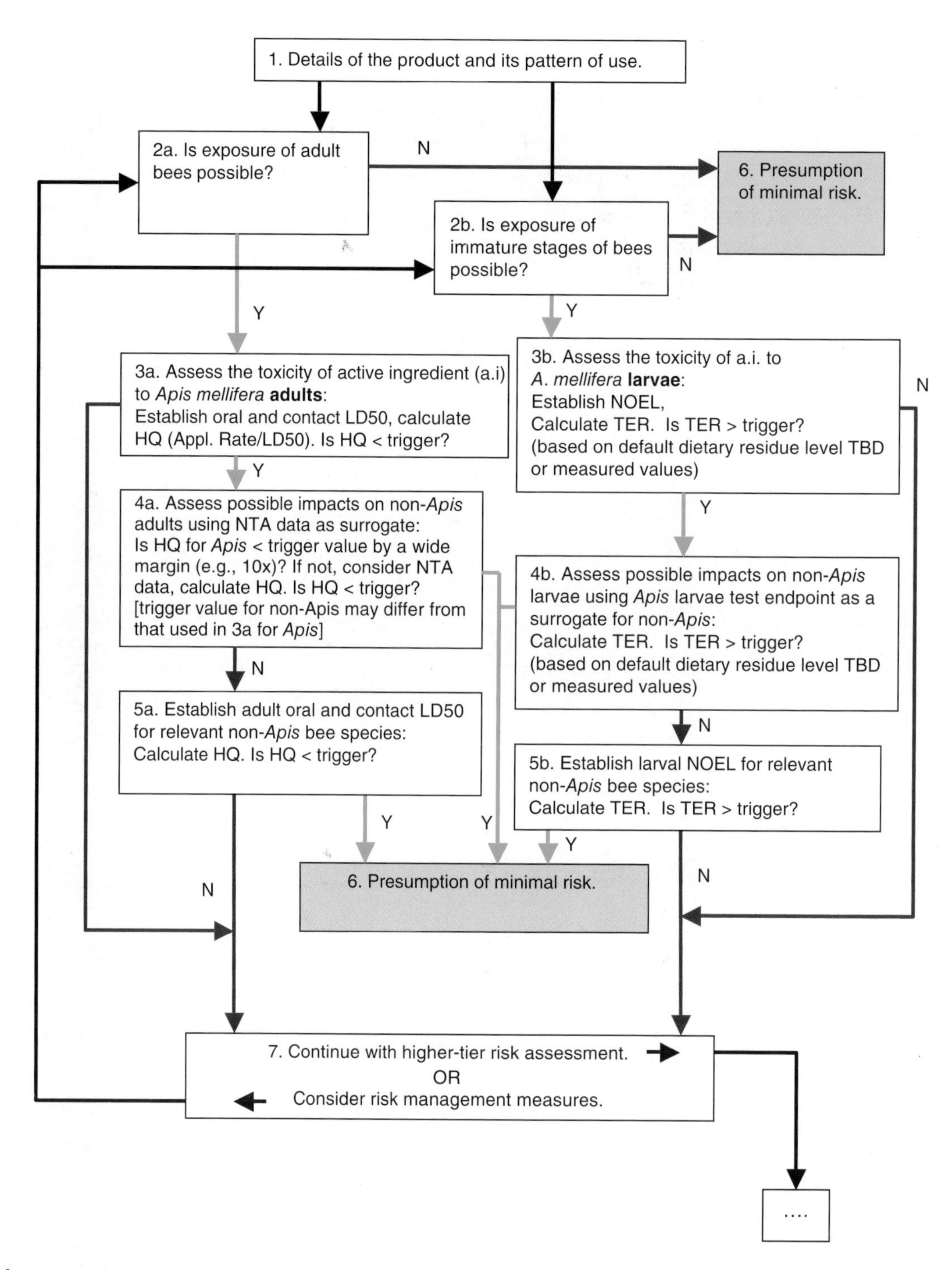

Plate 10.2 Insect pollinator screening-level risk assessment process for foliarly applied pesticides.

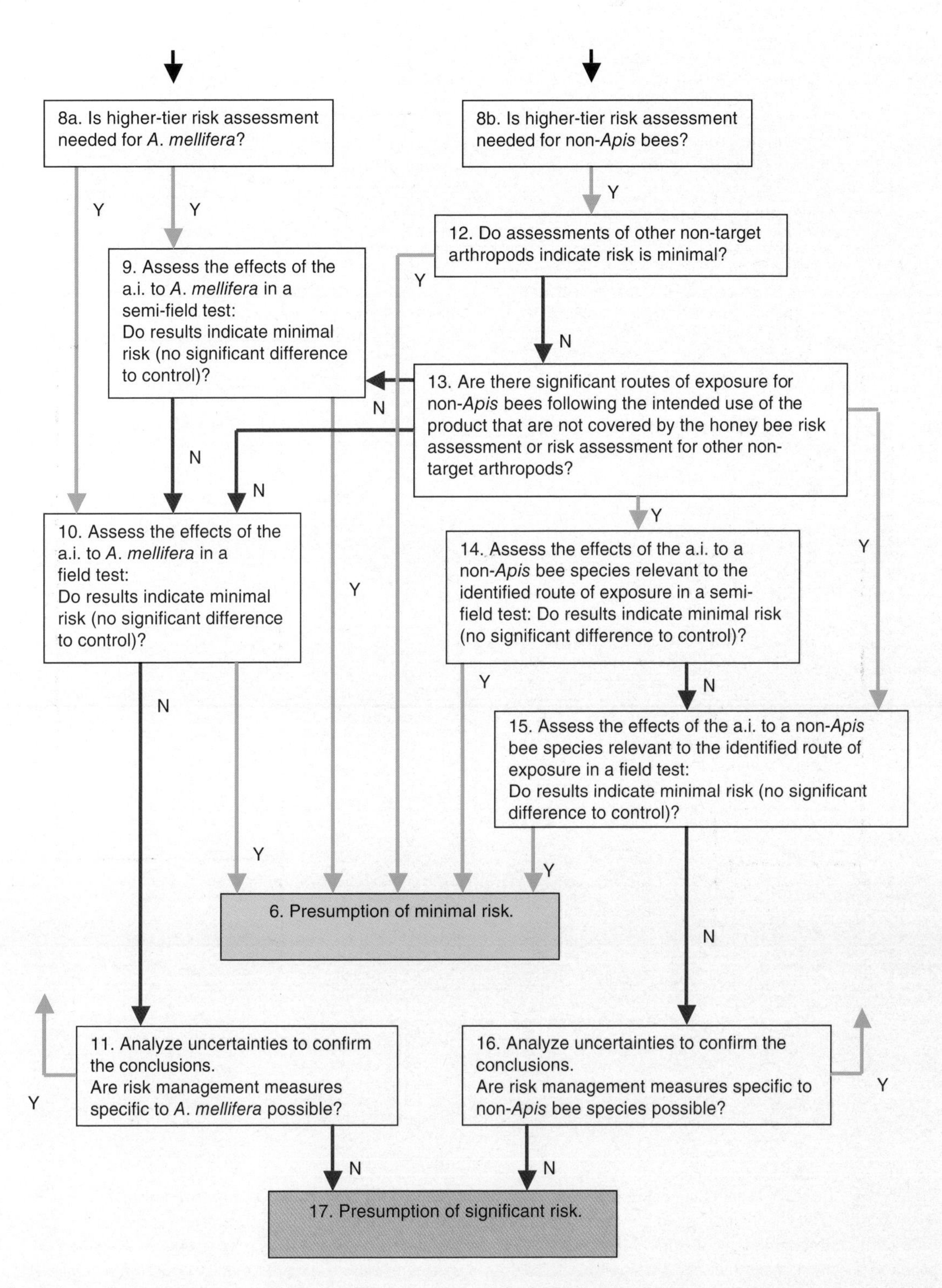

Plate 10.3 Higher tier (refined) risk assessment process for foliarly applied pesticides.

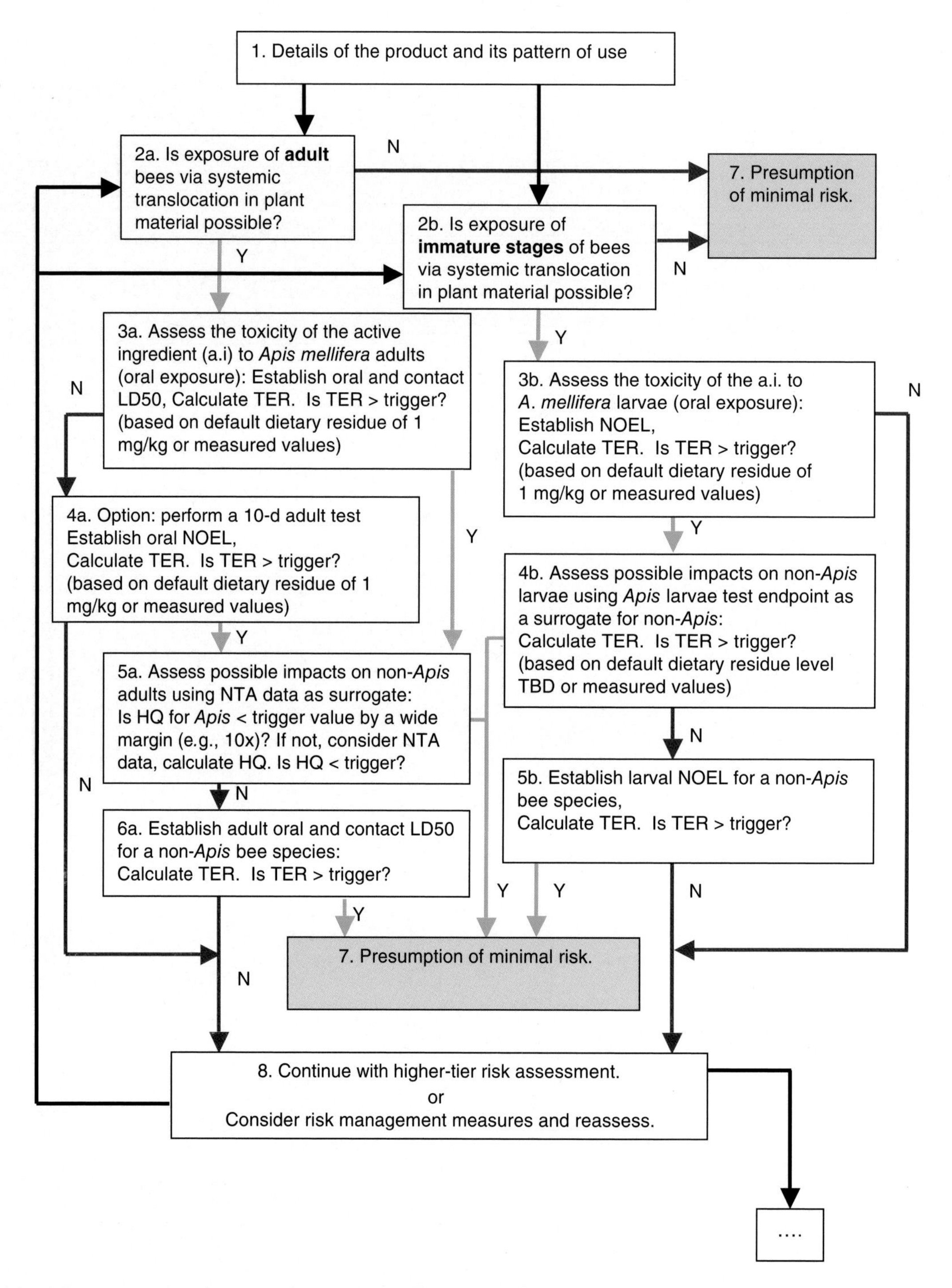

Plate 10.4 Insect pollinator screening-level risk assessment process for soil and seed treatment of systemic pesticides. Note that this flow chart may apply for trunk injection as well, as modalities of exposure of pollinators are similar as for soil or seed treatments. For trunk injection, however, further data are needed to appropriately describe the range of expected residue concentrations in nectar and pollen. As a consequence, no default value is currently available for a quantification of the risk (Boxes 3a and 3b). A compilation of available data could be made, with a particular attention to the corresponding injection protocols as it varies with the active substance involved and the tree.

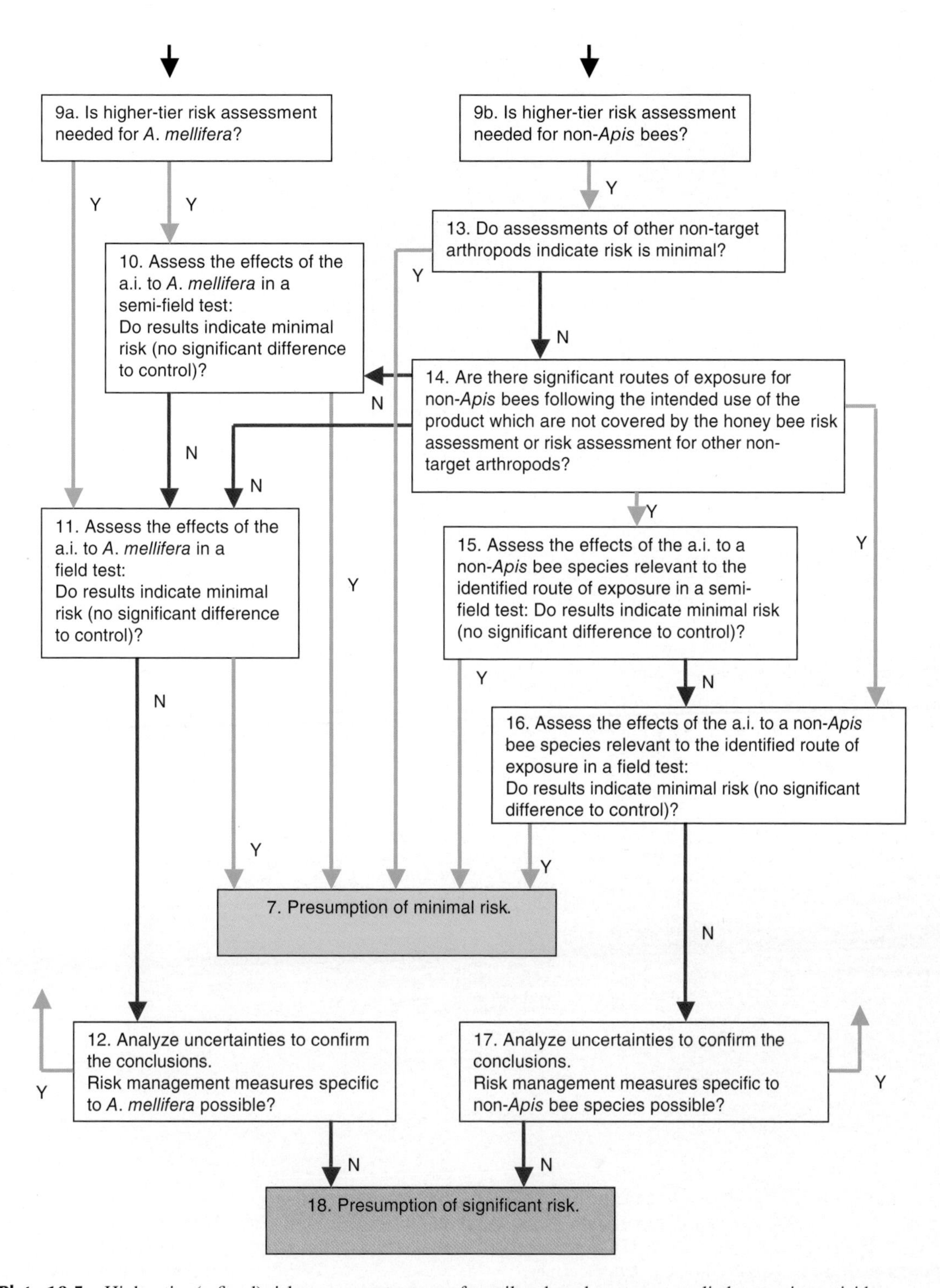

Plate 10.5 Higher tier (refined) risk assessment process for soil and seed treatment applied systemic pesticides.

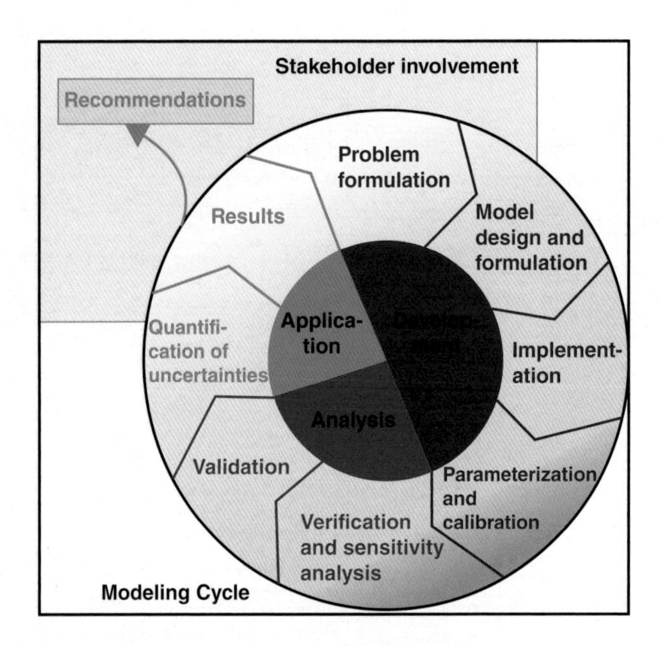

Plate 11.3 Tasks of the "Modeling Cycle," that is, of the iterative process of formulating, implementing, testing, and analyzing ecological models (after Schmolke et al. 2010b). Full cycles usually include a large number of subcycles, for example, verification leading to further effort for parameterization or reformulation of the model. The elements of this cycle are used to structure a new standard format for documenting model development, testing, analysis, and application for environmental decision making, TRACE (Schmolke et al. 2010b).

Plate 14.1 Guttation water on a strawberry leaf, photo by Mace Vaughan (Xerces Society for Invertebrate Conservation).

10 Overview of a Proposed Ecological Risk Assessment Process for Honey bees (*Apis mellifera*) and Non-*Apis* Bees

A. Alix, T. Steeger, C. Brittain, D. Fischer, R. Johnson, T. Moriarty, E. Johansen, F. Streissel, R. Fischer, M. Miles, C. Lee-Steere, M. Vaughan, B. Vaissiere, G. Maynard, M. Kasina, R.C.F. Nocelli, C. Scott-Dupree, M. Coulson, A. Dinter, and M. Fry

CONTENTS

Pesticide Risk Assessment for Pollinators, First Edition. Edited by David Fischer and Thomas Moriarty.

10.1 INTRODUCTION

Ecological risk assessments are intended to evaluate the likelihood that adverse ecological effects may occur as a result of exposure to one or more stressors (USEPA 1992). Typically, at the first tiers, risks are evaluated for individual taxonomic groups (e.g., freshwater fish, upland game birds, or terrestrial plants) using surrogate species. At higher levels of refinement, risks to individual taxa may be further integrated to determine whether there are effects to the community. However, risk assessments are typically conducted at the taxon level (USEPA 2004). The intent of this chapter is to describe a proposed method for estimating risk to honey bees (*Apis mellifera*) and non-*Apis* bees from pesticides that are applied via sprays (acting on contact) and via seed or soil treatments and tree trunk injections (acting systemically).

In general, a pesticide risk assessment process is used for evaluating new compounds or new products entering the market or those compounds undergoing re-evaluation, as in the 10-year process of re-evaluation in the European Union (EU) or in North America where chemicals are re-evaluated every 15 years. As with risk assessments for other taxonomic groups, the proposed risk assessment method described in this document makes use of surrogate species. The ecological risk assessment process described consists of a series of steps or phases, which are intended to be iterative where information gathered at each step is evaluated against the protection goals. The risk assessment process consists of a *problem formulation* (Phase 1), *analysis* (Phase 2), and *risk characterization* (Phase 3). This generic process is depicted in Figure 10.1. In Phase 1, *problem formulation*, measurement endpoints are identified in relation to protection goals and corresponding assessment endpoints, a conceptual model is prepared, and an analysis plan is developed. Based on the conceptual model and its associated risk hypothesis, the analysis plan articulates how the risk hypothesis will be tested. In Phase 2, *analysis*, available measures of exposure and measures of effect are evaluated. Through environmental fate data, the movement of a stressor (i.e., the pesticide and relevant transformation and breakdown products) in the environment is characterized; this is frequently termed the exposure characterization or exposure profile. Similarly, the potential acute and chronic effects of a chemical are characterized in what is frequently termed the stressor-response profile. Additionally, the proposed or existing uses of a compound are characterized and, based on these uses and the environmental fate of the compound, predicted or estimated environmental concentrations (PEC or EEC) are derived.

Once effects and exposure are characterized, the risk assessment proceeds to Phase 3, *risk characterization*. Typically, the risk characterization consists of two steps, that is, risk estimation and risk discussion (evaluation). In the risk estimation step, the measures of exposure (e.g., EECs or PECs) and measures of effect are integrated to develop risk estimates. These risk estimates may be based on point estimates of exposure and a point estimate of effect, for example, for Tier 1, exposure is based on application parameters assumed to result in the highest exposure for a particular use, and point estimates of effect, for example, the acute median lethal dose to 50% of the species tested (LD50). If initial values for potential exposure and effects result in risk estimates that exceed regulatory triggers, then these point estimates can be refined through higher tier testing with regard to both potential exposure or potential effects. Possible refinements in exposure estimates are discussed in Chapter 7 while possible refinements in effects are discussed in Chapter 8 (laboratory studies) and Chapter 9 (semi-field/full-field studies). As ecological risk assessment methodologies evolve, refined estimates could be based on distribution-based estimates of either exposure (e.g. residue concentrations in pollen from field monitoring studies based on application rate reflecting the worst case for a particular use), or effects (e.g., species sensitivity distribution using LD50 values for non-*Apis* species).

Regardless of whether point estimates or distribution-based estimates are used, the integration of exposure and effects data is typically expressed as a ratio (quotient), and it is this ratio that is considered to be the "risk estimate." If point estimates of exposure and effects are used as inputs, the risk quotient (RQ) is a deterministic point estimate of risk. If the exposure or effects inputs are probability distributions of possible values, the risk estimate is itself a "joint" probability distribution and represents a probabilistic estimate. Deterministic

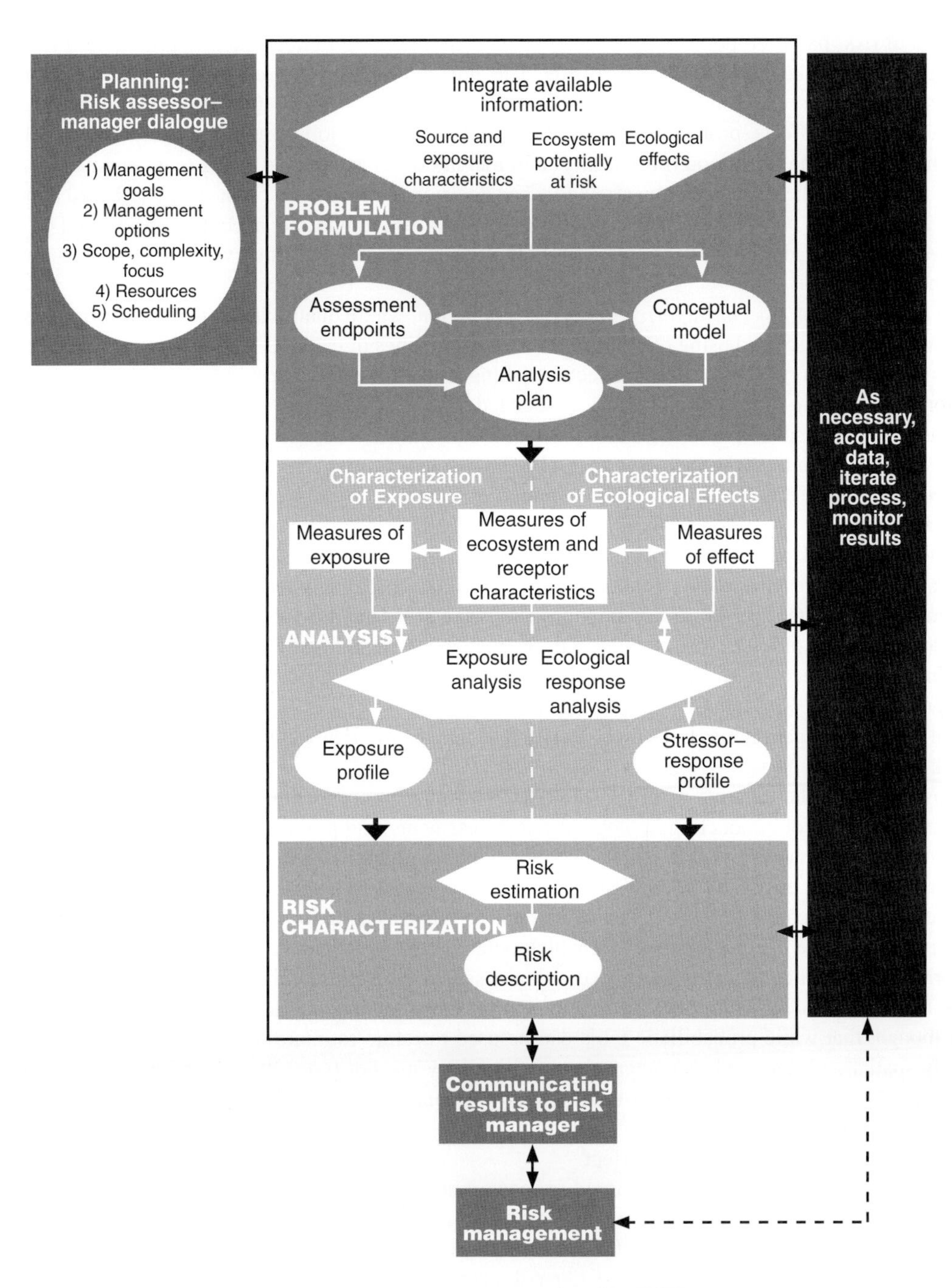

FIGURE 10.1 Diagram of ecological risk assessment process employed by USEPA.

estimates of risk, based on point estimates of exposure and effects, do not typically provide information on the magnitude and likelihood of adverse effects. This uncertainty is in most cases accounted for with the use of assessment factors. In refining the risk assessment on the basis of distribution-based estimates of either or both exposure and effects, probability distributions and particularly joint-probability distributions allow the estimation of both the likelihood (probability) and magnitude of an adverse effect (e.g., estimates of a 40% chance that 60% of the species will be affected). The decision to move from point-estimate-based approaches to distribution-based approaches[1] that may also be spatially and temporally specific is predicated on the risk manager's need for additional information to support their decision and the availability of data to support such approaches.

The second part of *risk characterization* is risk evaluation, where quantitative estimates of risks are, when necessary, further described using qualitative data. Multiple lines of evidence are used to more fully describe what is known about potential adverse effects resulting from the use of a pesticide. Risk evaluations include additional discussion about the variability associated with the measured endpoints along with associated uncertainties, that is, attempts to characterize what is not known. When necessary or possible, the intended effects of relevant mitigation measures may also be discussed. Any incident data, that is, adverse effects reported involving the actual use of the compound in the field, are also discussed to further characterize potential effects.

Although the risk assessment process is depicted as three distinct phases, each phase is intended to be iterative. As more information (data) becomes available, the outcome of the process should evolve to accommodate the data. The risk assessment process is therefore intended to take advantage of multiple lines of evidence and the problem formulation with its conceptual model and risk hypothesis may be refined as more information becomes available. A critical component to this iterative process is clear communication between the risk assessor and the risk manager to ensure that protection goals are adequately articulated and that the relevant mitigation measures on risk estimates may be implemented and potentially evaluated within the risk assessment.

Consistent with the iterative nature of the risk assessment process, regulatory authorities typically rely on a tiered process for conducting ecological risk assessments; the preliminary, or screening-level (Tier 1) assessments are intended to screen substances for which a potential risk cannot be excluded. Higher tiers attempt to refine risk estimates to (1) identify whether a potential risk will likely be encountered under more realistic assessment conditions, that is, using less conservative assumptions regarding potential exposure and effects; (2) determine the conditions under which potential risks may occur; and, (3) identify the spatial and temporal characteristics of risks. The tiered risk assessment process identifies those chemicals for which a higher level of resources should be devoted to support more refined and detailed assessments. It should be noted though, that while probabilistic tools can be used to refine estimates of exposure and effects, and to quantify spatial and temporal characteristics of risks, they are not typically applicable to determining the conditions of occurrence for risk. Additionally, such refinements are typically focused on specific uses which have exceeded trigger values and which require a more detailed understanding of the potential magnitude, likelihood, or duration of a particular effect.

Decision criteria are used within a tiered framework as a basis for discriminating potential risks among substances. Screening-level risk assessments may have predetermined decision criteria to answer whether potential risks exist, as for example in the EU where decision-making criteria are defined for all groups of organisms (EC 1997). Conversely, higher tier risk assessments may not have predetermined or uniformly defined decision criteria since the management decision may change from yes/no to questions regarding

[1] Species sensitivity distributions are an option to refine the evaluation of effects for risk assessment performed for a group of organisms and not at the level of a species, for example, the honey bee.

"what, where, and how great is the risk," as for example in the United States (USEPA 1998) and may also be associated with restrictions/conditions intended to limit the risk (which is the case in both the EU and United States).

In the following sections, the risk assessment process for honey bees and non-*Apis* bees is described. Consistent with the tiered process discussed in the preceding sections, the following sections propose risk assessment flowcharts discussed during the Workshop and are intended to illustrate the different steps mentioned above. Each step of these risk assessment processes are then discussed in greater detail, starting with screening-level assessments (Tier 1) and proposed refinements that incorporate additional data on potential exposure and effects to both *Apis* and non-*Apis* bees. The proposed process is delineated for pesticides that are applied foliarly and act on contact with or ingestion by insects. A different risk assessment process is articulated for pesticides that are applied to soil or as a seed treatment. For soil and seed treatments that are systemic, the chemical is taken up by the plant and distributed either through the xylem (i.e., translocation through the plant in the direction of xylem flow (acropetal[2]) or through the plant phloem (i.e., translocation through the plant in the direction of phloem stream (basipetal[3] and acropetal). The route of exposure to systemic compounds applied as soil, seed, or tree trunk injections is primarily through ingestion of residues in pollen or nectar.

10.2 PROTECTION GOALS, ASSESSMENT AND MEASUREMENT ENDPOINTS, TRIGGER VALUES FOR TRANSITIONING TO HIGHER LEVELS OF REFINEMENT, AND RISK ASSESSMENT TERMINOLOGY

As previously discussed, the initial phase of a risk assessment process is problem formulation. Problem formulation articulates the intent of the risk assessment and is predicated on particular protection goals for which the regulatory authority is responsible. In order to build a proposed risk assessment process for pollinators, the participants of the Workshop identified plausible, surrogate protection goals, that included:

i. protection of pollination services provided by *Apis* and non-*Apis* species
ii. protection of honey production and other hive products; and
iii. protection of pollinator biodiversity,

In order to structure an assessment that allows addressing risk management concerns, that is, realize protection goals, it is important to define assessment endpoints. Assessment endpoints are intended to be explicit expressions of the actual environmental value that is to be protected and are operationally defined by an ecological entity and its attributes (USEPA 1998). For assessing potential risks to *Apis* and non-*Apis* bees the ecological entities would be the organisms themselves (e.g., larval and adult honey bees and bumble bees) and potential attributes would consist of survival, development, and reproduction. The ability of assessment endpoints to support risk management decisions depends on the extent to which they target susceptible ecological entities and measurable ecosystem characteristics (USEPA 1998). Protection of the growth, reproduction, and survival at the colony or population level of these species will conserve pollination services, biodiversity contributed by pollinators, and availability of hive products (e.g., honey production). The conventional assessment endpoints of survival, development, and reproduction can be articulated for

[2] Acropetal refers to the direction of movement and is typically intended to denote movement from the base of a plant (e.g., roots) toward its apex.

[3] Basipetal refers to the direction of movement and is typically intended to denote movement from the apex of a plant toward its base.

Apis and non-*Apis* bees to include colony size and survival for honey bees, and population size and survival for non-*Apis* bees.

Assessment endpoints, growth, reproduction, or survival, are stand-alone objective measures. Measurement endpoints are attributes that are examined at the study level which, taken either individually or together, allow one to evaluate the consequences to the ecological receptor of concern regarding assessment endpoints. In initial (screening level) laboratory studies, it is practical to measure endpoints such as individual survival, toxicity to and developmental effects on larvae (brood), and behavioral effects (e.g., effects that become manifest in adults due to exposure as larvae). These measurement endpoints are relevant because they provide the body of data that is used to determine if there is a consequential effect at the colony or population level and can be indicative of the ability of a colony to grow, reproduce, or survive. In higher tier tests, it may be possible to directly measure effects on colony or population size and viability. However, as noted in previous chapters, further research is required to ascertain which, and at what level (sublethal) effects is indicative of a colony-level, or population-level effect. The linkage between protection goals, assessment endpoints, and possible measurement endpoints are critical for an effective evaluation of risk (Table 10.1).

The terminology of risk assessment can be confusing due to the differences among regulatory authorities. Many parts of the processes outlined in this document make reference to the European and Mediterranean Plant Protection Organization (EPPO) methodology, and the testing methods for non-target terrestrial arthropods thereof (Table 10.2).

Note that in Tier 3 analysis, where a field study is performed, neither a hazard quotient (HQ) or RQ, nor a toxicity exposure ratio (TER) is calculated. Rather, effects are characterized, statistically significant or not, in the context of actual exposure conditions and in the context of whole hive biology.

TABLE 10.1

Linkage of Protection Goals, Assessment Endpoints, and Measurement Endpoints for Social Bees (Including *Apis*) and Solitary (Non-*Apis*) Bees. Initials (L) and (F) Designate Endpoints Most Applicable to Laboratory (L) Studies and Field (F)

Protection Goal	Assessment Endpoints	Measurement Endpoints Population Level or Higher	Measurement Endpoints Individual Level
Pollination services	Population size and stability within the crop or its boundaries	Social bees: colony survival (F), colony strength (F) Solitary bees: population size (F) and persistence (F) over time	Social bees: individual survival (L, F), fecundity (F), brood success (L, F), behavior (L, F) Solitary bees: individual survival (L, F), reproduction (F), behavior (L, F)
Hive products (honey, etc.)	Production of hive products	Production of hive products (F)	Individual survival (L, F), brood success (L, F), behavior (L, F)
Pollinator biodiversity	Species richness and abundance on the crop/in the boundaries	Colony survival (F), colony strength (F), brood success (F), behavior (F) Species richness and abundance (F)	Individual survival (L, F), brood success (L, F), behavior (L, F)

TABLE 10.2
Risk Estimates and Their Components Used by Regulatory Authorities

Ecological Risk Estimate	Effects Component	Exposure Component	Comment	Where and How it is Used
Hazard quotient (HQ): effects/ exposure	LD50 measured as μg/bee	Dermal exposure concentration or oral dosing concentration as g/ha	Numerator and denominator are expressed in dissimilar measurement units	Used in European assessments Used in Tier 1 analysis
Risk quotient (RQ): exposure/ effects	LD50 measured as μg/bee	Contact exposure concentration, or oral dose concentration	Numerator and denominator are expressed in same measurement units	Used in North American assessments Used in Tier 1 analysis
	No observed adverse effect level (NOAEL) measured as μg/bee	Oral feeding concentration (solution) or dietary intake (pollen or nectar)	Numerator and denominator are expressed in same measurement units	Used in North American assessments Can be used in Tier 1 and Tier 2, analyses
Toxicity exposure ratio (TER): exposure/ effects	LD50 or the no observed adverse effect level (NOAEL) measured as μg/bee	Oral feeding concentration (solution) or dietary intake (pollen or nectar)	Numerator and denominator are expressed in same measurement units	Used in Tier 1 analysis (for larvae) and Tier 2 analysis

10.3 RISK ASSESSMENT FLOWCHARTS

This section illustrates the proposed risk assessment process identified by the participants of the 2011 SETAC Workshop on Pesticide Risk Assessment for Pollinators. The decision process is described and depicted in flowcharts to better highlight the progression of events through the tiers. Risk assessment starts with a preliminary verification that a risk assessment is warranted by first describing the routes of exposure that are considered likely and will trigger further evaluation. This leads to screening steps intended to exclude situations where the potential for adverse effects is considered low and with a sufficient margin of safety to conclude no further analysis is necessary. The process then focuses on uses for which further characterization of the risks is necessary and guides the assessor in efforts to identify the necessary data to enable the estimation of effects and exposure levels needed to assess potential risks from these scenarios.

An overview of each step in the problem formulation and risk assessment process, that is, screening-level assessment to more refined evaluation of effects and exposure based on laboratory data, to higher tiered assessments involving semi-field and field studies can be found in Chapters 5 and 6. Efforts to refine risk estimates are typically predicated on refining estimates of potential exposure and effects. For detailed descriptions of the studies to be undertaken to generate these data, refer to Chapter 7 (assessing exposure), Chapter 8 (laboratory-based effect studies), and Chapter 9 (field-based effect studies).

The flowcharts below are used to depict a generic risk assessment process that was developed during the Workshop. Two proposed processes distinguish between compounds applied as spray for which the worst case exposure may be expected through direct contact of pollinators with spray droplets during the flowering period (Figures 10.2 and 10.3) and, products used as soil or seed treatments for which an exposure may occur as a result of the systemic properties of the compound or its degradates (Figures 10.4 and 10.5). It is important

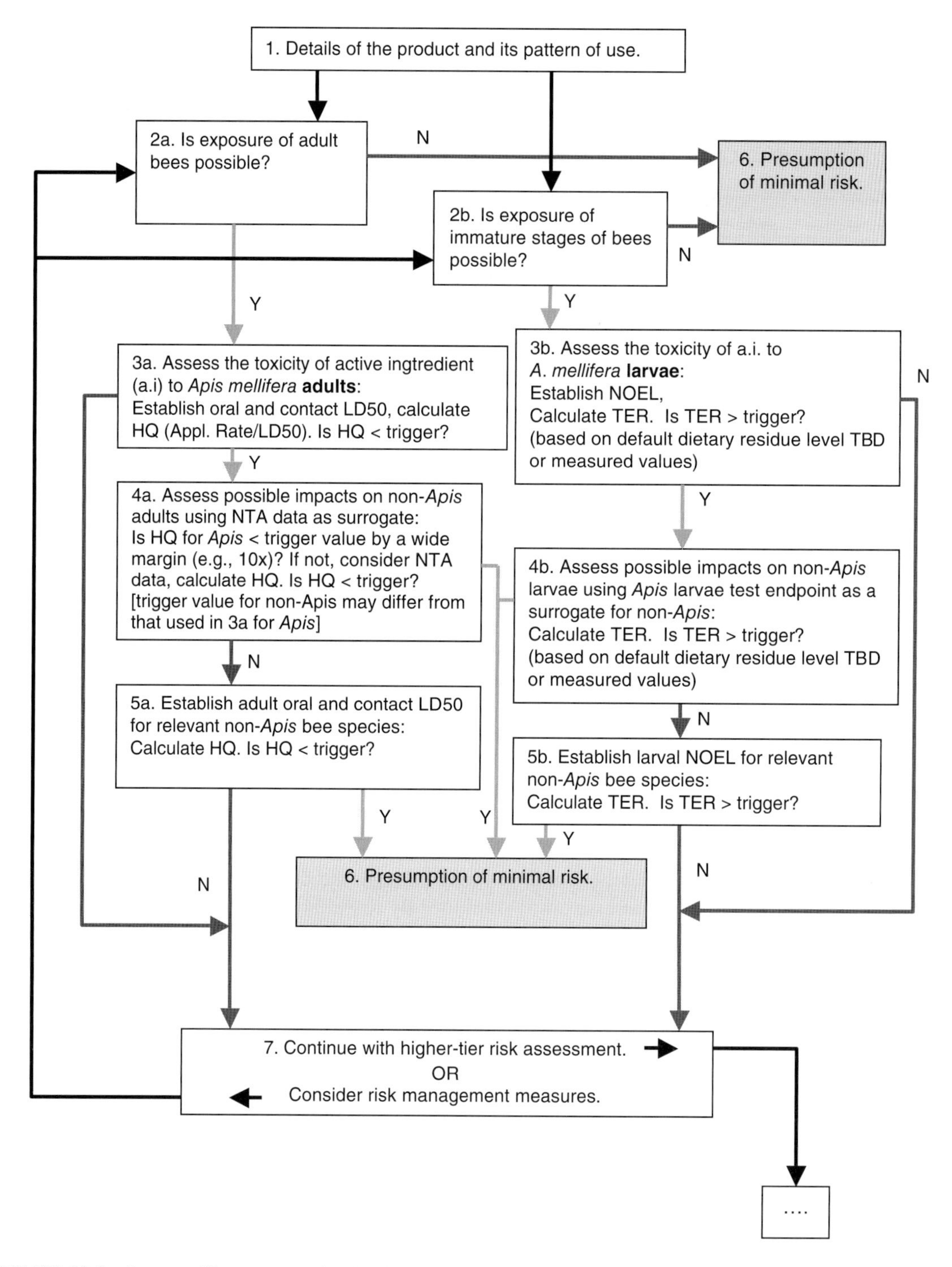

FIGURE 10.2 Insect pollinator screening-level risk assessment process for foliarly applied pesticides. (For a color version, see the color plate section.)

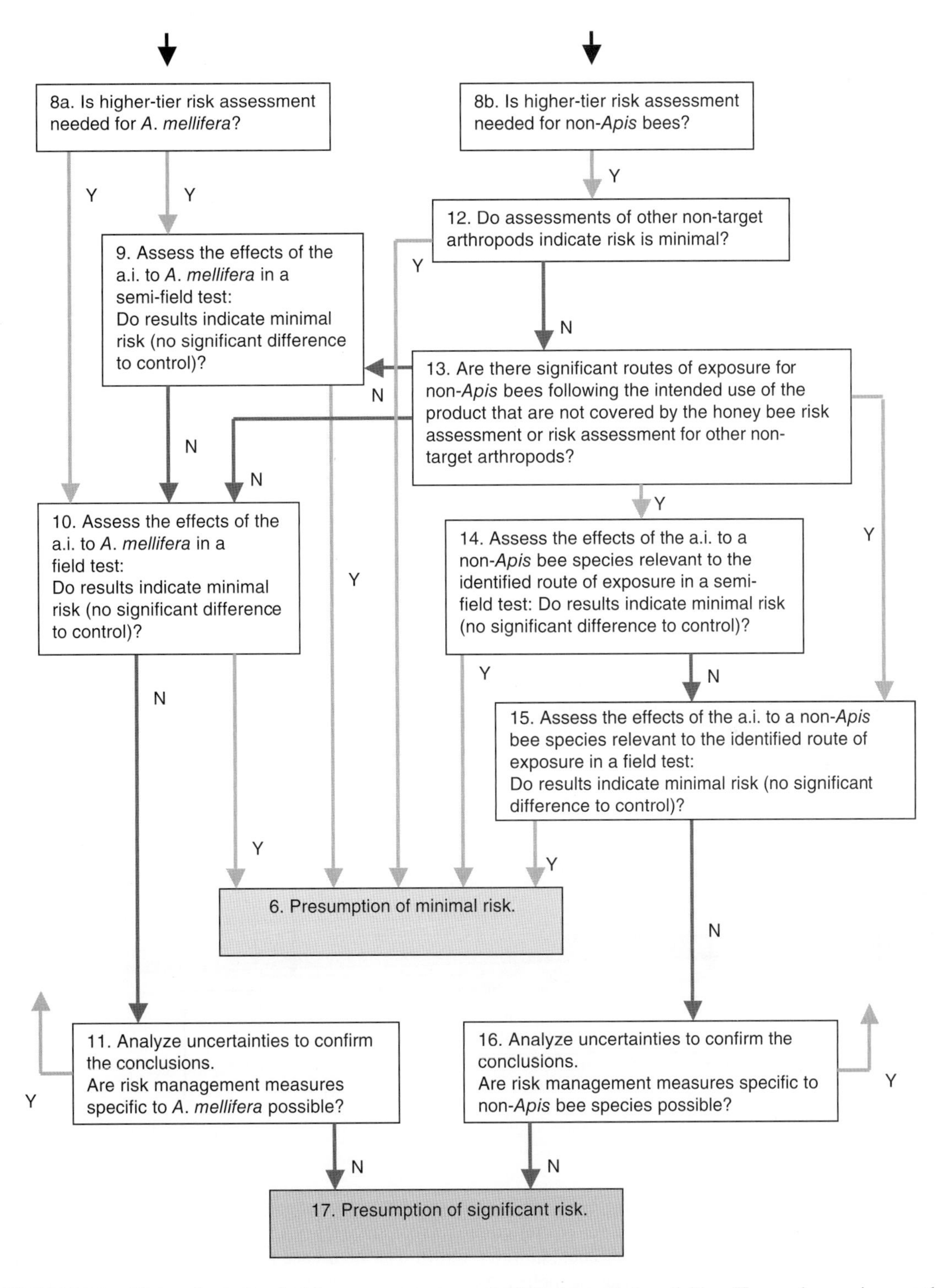

FIGURE 10.3 Higher tier (refined) risk assessment process for foliarly applied pesticides. (For a color version, see the color plate section.)

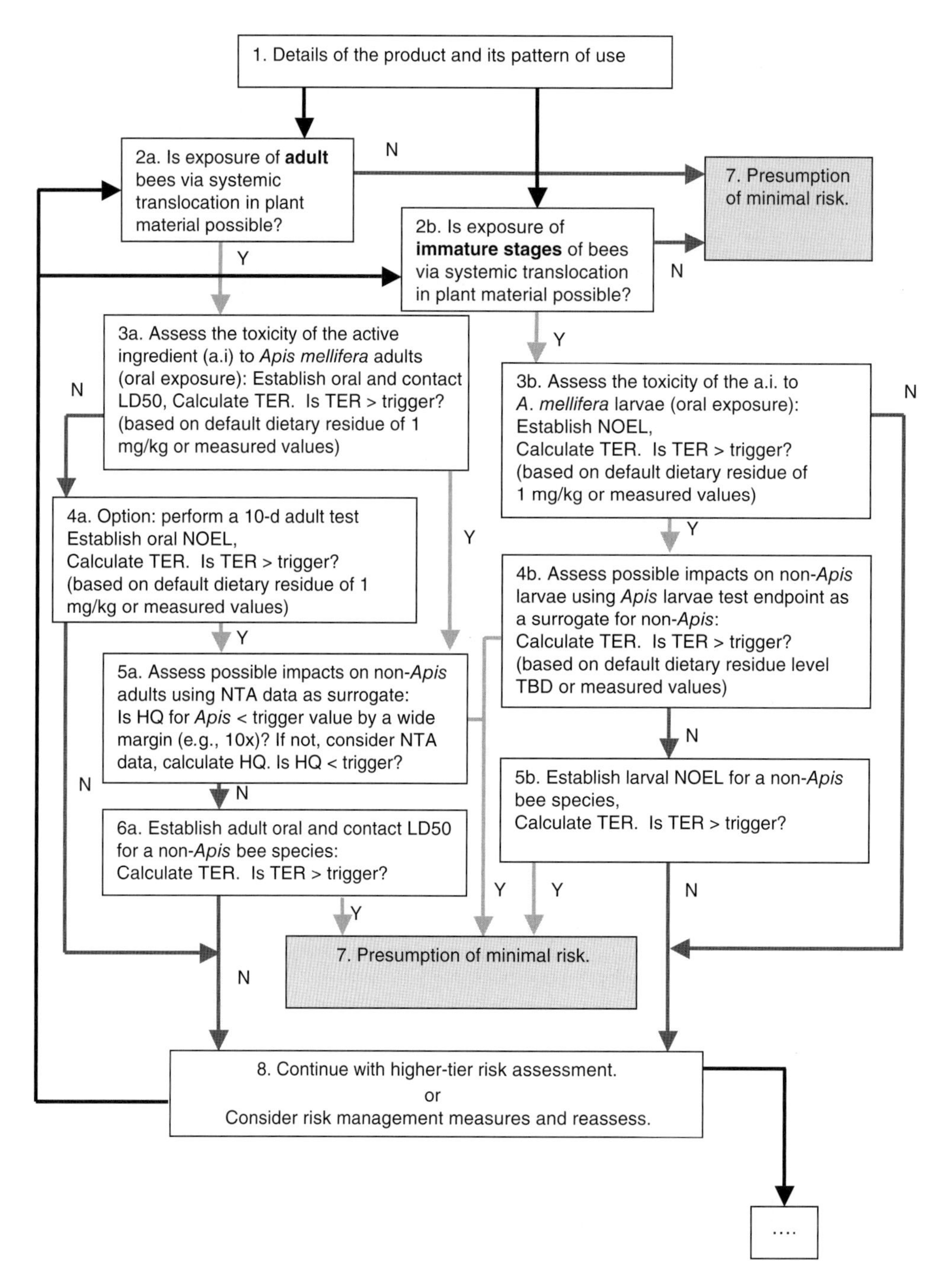

FIGURE 10.4 Insect pollinator screening-level risk assessment process for soil and seed treatment of systemic pesticides. Note that this flow chart may apply for trunk injection as well, as modalities of exposure of pollinators are similar as for soil or seed treatments. For trunk injection, however, further data are needed to appropriately describe the range of expected residue concentrations in nectar and pollen. As a consequence, no default value is currently available for a quantification of the risk (Boxes 3a and 3b). A compilation of available data could be made, with a particular attention to the corresponding injection protocols as it varies with the active substance involved and the tree. (For a color version, see the color plate section.)

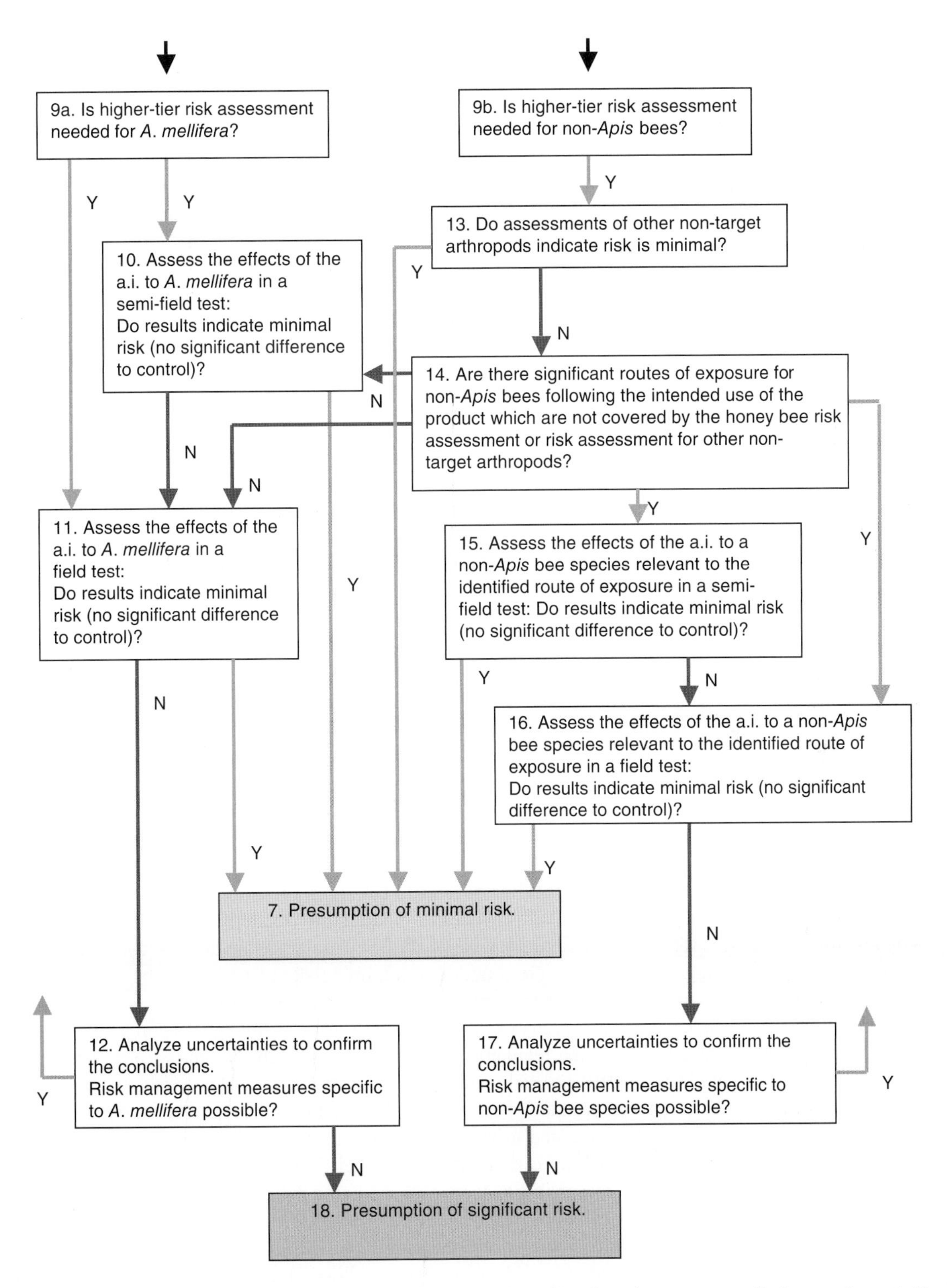

FIGURE 10.5 Higher tier (refined) risk assessment process for soil and seed treatment applied systemic pesticides. (For a color version, see the color plate section.)

to note that contact exposure to a systemic compound is also possible if it is applied as a spray application around or during the flowering period, for example, in the case of pre-bloom application. In this case, the reader may also find useful recommendations in the flowchart for soil or seed treatments.

Each box of these flowcharts is numbered and the nature of the data and reasoning behind each step of the process is provided below. As noted earlier, suitable level of concern (LOC) values (i.e., trigger values) for transitioning to higher levels of refinement are linked to risk management decisions and protection goals of individual regulatory authorities. The trigger values depicted in Figures 10.2 through 10.5 are generic. However, the more detailed but related risk assessment scheme in Appendix 6, which modifies the EPPO guidance (EPPO 2010), contains some trigger values currently used in the European regulatory process (EC 2010). As stated in other parts of this document, it is not the intent of this document, or SETAC, to recommend or support any particular trigger criteria but rather to emphasize the role that these values play in an efficient risk assessment process.

10.4 SPRAY APPLICATIONS

The risk assessment process for insect pollinators following the use of spray products (Figures 10.2 and 10.3) is depicted in the flow chart with numbered boxes and arrows in the direction that should be followed in response to a "yes" or "no" answer. More details regarding each of the steps are provided below.

The risk assessment process begins by asking whether exposure is possible (Box 2a); if exposure is not possible, then there is a presumption of minimal risk (Box 6). For sprayed applications, the screening level considers the worse case exposure assumption of a direct overspray to plants where bees are actively foraging. Potential effects of the chemical thus result from the overall effects of the direct spray on foraging bees.

As depicted on the left-hand side of Figure 10.2, at the screening level, the potential risk to adult honey bees from spray applications is assessed through calculation of an HQ (Box 3a). The assessor calculates an HQ by dividing the theoretical exposure, that is, the application rate expressed in terms of weight per unit area (e.g., grams active ingredient/hectare) by the most sensitive acute median lethal dose to 50% of the organisms tested, that is, the (contact) LD50 value, derived from laboratory studies. If the HQ value passes a regulatory trigger value, then there may be a presumption of minimal risk to adult honey bees and the reviewer proceeds to assess possible impacts to non-*Apis* adults (Box 4a).

To evaluate potential risk to *larval* honey bees, the assessor calculates a TER by dividing the most sensitive No-Observed-Effect level (NOEL) from the honey bee larval toxicity test by the theoretical maximum concentration in pollen and nectar (Box 3b). While several test designs currently exist to assess effects to larvae, adoption of this step in a formal, regulatory process would require standardization of a particular test design. Possible test designs for lower tier laboratory-based studies with larvae are discussed in Chapter 8. If the TER value passes the trigger value, then a presumption of minimal risk to larval honey bees can be made and the reviewer proceeds to evaluate possible impacts on non-*Apis* larvae (Box 4b).

Default Exposure Estimates for Screening Level Analysis for Apis Larvae: Although a theoretical maximum concentration has been established by some regulatory authorities for systemic products (e.g., 1 mg/kg or ppm, EPPO 2010; Alix and Lewis 2010) no such exposure model or theoretical maximum concentration level has been formally set for sprayed products. Pesticide residues resulting from direct overspray on food items for birds and mammals can be estimated using a residue per unit dose (RUD) approach favored by Hoerger and Kenaga (1972). The USEPA terrestrial exposure model (T-REX) has been revised to include insect residue data that could represent reasonably conservative screening values (USEPA 2012b). In the most recent guidance produced by the European Food Safety Authority (EFSA 2009), a range of RUD values have been developed for different crops and food sources. Furthermore, the USEPA toxicity of residues on foliage test may provide insight on the magnitude of residues on foliage following a particular application rate and the period of time these residues remain toxic (USEPA 2012a). Further research is necessary to both validate current screening exposure values used by regulatory authorities, as well as to develop RUD values, or other (screening) exposure models specific to pollinators.

The proposed risk assessment scheme also considers potential risks to non-*Apis* bees. At the screening level, risk to non-*Apis* bees is evaluated by employing effects data from honey bee acute oral or contact (LD50) studies (Box 4a depicting the calculation of an HQ for non-*Apis* adults), and chronic larval honey bee toxicity (NOEL) test data (Box 4b depicting the calculation of a TER for non-*Apis* larvae). In cases where Tier 1 (screening-level) data on *Apis* bees are not sufficient to conclude low risks to non-*Apis* bees (i.e., using a trigger value for *Apis* species modified with an appropriate safety factor to account for interspecies variation), then it may be concluded that the substance does not pass the screening step. In this case, data from non-target arthropods (NTA), typically required in the European registration process, could be considered (Box 4a and 4b) as they may provide useful information on the choice of non-*Apis* species to be further tested if potential risk cannot be excluded upon examination of the available NTA data. Participants in the Workshop agreed that NTA data could be utilized as it typically includes toxicity estimates for the predatory mite (*Typhlodromus pyri*) and the parasitic wasp (*Aphidius rhopalosiphi*). Refined risk estimates for non-*Apis* bees would then require development of adult oral and/or contact LD50 values for the relevant non-*Apis* species and an HQ (i.e, application rate/LD50) developed for adult bees (Box 5a). Similarly, where risk estimates do not meet trigger criteria for non-*Apis* bee larvae, then a NOEL for relevant non-*Apis* bees is necessary (Box 5b) to calculate a TER. As with toxicity estimates for adult non-*Apis* bees, toxicity test methods would have to be developed for larvae of relevant non-*Apis* bees. If risk estimates for either adult or larval non-*Apis* bees are within regulatory criteria, then minimal risk is presumed (Box 6); however, if not, then the reviewer should proceed to higher tier (refined) assessment methods depicted in Figure 10.3 or consider risk mitigation measures intended to reduce exposure (Box 7). Where risk mitigation measures are imposed, the reviewer should then re-evaluate whether exposure to adults (Box 2a) or larvae (Box 2b) has been sufficiently reduced to presume minimal risk (Figure 10.2). Again, if minimal risk cannot be presumed, the reviewer should proceed through the screen using the revised exposure numbers based on the proposed mitigation.

The proposed refined risk assessment for sprayed products (Figure 10.3) begins by asking whether higher tier risk assessment is needed for honey bees (Box 8a) or for non-*Apis* bees (Box 8b). The screening level risk assessment is typically based on effects data on individual bees collected through laboratory studies. However, in refined risk assessments, the reviewer considers the results of semi-field and full-field tests, which are typically conducted at the colony level rather than at the level of the individual bee. The refined risk assessment process therefore attempts to capture more realistic effects data as well as incorporating more refined estimates of exposure. For honey bees, effect estimates from semi-field studies (Box 9) or full-field studies (Box 10) are used to determine whether maximum application rates result in effects. If minimal risk cannot be presumed from the results of semi-field studies, then the reviewer should consider full-field studies where such studies can determine effects under more realistic test conditions (Box 10). In cases where full-field studies do not result in risk estimates that are consistent with protection goals, then the reviewer should conduct an analysis of uncertainties associated with the review process and determine whether possible mitigation specific to honey bees has been adequately considered (Box 11). As in the screening-level assessment, the impact of mitigation measures should be considered through the refined risk assessment process to ensure that their result is inconsistent with protection goals. After such an analysis, if risk estimates still do not meet regulatory criteria, then there is a presumption of significant risks (Box 17).

In the case of non-*Apis* bees, the reviewer assesses potential risks via data on NTAs (Box 12) and determines whether there are actual significant routes of exposure which are not accounted for by the higher tier tests conducted using honey bees (Box 13) such as from contaminated nest material. If risk concerns to non-*Apis* bees cannot be minimized, higher tier effects testing discussed in Chapter 9 using non-*Apis* bees relevant to the specific potential route of exposure are then considered, possibly first through a semi-field test (Box 14) with the option to extend the investigation to the full-field level (Box 15). As with honey bees, the process and underlying assumptions or uncertainties associated with risk estimates should be carefully analyzed (Box 16) and the reviewer should consider possible mitigation measures specific to non-*Apis* bees. The potential effects of mitigation options must be considered at each of the steps within the refined process whether it is an *Apis* or non-*Apis* analysis. If after this analysis, estimates are considered reasonable and potential mitigation measures cannot reduce potential exposure and potential risks, then the reviewer must presume significant risk to the non-*Apis* species, under the proposed conditions of use.

10.5 SOIL AND SEED TREATMENT APPLICATIONS FOR SYSTEMIC SUBSTANCES

The screening-level and refined risk assessment processes for soil or seed treatment-applied pesticides incorporate different degrees of ecological realism (Figure 10.4; Figure 10.5). Pesticides used as seed treatment or soil treatments are conservatively assumed to be systemically distributed to plant tissues as the plant develops. Each step (box) depicted in the flow chart is numbered and arrows depict the direction that should be followed in response to a yes or no answer. More detail regarding each of the steps is provided below.

When evaluating potential acute risk to adult honey bees from soil or seed treatments[4] with systemic compounds, the assessor first asks whether exposure is possible to the adult (Box 2a) or immature stages (Box 2b) via systemic translocation of residues in plant material. If exposure to honey bee adults is considered likely, the review calculates a TER (Box 3a) using either an acute oral or contact LD50 value for honey bee adults. In Europe, a Tier 1 TER is estimated by dividing a screening exposure estimate by the screening level hazard value. Currently, EPPO has a proposed conservative default exposure value of 1 mg a.i./kg, which relies on the default maximum concentration estimated in pollen or nectar from residues in whole plants for use with soil and seed treatments. If the risk estimate for the adult honey bees does not meet the regulatory criterion for low risk, then the reviewer should proceed to higher tier risk assessment (options to proceed with a 10-day adult test (Box 4a), or more refined studies) or consider risk mitigation measures and reassess (Box 8). If the TER value for the adult honey bees meets the regulatory criterion for low risk, then the reviewer proceeds to evaluate potential impacts on non-*Apis* adults (Box 5a). Here the assessor may consider data on NTAs. Where risk assessments for non-*Apis* bees do not meet the regulatory criterion for low risk (i.e., meets the regulatory criterion for low risk to *Apis* by a wide margin), then acute oral/contact LD50 values should be developed for non-*Apis* bees and a TER calculated (Box 6a). As with honey bees, if the risk estimate does meet the regulatory criterion for low risk, then the reviewer should proceed to higher tier (refined) risk assessment (semi-field or field study) or consider risk mitigation measures and reassess (Box 8).

For larval assessments, the same process as that discussed for spray applications is followed (Boxes 3b, 4b, and 5b of Figure 10.4). Additionally, the same process for higher tier (refined) risk assessment is used as discussed for spray applications. Participants of the Workshop noted the lack of information on potential exposure (nectar and pollen) related to trunk injection; and that further data are needed in this area (see Chapter 13). In the meantime, participants of the Workshop recommended that potential (screening) risks from trunk injection be estimated in the same manner as soil and seed scenarios.

As discussed previously, risk assessment is intended to be an iterative process. At a screening level, when risk estimates do not meet decision criteria (i.e., where a presumption of minimal risk cannot be made), the conditions under which the estimated risks occur should be more closely examined. More detailed fate considerations (such as degradation), or use considerations (such as timing of application, or application intervals) should be considered before additional testing is required.

[4] Although not specifically discussed at the Workshop, treatments with systemic compounds can include tree trunk injections as well.

10.6 SCREENING-LEVEL RISK ASSESSMENTS (TIER 1)

As noted, ecological risk assessments typically follow a tiered process (depicted in Figure 10.1). Substances move through lower tiers to higher tiers when the information indicates potential risk cannot be excluded. The first tier of that process is the screening-level assessment, which is intended to effectively and rapidly:

- exclude substances of low risk concern from entering into resource intensive higher tier risk assessment; and
- identify substances for which a potential risk to bees cannot be excluded and for which a higher tier risk assessment is needed.

The screening-level assessment should allow for the most efficient allocation of resources and minimize the number of substances forwarded for higher tier evaluation while still identifying substances of potential risk to bees. An efficient screening step in the risk assessment process is essential as it optimizes the success in achieving protection goals. At a screening level, the intent is then to use an appropriately sensitive species that is suitable to ensure that protection goals will be met. In this context, in designing the risk assessment process, participants proposed the *honey bee* as a reasonable surrogate for both *Apis* and non-*Apis* bees at a screening level for evaluating acute toxicity to adults. The reasons for this are:

- the biology and availability of *A. mellifera* makes it well-suited and lends itself to testing and analysis;
- the relative sensitivity of the honey bee compared to non-*Apis* species (based on available data)
- tiered toxicity test guidelines are widely available for *A. mellifera*; and
- conducting and interpreting the results of these tests does not require specialized backgrounds or conditions.

As illustrated in the flow chart (Figure 10.1), the screening step most often relies on the calculation of risk estimates, termed RQ, HQ, or TER. These risk estimates are compared to numerical regulatory decision criteria, termed an "LOC" or "trigger criterion." An LOC is a value against which a risk estimate is compared. It is intended to be protective in that it typically accounts for uncertainties related to intra- and inter-species variation in sensitivity, extrapolation of short-term toxicity to long-term effects, and extrapolation of laboratory results to the field.

Depending upon the type of risk estimate used (RQ or TER), if the estimate is above or below the LOC, then a determination of minimal risk is presumed, or whether additional refinements are necessary. For example, if screening-level risk estimate results in a TER (where the effects estimate is divided by the exposure estimate) that exceeds the LOC, then minimal risk is presumed (i.e., if TER > LOC = minimal risk is presumed); conversely, if the TER does not exceed the trigger value, then minimal risk cannot be presumed, and a higher tier risk assessment may be needed. The RQ is the reciprocal of the TER in that the exposure estimate is divided by the effects estimate; therefore, the RQ value is interpreted opposite to the way in which the TER is interpreted, that is, if the RQ exceeds a trigger value, then minimal risk is not presumed and a higher tiered risk assessment may be needed. If the RQ value is greater than the LOC (or trigger value), then minimal risk cannot be presumed.

10.7 FACTORS LIMITING CERTAINTY IN SCREENING ASSESSMENTS

Screening-level assessments are typically based on conservative assumptions regarding both exposure and effects. For example, at a screening level assessment for honey bees, the EPPO system does not account for good practices such as avoiding spray application during foraging times but conversely, not all routes of potential exposure are reflected. Given all the potential variables to consider, the participants of the Workshop believed that the proposed screening level analysis is conservative and protective for other potential routes of exposure.

Similarly, although mortality is the primary effect reported and used to generate LD50 values in acute toxicity tests, adverse effects on behavior are also reported. As discussed in earlier chapters, the extent to which sublethal effects occur and whether they ultimately affect assessment endpoints such as impaired survival, growth, and reproduction at the colony level remains an uncertainty for many compounds. However, since effects on behavior are frequently, but not exclusively, associated with insecticides or acaricides which will also potentially affect acute survival, the majority of these compounds will be subject to higher tier risk assessment where the sublethal effects will be more thoroughly evaluated. In addition, other information presented in the data profile of a compound (such as mode of action, route of uptake, toxicity, and effects on other types of terrestrial arthropods) should always be examined (EPPO 2010), and integrated with the findings of the screening step as part of the overall risk assessment for honey and non-*Apis* bees.

The capacity of the screening-level assessment to properly screen substances of low likelihood of adverse effects from substances for which further assessment is necessary has been evaluated through a review of the honey bee kill incidents recorded in the United Kingdom survey network Wildlife Incident Investigation Scheme (WIIS) (Mineau et al. 2008). The Mineau et al. 2008 analysis supports the utility and efficacy of the Tier 1 screening methodology, provided that considerations on the mode of action and use patterns are also kept in mind, as for any risk assessment process.

10.8 REFINEMENT OPTIONS FOR SCREENING-LEVEL RISK ASSESSMENT

If the results of a screening-level assessment indicate that a minimal risk cannot be concluded, the process moves to a series of refinements in exposure or effects data (see Figures 10.2 through 10.5). There are a number of options to further refine a risk assessment through a more in-depth description or characterization of exposure or of effects. These options are described, regarding their possible methodologies, in previous chapters. As refinements progress, different TERs and RQs are developed.

In the deterministic risk assessment approach, the primary outcome of the (Tier 1) risk characterization is the calculation of the RQ, or the TER depending on the country or region where the assessment is being performed. Both the RQ and the TER are single number (point) risk estimates. In reality, risk is more complex and therefore, a single point estimate can be misleading. As a consequence, the assessor should characterize the RQ or TER with a description of the uncertainties, assumptions, strengths, and limitations associated with the risk estimate. These sources of variability and uncertainty should be discussed during characterization of the exposure and effects and should include refinement options used in ultimately determining the RQ or TER. At the higher levels of refinement (e.g., semi-field and field tests), the level of impact is directly measured in experiments that are intended to reproduce the operational conditions of the subject pesticide product. In this case, TER and RQ values are no longer calculated.

Exposure is the first component of the risk to be examined to determine whether a risk assessment is needed, and the first to be explored to refine a potential risk. The relative importance of different exposure routes for *Apis* and non-*Apis* bees provides a guide for proceeding through the levels of refinement (Table 10.3). The main exposure routes identified for evaluation in the screening-level assessment are oral intake of nectar and pollen, and contact exposure. While not all exposure routes are included in the screening-level (Tier 1) risk assessment (e.g., wax and drinking water are not evaluated at Tier 1); and, direct overspray is considered as the worst case (high-end) exposure, it is important for the assessor to consider additional exposure routes for higher tier risk assessment purposes (Table 10.3 presents potential exposure routes for different bees).

TABLE 10.3
Likelihood of Exposure to *Apis* and Non-*Apis* Bees from Various Routes

	Apis		Non-*Apis*	
Exposure	**Adults**	**Larvae**	**Adults**	**Larvae**
Nectar	+ + +[1]	+	+ to + + +	+
Pollen	+ to + + +	+ +[2]	+ to + + +	+ + to + + +
Water[a]	+ to + +	+[3]	+	+
Nesting material	+[4]	+[4]	+ to + + +[4,5]	+ to + + +[6,7,8]
Exposure to soil	±	–	– to + + +	– to + + +
Foliar residues				
(Contact and direct spray)	+ + +	–	+ + +	– to + + +
Direct spray	+ + +[9]	–	+ + +[9]	–

[a]Collect water for cooling (evaporative cooling; take up into crop, regurgitate it, and flap wings to distribute) and honey production.
[1]Particularly for nurse bees; [2]bee bread; [3]provided by nurse bees; [4]wax; [5]leaves and soil for cement; [6]leafcutting bees; [7]soil used to cap cells; [8]exposure to soil; [9]at flowering.

Other insects may experience these exposure routes and testing methods are available for these species and field data may be available. As an example, parasitoid species also feed on nectar, such as the predatory mite *T. pyri,* or the ladybird beetle *Coccinella septempuctata* feeds on pollen. The predatory and parasitoid coleopteran *Aleochara bilineata* is a soil dweller at the adult stage. Therefore, review of these data when available may be useful in determining the major exposure routes to be investigated in a risk assessment for pollinators.

10.8.1 Refinement Options for Spray Applications

10.8.1.1 *Apis* Adults

If the HQ for adult *Apis* exceeds the LOC in the screening-level (Tier 1) assessment, then further information is required. Refinements can be made for exposure or effects, depending upon the profile of the active substance and its residues.

For spray application, an option for refining exposure estimates is to move from the screening-level default values to product-specific field modeling or measurement data to better quantify exposure. If an application during flowering cannot be excluded, this option may have several levels of refinement such as consideration of the interval between application and flowering and the expected level of residues to which bees could be exposed, for either modeled or measured estimates of refined exposure. Measurements of actual exposure may be achieved by the use of existing residue data, for example, magnitude of residue, or by implementing tunnel or field residue studies to estimate the level of exposure in treated crops and considering different modalities for the period of treatment.

While most semi- and full-field toxicity tests generate data on both exposure and effects, they may also be pursued with an exclusive aim of providing realistic exposure estimates. In this case, it is important that data generated from the field test is recorded so that it may be directly compared to the ecotoxicity data (i.e., the results and endpoints are expressed in the same units and represent comparable measures of exposure).

With respect to residue concentrations in nectar and pollen (or foliage where appropriate), the reviewer should consider the 90th percentile of measured concentrations as a conservative measure of exposure. However the decision to use a 90th percentile or other value ultimately depends on the dataset. If data are derived from only a single test on one crop, then a specified percentile, for example, 90th percentile, should be sufficiently vetted to reflect the uncertainty and variability as is frequently done in support of probabilistic approaches. If several trials have been undertaken, or data are derived for several crops, then a mean or a lower percentile may be more appropriate and would achieve the same level of protection. The selection of a particular crop for the evaluation of residues must consider whether the resulting data are sufficiently conservative to enable those data to serve as a surrogate for other uses.

The initial test to measure the effect of a compound is a lethality test consistent with relevant life stage and exposure route (e.g., oral LD50, or larval toxicity test). As effects tests become more refined, they incorporate more environmentally realistic conditions and begin to reflect both intrinsic toxicity and potential-enhancing or compensatory effects, related to environmental conditions.

To further refine the toxicity endpoint, additional *Apis* studies that could be relevant for the adult life stage include:

- 10-day feeding study (adult survival);
- toxicity of residues on foliage study;
- semi-field data; and
- field data.

A description of the studies that may be appropriate is found in Chapter 9; these studies are discussed briefly below.

The 10-day adult study is an extension of the standard laboratory oral exposure method (OECD 215). The test exposes adult bees for a period of 10 days and measures lethal effects after ingestion of product over the entire test duration. A NOEL is derived that may be used similarly as a LD50 in RQ calculations. Because this test only addresses oral exposure, it is not sufficient to address the uncertainties associated with sprayed compounds and is actually considered to be useful when refining estimates of effects for systemic soil or seed treatments. Currently there is no internationally recognized guideline for the 10-day feeding study nor for the larval toxicity testing in the laboratory; therefore, these tests need to be developed and validated before formal inclusion into a regulatory risk assessment scheme. The endpoint from a 10-day feeding study could be compared to either the default (screening-level) exposure concentration, or to refined exposure concentrations based on field measurements, both expressed in mg a.i./kg.

The USEPA foliar residue toxicity study is more representative of the conditions of exposure for bees after a spray event. This study is designed to evaluate the effects from exposure to dry and aged residues (3, 6 and 24 hours) and thus provide information on the level of bioavailability and length of residual hazard of the substance.

As discussed in Chapter 9, semi-field, and field studies reproduce more closely the conditions of exposure of bees in a treated crop. The test provides information on colony health based on bee survival and development related to actual field application parameters. Semi-field and field tests can be undertaken with pollinator-attractive crops treated at flowering (e.g., *Phacelia*), or pursued with the actual target crop when a treatment at flowering cannot be excluded. Semi-field and field tests can also provide additional information to refine an assessment such as information on potential exposure outside the flowering period of the crop, or through spray drift onto flowers in vegetated areas, or onto flowering weeds within the crop (e.g., in orchards). Finally semi-field and field tests may allow the evaluation of the efficacy of certain risk mitigation measures to limit exposure such as reduced application rates, or modifying application intervals.

10.8.1.2 *Apis* Larvae

As for the adults, an option for refinement of exposure is to move from the screening-level default values (e.g., application rate or default consumption rate), to product-specific field modeling or actual measured residues (e.g., in pollen and nectar, honey or bee bread) to better quantify exposure of larvae. The same considerations with regard to the generation and use of these data apply.

Additional *Apis* studies that could be relevant for the larval or immature life stages include:

- brood-feeding studies (brood development[5]);
- semi-field studies; and
- field studies.

The brood-feeding study aims at evaluating the effects on the development of the honey bee to derive a no-observed-effect concentration (NOEC). This NOEC can then be compared to either default (screening-level) concentration estimates or to refined concentrations based on field measurements.

The semi-field and field tests are similar with respect to measurement of effects on adults and both can provide information on colony health and brood development. As discussed elsewhere, field studies typically do not lend themselves to producing a dose–response relationship (i.e., a NOEC or LOEC) due to scale and logistical reasons. Consequently, the assessor must evaluate whether the study results indicate that a minimum level of risk exists (for example, no significant (or limited) difference between test and control plots). Increased levels of refinement toward characterizing effects beyond the laboratory and semi-field may involve assessing impacts of the formulated product in full-field tests. Further discussion and guidance on semi-field, and field tests can be found in Chapter 9, and discussion and guidance on brood tests can be found in Chapter 8.

10.8.1.3 Non-*Apis* Adults

Non-*Apis* bees may differ from honey bees in their exposure and sensitivity to plant protection products (Miles and Alix 2011; Devillers et al. 2003). Most non-*Apis* bees are solitary, with single females that forage for pollen and nectar to feed their offspring, construct their nests, and lay eggs (see introduction to non-*Apis* biology). The death of a foraging female implies the cessation of her reproduction (Tasei 2002). In comparison, when a (honey bee) colony loses female workers, the loss may be compensated by the colony, for example, by engaging inactive workers (Robinson 1992) or through reduced foraging age (Winston and Fergusson 1985), so the colony may continue to develop as a viable unit. For bumble bees some colony recovery is also possible (Schmid-Hempel and Heeb 1991). However, the death of the bumble bee queen in the spring signifies the death of the potential colony that would be formed (Thompson and Hunt 1999).

In comparison to honey bees, the life-history traits of non-*Apis* bees such as sociality and nesting behavior result in a greater importance of certain exposure routes. For example, alfalfa leafcutting bees (*Megachile rotundata*) may be more exposed to foliar residues (George and Rincker 1982), ground-nesting bees to soil residues and larvae to pollen residues. These differences mean that representatives of the main non-*Apis* groups for which we have sufficient knowledge should be considered for higher tier testing of a plant protection product for bees when a risk cannot be excluded. Where non-*Apis* species are chosen for higher tier evaluation they should be amenable to experimentation, provide reliable and reproducible results, and the methods should comply with internationally recognized and validated guidelines (e.g., OECD test guidelines). The exact choice of species may be based on the proposed use of the product and on regional

[5] For example the method of Oomen et al. (1992).

(species) considerations; however, it should be possible to extrapolate from "standard" species (e.g., *Bombus* sp.) to reduce the need for unnecessary testing.

Participants of the Workshop proposed that higher tier testing could be conducted with social non-*Apis* bees from the tribes Bombini and Meliponini and solitary bees that are ground-nesting and cavity-nesting (Table 10.5). While techniques exist for both laboratory and field or semi-field tests for Bombini spp. (*B. terrestris* and *B. impatiens*) standardization is needed (for review on *Bombus* spp. see Van der Steen 2001). Similar tests are in development for Meliponini species. Sufficient knowledge exists of the ecology of the Bombini and Meliponini tribes to be able to predict the main exposure routes (see Chapter 7). For cavity-nesting solitary bees (*Osmia lignaria* and *M. rotundata*), laboratory and field or semi-field tests have already been successfully developed (Alston et al. 2007; Abbott et al. 2008; Ladurner et al. 2008). For ground-nesting bees, while primary exposure routes can be predicted, there are not yet the techniques to perform standardized tests on them in the laboratory or the field. Until such techniques are available, the solitary cavity-nesting bees may sufficiently represent "solitary non-*Apis*" as a group, taking into account that for ground-nesting species, soil residues may play a more important route of exposure. Note, however, that even for Bombini and Meliponini tribes no validated or internationally recognized test protocols exist which currently limits their inclusion into a risk assessment scheme at this point in time and further research is needed.

Exposure Similar to the refinement process for adult honey bees, the option for refinement of exposure to adult non-*Apis* bees is to move from the screening-level default values to product-specific field modeling or measurement data to better quantify exposure of non-*Apis* larvae (Table 10.3). The same considerations with regard to the generation and use of these data apply.

Effects As discussed previously, at a screening level in the proposed risk assessment scheme, the adult *A. mellifera* is used as a surrogate for non-*Apis* species. To take into account interspecies variation and the different life-history characteristics a safety factor may be built into the LOC for *Apis* to non-*Apis* (participants of the Workshop considered a 10× factor conservative). Then, as illustrated in the flow chart, if the HQ is less than the adjusted non-*Apis* LOC, then risk is presumed to be low for non-*Apis* species; and, where it is not, further refinement of the ecotoxicity data may be undertaken.

When available, NTA data may be considered at this stage, as it may provide relevant information on effects (and route-specific exposure) to non-*Apis* species (see Table 10.4).

The nectar-feeding parasitoid *A. rhopalosiphi* and the soil-dwelling beetle *A. bilineata* are among the most sensitive of the NTAs tested under the European ESCORT scheme (Candolfi et al. 2001; Barrett et al. 1994). Adult parasitoids such as *Aphidius* also feed on nectar, which makes it a good NTA representative for exposure conditions of pollinating species. Similarly, approximately 70% of non-*Apis* bees are ground-nesting (Michener 2000) and the ground-dwelling beetle *A. bilineata*, is tested for sensitivity to plant protection products through sand or soil under the (Alix et al. 2011; Barrett et al. 1994), such that data from its contact toxicity tests may be considered informative for ground nesting bees. In the cases where a refined risk assessment has been triggered for non-*Apis* adults, the dataset developed in the European process may contain information on up to 8–10 species in the laboratory and more when semi-field or field testing have to be undertaken for refined risk assessment purposes (Candolfi et al. 2001) (Table 10.4). In these cases, inventories of the species identified in the crops tested may also be useful information in evaluating whether a particular concern is raised for non-*Apis* species which would need to be investigated further. Additional work is needed to understand the relative sensitivity of NTAs typically used in toxicity testing to non-*Apis* bees for which they may be used as surrogates.

If relevant NTA data cannot be found, then the assessor may consider selection of an appropriate non-*Apis* species for use in acute laboratory testing (Table 8.3, see Chapter 8) and data from residue studies and field measurements (i.e., pollen, nectar, foliage and soil) can inform study design with respect to exposure for

TABLE 10.4
Testing Methodologies Developed for the Risk Assessment to Non-Target Arthropods Developed in the European Process of Evaluation of Pesticides

Testing Scale	Species (and Stages Tested)
Tier 1 laboratory: artificial substrate	*Aphidius rhopalosiphi* (adults + life cycle[a]) *Typhlodromus pyri* (protonymphs + life cycle[a])
Tier 2 (extended) laboratory: natural substrate	*Aleochara bilineata* (adults + life cycle[a]) *A. rhopalosiphi* (adults + life cycle[a]) *Chrysoperla carnea* (larvae + life cycle[a]) *Coccinella septempunctata* (larvae + life cycle[a]) *Orius laevigatus* (nymphs + life cycle[a]) *Pardosa* sp. (adults) *Poecilus cupreus* (adults) *Trichogramma cacoeciae* (adults + life cycle[a])
Semi-field	*P. cupreus* (adults) Methods can be adapted for many species
Field	Arthropods (populations and communities)

Source: Data from Candolfi et al. 2001.
[a]Studies purporting to examine the life cycle of species may focus on a particular aspect of the life cycle and may not include the entire life cycle.

non-*Apis* (see also Chapter 7). For example, a plant protection product with high foliar residues would suggest that higher tier testing should be performed on alfalfa leafcutting bees (*M. rotundata*) if such bees will visit the crop to harvest nesting material and exposure may occur.

Alternatively, as shown in the flow charts (Figures 10.2, 10.3, 10.4, and 10.5), non-*Apis*-specific test data for adult contact or oral toxicity can be generated. These data are likely to be in the form of an LD50 (μg/bee), to be used in developing an HQ similar to that for adult *Apis*. For the assessment criteria to be met, the HQ must not exceed the LOC (trigger value); if it does not exceed a concern, the assessment need not proceed further. Issues of LOC (triggers) and safety factors (such as intra-species variation) may be further discussed by the respective regulatory authorities.

Refinement of effects data beyond the laboratory and semi-field or field may involve assessing impacts of the formulated product. Guidance on the types of tests may be found in Chapter 8. The field or semi-field tests will monitor behavior and quantify bee mortality and fecundity of one or several selected non-*Apis* species likely to be encountered in the crops to be treated with the product. (see Chapter 8, Hazard, Field for methods and advantages of field tests on non-*Apis* bees). Table 10.5 at the end of this section highlights the availability of laboratory and field tests for representative groups of social and solitary non-*Apis* bees.

Risk Characterization (Estimation) For both *Apis* and non-*Apis* assessments, when higher level field data are developed, the results are not expected to be applied in a TER or quotient context, but may be used directly in the risk assessment. Again, mitigation of potential risk remains as an important pathway to meeting protection goals whether at the screening or higher tier steps of the analysis. Risk characterization will depend upon the data generated and refinements therein. Below is a brief discussion of refinements to input studies.

TABLE 10.5
Available Laboratory and Field Tests with Representative Groups of Solitary and Social Non-*Apis* Bees

		Solitary		Social	
Study Type		**Tunnel-nesting (Tube, Wood)**	**Ground-nesting**	**Bombini (Bumble Bees)**	**Meliponini (Stingless Bees)**
Laboratory	Adult	Zone: temperate north *Megachile rotundata* (Huntzinger et al. 2008; Scott-Dupree et al. 2009) *Osmia lignaria* (Ladurner et al. 2005; Scott-Dupree et al. 2009) Zone: tropics *Xylocopa* spp. (tests in development)	Zone: temperate north *Nomia melanderi* (Johansen et al. 1984; Mayer et al. 1998)	Zone: temperate north *Bombus terrestris* (Thompson 2001) *Bombus impatiens* (Scott-Dupree et al. 2009; Gradish et al. 2011b[a])	Zone: tropics several species, and tests in development (Macieira and Hebling-Beraldo 1989; Valdovinos-Nunez et al. 2009)
	Larva	Zone: temperate north *M. rotundata* (Peach et al. 1995; Gradish et al. 2011a; Hodgson et al. 2011) *O. lignaria* (Abbott et al. 2008) Zone: tropics *Xylocopa* spp. (tests in development)		Zone: temperate north *B. terrestris* (Thompson 2001) *B. impatiens* Gradish et al. 2010; Gradish et al. 2011b[a])	Zone: tropics (tests in development)
Field	Semi-field	Zone: temperate north *M. rotundata* (Johansen et al. 1984; Tasei et al. 1988; Mayer and Lunden 1999) *Osmia bicornis* (Konrad et al. 2008) *O. lignaria* (Ladurner et al. 2008)		Zone: temperate north *B. terrestris* (Tasei et al. 2001) *B. impatiens* (Gels et al. 2002)	Zone: tropics (tests in development)
	Field	Zone: temperate north *M. rotundata* (Torchio 1983) *O. lignaria*	Limited availability of tested species *Nomia melanderi* (Mayer et al. 1998)	Zone: temperate north *B. terrestris* (Tasei et al. 2001) *B. impatiens*	Zone: tropics (tests in development)
Exposure (pollen, nectar, foliar, soil)		Can be developed. For pollen provisions in the field, see Abbott et al. 2008 For foliar residues, see George and Rincker 1982		For pollen, see Morandin et al. 2005	

[a]Needs standardized guidelines of currently used lab bioassay and microcolony assays.

10.8.1.4 Non-*Apis* Larvae

Exposure A general description of exposure sources for non-*Apis* species (immature stages) is provided in Table 10.3. Where honey bee larvae are exposed primarily in larval food (which is processed pollen) this should be considered when generating a refined (exposure) analysis for non-*Apis* species. For example, pollen sampled in the field or from loads taken at the hive entrance (pollen traps) or from forager bees directly may represent concentrations found in unprocessed food sources. Concentrations of residues from pollen sampled from within hive food stores or from larval cells could be more relevant to honey bee larvae.

Non-*Apis* larvae may also be exposed through contact with the pollen and nectar food provision in the nest. In addition, the larvae of ground-nesting bees and cavity-nesting bees which separate their nest cells with soil (e.g., *O. lignaria*) may come into contact with soil applied plant protection products. Similarly, the larvae of leafcutting bees may come into contact with a plant protection product through residues on the foliage used to construct its nest (see Chapter 7, Exposure). Non-*Apis* species thus have various sources of exposure (e.g., treated soil or nesting material). Refining potential exposure estimates to non-*Apis* bees to account for the different exposure sources would be difficult to achieve in a specific exposure test. In this case, it would be more appropriate to refine potential exposure and risk through a semi-field or field study (see Chapter 9).

Effects As discussed earlier, honey bee larvae are proposed as a surrogate for non-*Apis* larvae as there is currently no formal guideline established for testing non-*Apis* larvae.

As the assessor moves through the proposed process, they may consider NTA data, if available, which may provide relevant information to refine potential risk to non-*Apis* species (Candolfi et al. 2001). These tests measure a wide range of endpoints including juvenile and adult survival, fecundity, or larval development depending on the species being tested (Table 10.4). The NTA tests are frequently designed to detect relatively small changes in sublethal endpoints; therefore, an understanding of an application rate that may result in low impact on growth or fecundity or other sublethal parameters may be derived. Beyond laboratory tests, refining an understanding of potential effects to non-*Apis* larvae may involve field tests with formulated products (see Chapter 9). While field and semi-field tests have not been specifically developed for ground nesting bees, monitoring of cavity-nesting bees through field or semi-field tests may provide information on some of the larval exposure routes that are unique to non-*Apis* species. Table 10.5 at the end of this section highlights the availability of laboratory and field tests for representative groups of social and solitary non-*Apis* bees.

Risk Characterization (Estimation) If effects data on non-*Apis* larvae have been generated and provide a NOEC, then this value could be used as in TER calculation. Both default and refined exposure estimates may also be used in TER calculation. As noted in the flow charts, should this assessment indicate risks that are not consistent with protection goals, then, either mitigation measures may be considered or the assessment may proceed to further refinement.

Again, when data are generated from field tests, it is not expected that the results are conveyed in a TER (quotient-based) context, but rather incorporated directly into a risk assessment.

10.8.2 Soil or Seed Treatment Application for Systemic Substances (Also Including Trunk Injection)

10.8.2.1 Exposure—*Apis* and Non-*Apis*

While there are differences in the screening-level assessment for calculation of HQs/TERs between sprayed pesticides and systemic substances, the general approach to refining the risk assessment for systemic applications is largely similar to that for spray applications. The primary difference is that for systemic chemistries, exposure levels via contact are largely below that which may be encountered via an oral route of exposure. Exposure routes specific to non-*Apis* bees (Table 10.3) have important ecological consequences in risk assessment. For example, for systemic compounds, leafcutter bees may be exposed orally through the foliage used to build their nests. The most appropriate way to explore this further is through simulating exposure conditions in a semi-field or a field test (see Chapter 9).

As stated earlier, for trunk injection, further data are needed to appropriately describe the range of expected residue concentrations in nectar and pollen that may be used in a risk estimate for this application method. In the future, a compilation of available data could be made, with particular attention to the corresponding injection protocol as it varies with active ingredient and tree species.

10.8.2.2 Effect—Adult *Apis*

If risk cannot be excluded at the screening-level assessment, then a Tier 2 assessment, based on the 10-d NOEL for young adult honey bees, can be conducted. The 10-day test is an appropriate measure to refine the acute effects endpoint employed in the Tier 1 assessment (i.e., oral LD50). The 10-day test may be run based on the default maximum concentration estimated in pollen or nectar, or on refined measured values, if these are available (see section 10.8 for more details on the options). In this case, if the TER value exceeds triggers, then one may reach a presumption of low risk to adult honey bees from soil or seed applications. If viable exposure routes exists for the immature stages of either honey bees or non-*Apis* species, (e.g., through contaminated pollen or bee bread), then the approaches for refinement to soil or seed scenarios are similar as that for spray treatments (see Chapter 9). For higher tier testing (semi-field and field testing) protocols may be adapted to reflect crops grown from coated seeds or to products applied on or to soil, or for trunk injection. These tests may include monitoring of effects at sowing if measurements from potential exposure via seed dust (if it cannot be excluded or mitigated), or measurements of potential exposure to non-*Apis* species that might frequent the soil.

10.8.2.3 Risk Characterization (Estimation)

Similar principles as for spray application do apply for soil or seed treatments and trunk injection.

10.9 CONCLUSIONS ON THE RISKS AND RECOMMENDATIONS

Concluding a risk assessment is probably the step that best reflects how case-related the risk assessment process can be. Conclusions could be very brief and simply indicate that under the assessment that was conducted (i.e., whether it was screening level or a higher tiered assessment) the use of the product meets the protection goals of the respective regulatory authority. However, where a refined risk assessment was triggered, there is a need to clearly express the following information in the conclusions:

- what concerns were identified at the screening step;
- whether/what concerns were identified in higher tier assessments;
- whether results of the higher tier assessment, addressed potential risk concerns;
- whether/which mitigation measure were considered at different levels of analysis, and whether the mitigation measures reduced potential risks to an acceptable level;
- whether, despite higher tier analysis, all available lines of evidence, and consideration of mitigation measures, potential risks remain; and
- remaining uncertainties (if any) in the risk assessment.

Risk assessment conclusions should give particular emphasis to the four following areas which are essential in providing appropriate information to risk managers for decision making. These are:

- the appropriateness of the available data to assess potential risks posed by the subject compound, or product;
- defining the use parameters required in order that the protection goals can be met;
- characterization of any potential risks, including remaining uncertainties resulting from a lack of data or deficiencies in the existing data; and
- where refined risk analysis indicates risk, characterization should be provided regarding the growth, reproduction or survival of the organism (colony or population) and possible interactions with plants and ultimately with stated protection goals.

Risk assessment conclusions should characterize the possibility of risk based on the available lines of information (data, monitoring information, incidents, etc.). Characterization should include discussion of potential risk to any specific life stages or castes. In certain cases, exposure considerations should focus on gathering more refined data such as:

- characterizing spray drift onto adjacent crops or vegetation that are attractive to bees; and
- characterizing exposure to residues that could reach pollen or nectar of the crop for pre-flowering applications of systemic compounds, and of mobilization of soil residues in rotational crops (where relevant).

The risk assessment should be able to address the meaning of effects, for example, a temporary increase in the mortality of foragers, avoidance of a treated crop over the first days post treatment, etc. Field, and in some cases semi-field, studies may allow for the monitoring of colonies/populations over long periods and measurement endpoints may be available to address these concerns. Unresolved issues regarding time scale (temporal) or spatial scale could also be addressed through modeling tools when sufficiently developed.[6] Where uncertainties are related to "borderline" or "minor" effects and do not strictly compromise the protection goals, they may be appropriately addressed by implementing a monitoring study. The advantage of monitoring in this respect is to verify that protection goals will be met under conditions of agricultural practice in the real environment without any effort to control other stress factors.

If a decision is made not to authorize a use, then it must be based on the evidence that protection goals for a particular product cannot be met. The inability to meet protection goals implies that, based on the available lines of evidence and higher tiered analysis, neither exposure nor hazard can be reduced or avoided, and resulting risks may compromise protection goals. It is the responsibility of both the risk assessor and risk manager to discuss the conditions of the assessment and explore mitigation options, if these are warranted. Both the assessor and manager should consider whether information exists that would determine whether all options to refine or mitigate potential risks have been explored before a final decision is reached.

10.10 RECOMMENDING RISK MITIGATION MEASURES

Risk mitigation measures mainly aim at reducing the risks through a reduction of exposure. In principle, mitigation may be considered at any stage of the assessment process, such as prohibiting application during bloom. However, certain measures aimed at reducing the level of exposure or residues in relation to effect

[6] Modeling tools have been successfully developed in other areas of ecotoxicology for that purpose.

measure, are more effectively considered during higher tier testing, such as reduced application rates or increased application intervals. Dedicated field testing may be useful when dealing with the product-specific measures. The decision to consider mitigation measures at any step of the process involves issues of product efficacy, as well as national policies. A fuller address of mitigation measures is found in Chapter 13.

10.11 ADDITIONAL TOOLS IN SUPPORT OF RISK ASSESSMENT AND TO INFORM RISK MANAGEMENT

Tools that may help to better interpret data (e.g., statistical and mathematical tools) should be used, particularly when higher tier data have been generated. In addition to these tools which now often enter into the usual package of risk assessment, modeling, and landscape management approaches are possibly the most promising ones to further support both risk assessment and risk mitigation provided these tools are sufficiently vetted and validated against measured data.

10.11.1 Modeling Tools

Modeling tools may provide insight on uncertainties identified in risk analyses that cannot be readily addressed by laboratory and/or field studies. Modeling population dynamics may be used to simulate the fate of the population or colony over years of exposure to the product, or at a wider scale than the field, and may have the potential to address generic questions such as colony-level implications from individual-level effects. Development of models for honey bees and non-*Apis* bees could thus address general questions such as:

- What level of mortality or brood loss is of minimal consequence at the colony or population level?
- What magnitude and frequency of effects on adult survival and brood success are required to put the viability of a honey bee colony at risk?
- How do these thresholds vary according to season?

Answers to these generic issues are of great interest in conducting and interpreting risk assessments but also in support of decision making. The potential usefulness of modeling tools is discussed in more detail in Chapter 11.

REFERENCES

Abbott VA, Nadeau JL, Higo HA, Winston ML. 2008. Lethal and sublethal effects of imidacloprid on *Osmia lignaria* and clothianidin on *Megachile rotundata* (Hymenoptera: Megachilidae). *J. Econ. Entomol.* 101:784–796.

Alix A, Lewis G. 2010. Guidance for the assessment of risks to bees from the use of plant protection products under the framework of Council Directive 91–414 and Regulation 1107–2009. *OEPP/EPPO Bull.* 40:196–203.

Alix A, Bakker F, Barrett K, Brühl CA, Coulson M, Hoy S, Jansen JP, Jepson P, Lewis, Neumann P, Süßenbach D, van Vliet P. 2011. ESCORT 3: Linking Non-Target Arthropod Testing and Risk Assessment with Protection Goals. Proceedings of the European Standard Characteristics of Non-Target Arthropod Regulatory Testing Workshop ESCORT 2, Zuiderduin, The Netherlands, March 21–23, 2000. March 8–11, 2010. SETAC Press, 136pp.

Alston DG, Tepedino VJ, Bradley BA, Toler TR, Griswold TL, Messinger SM. 2007. Effects of the insecticide phosmet on solitary bee foraging and nesting in orchards of Captiol Reef National Park, Utah. *Environ. Entomol.* 36:811–816.

Barrett K, Grandy N, Harrisson EG, Hassan S, Oomen P. (eds). 1994. Guidance document on regulatory testing procedures for pesticides with non-target arthropods. *ESCORT Workshop (European Standard Characteristics of Non-Target Arthropod Regulatory Testing).* SETAC Publication, Wageningen, p. 53.

Candolfi MP, Barrett KL, Campbell PJ, Forster R, Grandy N, Huet MC, Lewis G, Oomen PA, Schmuck R, Vogt H. 2001. Guidance document on regulatory testing and risk assessment procedures for plant protection products with non-target arthropods. *ESCORT 2 Workshop (European Standard Characteristics of Non-Target Arthropod Regulatory Testing)*. SETAC Publication, Wageningen, p. 46.

Devillers J, Decourtye A, Budzinski H, Pham-Delègue MH, Cluzeau S, Maurin G. 2003. Comparative toxicity and hazards of pesticides to *Apis* and non-*Apis* bees. A chemometrical study. *SAR QSAR Environ. Res.* 14:389–403.

EC. 1997. Council Directive 97/57/EC of 22 September 1997 establishing Annex VI to Directive 91/414/EEC concerning the placing of plant protection products on the market. *OJL*. 265:87–109.

EC. 2010. Council Directive of 15 July 1991 concerning the placing of plant protection products on the market (Directive 91/414/EEC).

European Food Safety Authority [EFSA]. 2009. Guidance Document on Risk Assessment for Birds and Mammals on request from EFSA. *EFSA J.* 7(12):1438. doi:10.2903/j.efsa.2009.1438

EPPO. 2010. Environmental risk assessment scheme for plant protection products. *OEPP/EPPO Bull.* 40:1–9.

Gels JA, Held DW, Potter DA. 2002. Hazards of insecticides to the bumble bees *Bombus impatiens* (Hymenoptera: Apidae) foraging on flowering white clover in turf. *J. Econ. Entomol.* 95:722–728.

George DA, Rincker CM. 1982. Residues of commercially used insecticides in the environment of *Megachile rotundata. J. Econ. Entomol.* 75:319–323.

Gradish AE, Scott-Dupree CD, Cutler GC. 2011a. Susceptibility of *Megachile rotundata* to insecticides used in wild blueberry production in Atlantic Canada. *J. Pest. Sci.* 85(1):133–140.

Gradish AE, Scott-Dupree CD, Frewin AJ, Cutler GC. 2011b. Lethal and sub-lethal effects of some insecticides recommended for wild blueberry on the pollinator *Bombus impatiens*. *Can. Entomol.* 144:478–486.

Gradish AE, Scott-Dupree CD, Shipp L, Harris CR, Ferguson G. 2010. Effect of reduced risk pesticides for use in greenhouse vegetable production on *Bombus impatiens* (Hymenoptera: Apidae). *Pest. Manag. Sci.* 66(2): 142–146.

Hoerger F, Kenaga EE. 1972. Pesticide residues on plants: correlation of representative data as a basis for estimation of their magnitude in the environment. In: Coulston F, Korte F (eds), *Environmental Quality and Safety: Chemistry, Toxicology, and Technology*. Georg Thieme, Stuttgart, pp. 9–28.

Hodgson EW, Pitts-Singer TL, Barbour JD. 2011. Effects of the insect growth regulator, novaluron on immature alfalfa leafcutting bees, *Megachile rotundata*. *J. Insect Sci.* 11:43.

Huntzinger C, James RR, Bosch J, Kemp WP. 2008. Fungicide tests on adult alfalfa leafcutting bees *Megachile rotundata* (F.) (Hymenoptera: Megachilidae). *J. Econ. Entomol.* 101:1088–1094.

Johansen CA, Rincker CM, George DA, Mayer DF, Kious CW. 1984. Effects of aldicarb and its biologically active metabolites on bees. *Environ. Entomol.* 13:1386–1398.

Konrad R, Ferry N, Gatehouse AM, Babendrier D. 2008. Potential effects of oilseed rape expressing oryzacystatin-1 (OC-1) and purified insecticidal proteins on larvae of the solitary bee *Osmia bicornis*. *PloS ONE.* 3(7):e2664. doi:10.1371/journal.pone.0002664

Ladurner E, Bosch J, Kemp WP, Maini S. 2005. Assessing delayed and acute toxicity of five formulated fungicides to *Osmia lignaria* Say and *Apis mellifera*. *Apidologie* 36:449–460.

Ladurner E, Bosch J, Kemp WP, Maini S. 2008. Foraging and nesting behavior of *Osmia lignaria* (Hymenoptera: Megachilidae) in the presence of fungicides: cage studies. *J. Econ. Entomol.* 101:647–653.

Macieira OJD, Hebling-Beraldo MJA. 1989. Laboratory toxicity of insecticides to workers of *Trigona spinipes* (F. 1793) (Hymenoptera: Apidae). *J. Apic. Res.* 28:3–6.

Mayer DF, Lunden JD. 1999. Field and laboratory tests of the effects of fipronil on adult female bees of *Apis mellifera, Megachile rotundata* and *Nomia melanderi*. *J. Apic. Res.* 38:191–197.

Mayer DF, Kovacs G, Lunden JD 1998. Field and laboratory tests on the effects of cyhalothrin on adults of *Apis mellifera, Megachile rotundata* and *Nomia melanderi*. *J. Apic. Res.* 37:33–37.

Michener CD. 2000. *The Bees of the World*. The John Hopkins University Press, Baltimore and London.

Miles MJ, Alix A. 2011. Assessing the comparative risk of plant protection products to honey bees, non-target arthropods and non-*Apis* bees. In: Proceedings of the 11th ICPBR Meeting, Wageningen, The Netherlands, November 2011. Morandin LA, Winston ML, Franklin MT, Abbott VA. 2005. Lethal and sub-lethal effects of spinosad on bumble bees (*Bombus impatiens* Cresson). *Pest. Manag. Sci.* 61:619–626.

Mineau P, Harding KM, Whiteside M, Fletcher MR, Garthwaite D, Knopper LD. 2008. Using reports of honey bee mortality in the field to calibrate laboratory derived pesticide risk indices. *Environ. Entomol.* 37(2):546–554.

Oomen PA, de Ruijter A, Van der Steen J. 1992. Method for honeybee brood feeding tests with insect growth-regulating insecticides. *EPPO Bull.* 22:613–616.

Peach ML, Alston DG, Tepedino VJ. 1995. Sublethal effects of carbaryl bran bait on nesting performance, parental investment, and offspring size and sex ratio of the alfalfa leafcutting bee (Hymenoptera: Megachilidae). *Environ. Entomol.* 24:34–39.

Robinson GE. 1992. Regulation of division of labor in insect societies. *Annu. Rev. Entomol.* 37:637–665.

Schmid-Hempel P, Heeb D. 1991. Worker mortality and colony development in bumblebees, *Bombus lucorum* (L.) (Hymenoptera, Apidae). *Bull. Soc. Entomol. Suisse.* 64:93–108.

Scott-Dupree CD, Conroy L, Harris CR. 2009. Impact of currently used or potentially useful insecticides for canola agroecosystems on *Bombus impatiens* (Hymenoptera: Apidae), *Megachile rotundata* (Hymenoptera: Megachilidae), and *Osmia lignaria* (Hymenoptera: Megachilidae). *J. Econ. Entomol.* 102:177–182.

Tasei JN. 2002. Impact of agrochemicals on non-*Apis* bees. In: Devillers J, Pham-Delegue M-H (eds), *Honey Bees: Estimating the Environmental Impact of Chemicals.* Taylor & Francis, New York.

Tasei JN, Carre S, Moscatelli B, Grondeau C. 1988. Recherche de la D.L. 50 de la deltamethrine (Decis) chez *Megachile rotundata* F. Abeille pollinistatrice de la luzerne (Medicago sativa L.) et des effets de doses infralethales sur les adultes et les larves. *Apidologie* 19(3):291–306.

Tasei JN, Ripault G, Rivault E. 2001. Hazards of imidacloprid seed coating to *Bombus terrestris* (Hymenoptera: Apidae) when applied to sunflower. *J. Econ. Entomol.* 94:623–627.

Thompson H. 2001. Assessing the exposure and toxicity of pesticides to bumblebees (*Bombus* sp.). *Apidologie* 32:305–321.

Thompson HM, Hunt LV. 1999. Extrapolating from honeybees to bumblebees in pesticide risk assessment. *Ecotoxicology.* 8:147–166.

Torchio PF. 1983. The effects of field applications of naled and trichlorfon on the alfalfa leafcutting bee, *Megachile rotundata* (Fabricius). *J. Kans. Entomol. Soc.* 56:62–68.

U.S. Environmental Protection Agency [USEPA]. 1992. Framework for Ecological Risk Assessment [EPA/630/R-92/001]. Risk Assessment Forum, Washington, DC.

U.S. Environmental Protection Agency [USEPA]. 1998. Guidelines for Ecological Risk Assessment [EPA/630/R-95/002F]. Risk Assessment Forum, Washington, DC.

U.S. Environmental Protection Agency [USEPA]. 2004. Overview of the Ecological Risk Assessment Process in the Office of Pesticide Programs, U.S. Environmental Protection Agency Endangered and Threatened Species Effects Determinations. Office of Chemical Safety and Pollution Prevention, Washington, DC. http://www.epa.gov/espp/consultation/ecorisk-overview.pdf (accessed February 6, 2012).

U.S. Environmental Protection Agency. 2012a. User's guide for T-REX version 1.5 (Terrestrial Residue EXposure model). Office of Pesticide Programs, Environmental Fate and Effects Division. Available online at: http://www.epa.gov/oppefed1/models/terrestrial/trex/t_rex_user_guide.htm (accessed February 6, 2012).

U.S. Environmental Protection Agency. 2012b. Ecological Effects Test Guidelines. OPPTS 850.3030 Honey Bee Toxicity of Residues on Foliage. EPA 712-C-018. January 2012. http://www.regulations.gov/#!documentDetail;D=EPA-HQ-OPPT-2009-0154-0017 (accessed February 6, 2012).

Valdovinos-Nunez GR, Quezada-Euan JJ, Ancona-Xiu P, Moo-Valle H, Carmona A, Sanchez ER. 2009. Comparative toxicity of pesticides to stingless bees (Hymenoptera: Apidae: Meliponini). *J. Econ. Entomol.* 102:1737–1742.

Van der Steen JJM. 2001. Review of methods to determine the hazard and toxicity of pesticides to bumblebees. *Apidologie* 32:399–406.

Vaughn M, Shepherd M, Kremen C, Black SH. 2007. *Farming for Bees.* The Xerces Society, Portland, OR. p. 44.

Winston M. 1987. *The Biology of the Honey Bee.* Harvard University Press, Cambridge, MA. ISBN 0-674-07409-2.

Winston ML, Fergusson LA. 1985. The effect of worker loss on temporal caste structure in colonies of the honey bee (*Apis mellifera L.*). *Can. J. Zool.* 63:777–780.

11 Ecological Modeling for Pesticide Risk Assessment for Honey Bees and Other Pollinators

V. Grimm, M.A. Becher, P. Kennedy, P. Thorbek, and J. Osborne

CONTENTS

11.1 INTRODUCTION

Current pesticide risk assessment for honey bees is based on laboratory tests and on semi-field and field studies. Risk assessment schemes focus on quotients of the hazard imposed by a compound and the predicted exposure to this compound in the field. Depending on this quotient, in a tiered approach, individual larvae and adults or entire experimental colonies are tested under confined or open field conditions. This scheme provides a wealth of important information for risk assessment. Test methods, experimental designs, standardization, and new and comprehensive endpoints are under continuous development and will help improve the efficiency and reliability of current risk assessment schemes. There are, however, a number of questions relevant for ecological risk assessment that cannot be fully answered with laboratory and field studies. Ecological risk assessment tries to determine unacceptable risk on populations but it remains unclear how to establish whether an effect is unacceptable or not (Hommen et al. 2010). Tests focusing on the individual organisms deliver information on mortality or sublethal effects under laboratory conditions, but leave uncertain what these effects mean at the population level, for example, whether or not they impair the ability of the entire colony to persist, to cope with other stressors, and to reliably provide services such as honey production and pollination.

To assess effects on natural populations in general, ecological factors such as adaptive behavior, population structure, density dependence, exposure patterns, landscape structure, and species interactions need to be taken into account (Forbes et al. 2009). In addition, for social insects such as honey bees, it needs to be considered that the reproductive unit is not the individual worker bee but the entire colony and its queen. The colony and its functioning can be considered as a complex net of buffer mechanisms that has evolved to increase the fitness of the queen. The loss of individual worker honey bees might thus be less significant than in solitary species; beekeepers may see it differently if honey harvest is impaired. On the other hand, buffer mechanisms

Pesticide Risk Assessment for Pollinators, First Edition. Edited by David Fischer and Thomas Moriarty.

have only certain capacities. We cannot easily know these capacities and how they are affected by other stressors such as *Varroa* mites (*Varroa destructor*), viruses, changes in landscape, or beekeeping practices.

Semi-field and field studies allow inclusion and manipulation of some ecological factors, but certainly not all of them in all possible combinations within experimentally controlled conditions. They are expensive, time-consuming, require interpretation by experts, and may still be inconclusive as sufficiently controlled conditions are rarely achievable under field conditions. In addition, behavioral responses of colonies and foraging bees show large variations that can make it difficult to obtain any identifiable effects of a certain factor on honey bee populations.

Ecological models provide a tool to overcome limitations of empirical studies. They are widely used in theoretical and applied ecology because ecological systems are usually too complex, develop too slowly, and cover areas that are too large to be studied solely via controlled laboratory or field experiments. In the context of regulatory risk assessment, ecological models are often grouped with organism-level models addressing toxicokinetics and toxicodynamics (TK–TD) or dynamic energy budgets (DEB) to "mechanistic effect models" (Grimm et al. 2009). This terminology was introduced to distinguish these models, which simulate processes related to effects of pesticides on organisms and populations, from fate models which focus on the fate of pesticides in water and soil, and from statistical or empirical models, which establish correlative, but not causal relationships between factors. Ecological models can address all levels of organization beyond the individual, but ecological risk assessment usually focuses on populations (Galic et al. 2010; Schmolke et al. 2010a). In this chapter, we give a brief introduction into the rationale and approaches of ecological modeling of population dynamics. We present an example model to demonstrate the potential insights that can be gained from such ecological models, summarize current modeling practice, and describe recent attempts to establish good modeling practice (GMoP), which is needed to make mechanistic effect models applicable for regulatory risk assessment. We then provide an overview of existing models of honey bee colonies and give recommendations for the potential use of these models for pesticide risk assessment. Although this chapter focuses on honey bees, we will also briefly discuss how ecological modeling could support risk assessment of non-*Apis* pollinators. We will not discuss models addressing ecosystem services, which are important but belong to a different category of models and address different questions (Kevan et al. 1997; Williams et al. 2010).

11.2 EXAMPLE MODEL: COMMON SHREW

The following example model demonstrates how well-tested population models can be used to extrapolate the effects of toxicants observed at the individual level to the population level while considering different exposure patterns and landscape structures. Since such a demonstration does not yet exist for honey bees or other pollinators, we use a model of the common shrew (*Sorex araneus* L.). Wang and Grimm (2007) developed an individual-based population model of this species, which is a common insectivore. The purpose of the model was to explore the population-level consequences of acute mortality induced by pesticides.

The key behavior of the common shrew, which determines its response to heterogeneity in habitat quality and to the local density of conspecifics, is territoriality, that is, the aggressive defense of a certain area to secure resources and habitat. Therefore, the model is spatially explicit and represents each individual of the population, its life cycle, and its territorial behavior. The habitat consists of hexagonal units of 5 m diameter which are characterized by habitat type (e.g., grassland or hedge) and level of food resources on a given calendar day. Individuals are characterized by the variables age, gender, developmental stage (lactating offspring, subadult, adult), fertility (fertile, infertile; applies to females only), pregnancy, and home range. Home ranges are a set of habitat units occupied by a certain individual.

The processes of the model comprise development, mortality, reproduction, home range dynamics, dispersal, and mating. The model proceeds in daily time steps and covers an area of 25 ha. A full description of the model is given in Wang and Grimm (2007) using the standard format for describing individual-based

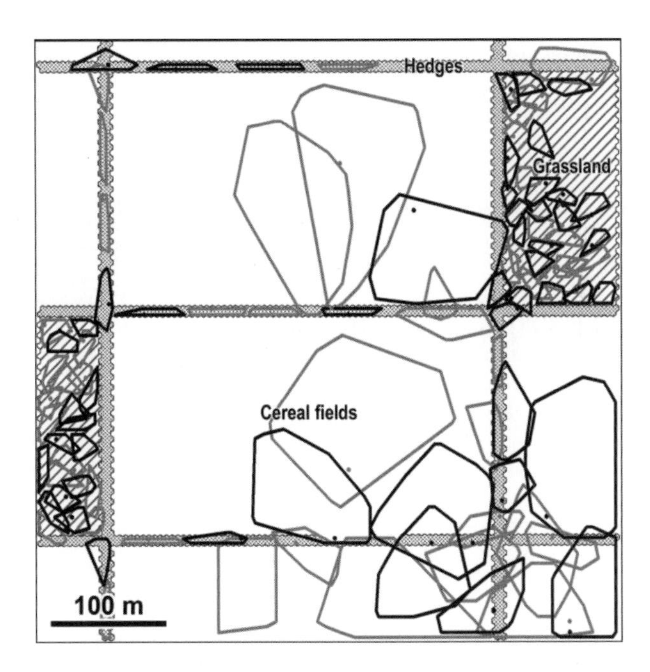

FIGURE 11.1 Output of an individual-based model of the common shrew (Wang and Grimm 2007) on a certain day of the simulation. Black lines delineate home ranges of males, gray lines of females. Home ranges in cereal fields need to be larger than in grassland or hedges because of lower resource levels. Home ranges are drawn as minimum convex polygons by connecting the outmost cells occupied by their owners (from Wang and Grimm 2007).

models (IBMs), ODD (Overview, Design concepts, Details; Grimm et al. 2006, 2010). The model allows the fate of each individual and its territory to be followed, day by day, in heterogeneous landscapes consisting of different habitat types (Figure 11.1).

Parameters affecting home range sizes were calibrated to match observations of a certain field study. Likewise, daily mortality was calibrated to obtain populations able to persist in good habitats. All other model parameters were taken from field studies. To make sure that the model captures important features of the internal organization of real populations of the common shrew, it was compared to multiple patterns observed in reality (Grimm et al. 2005; Grimm and Railsback 2005, 2012). Home range size and location varied with season, habitat type, and shrew density qualitatively similar to what is known from the field. Further patterns successfully tested were: proportion of pregnant and lactating females and the age distribution of juveniles and subadults. Thus, although the model certainly is not realistic in the sense that it takes into account all aspects of real populations, it is realistic enough to qualitatively predict the response of populations to additional mortality.

Accordingly, Wang and Grimm (2010) explored various hypothetical scenarios by applying pesticide-induced mortality on either April 1 or July 15; on that day, all individuals had an additional probability of 10% or 20% of dying. They contrasted orchards with and without 10% or 20% hedges, and compared different endpoints such as population size, daily population growth rate, recovery time, and extinction risk. They found that population size is more sensitive for detecting short-term effects than population growth rates; and that the landscape structure and timing of application had strong impacts on population recovery. For example, with 20% additional mortality on April 1, the population stabilized in orchards including 20% hedges, but continually declined in landscapes without hedges (Figure 11.2).

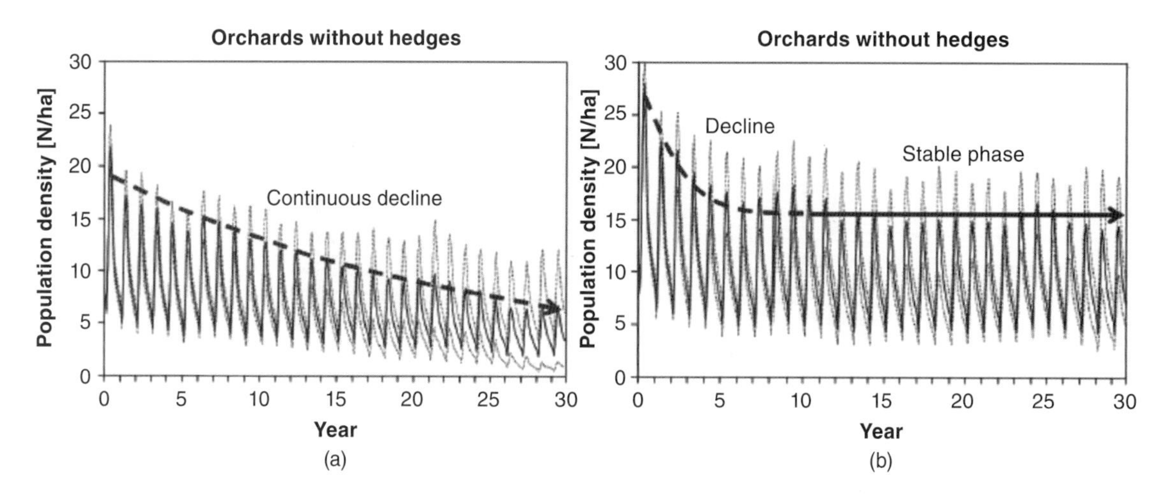

FIGURE 11.2 Population dynamics in orchards with and without 20% hedges with a yearly application of 20% additional mortality on April 1 (from Wang and Grimm 2010).

The model of Wang and Grimm (2007, 2010) can in principle be used for regulatory higher-tier risk assessments of small mammals. Its main limitation is that only few empirical studies exist that can be used for parameterizing, testing, and validating the model. But it clearly demonstrates the potential of well-tested ecological models for risk assessment of pesticides. A further exemplary demonstration of this potential can be found in Topping et al. (2009), who analyze, using much more detailed models, scenarios including skylarks, beetles, spiders, and field voles. Galic et al. (2010) give an overview of the types of insights for ecological risk assessment that can be gained from population models. Population models are all based on a model's ability to assess population status after integrating lethal and sublethal effects including behavioral changes, at the individual level.

11.3 RATIONALE AND APPROACHES OF MECHANISTIC EFFECT MODELING

Ecological models have to be based on conceptual models that reflect our current understanding of the system represented in the model. Conceptual models are usually formulated verbally or graphically, which by itself provide no means for testing whether they are consistent and complete. Modelers, therefore, use formal notations, based on mathematics and computer logics, to translate conceptual models into a framework that allows rigorous calculation of their consequences. Ecological models are thus, broadly speaking, tools for studying if–then scenarios: *if* we agree on a certain set of simplifying assumptions, *then* we have to accept the consequences predicted by the model.

At the beginning of modeling projects, we are usually unhappy with their consequences because they do not match observations, so we revise our assumptions. Model development is, therefore, an iterative process (Figure 11.3).

The "Modeling Cycle" (depicted in Figure 11.3) is relevant for any type of model, but many different types of model design and formulation exist (Schmolke et al. 2010a). Simple models, which are formulated via one or a few coupled differential equations, keep track of the processes causing changes in population size, such as mortality, reproduction, or disturbances. They are easy to communicate and understand but usually too poor in structure and mechanisms to be predictive and testable. Matrix models go beyond population size and consider the age, size, or stage structure of populations. They are frequently used to predict population growth

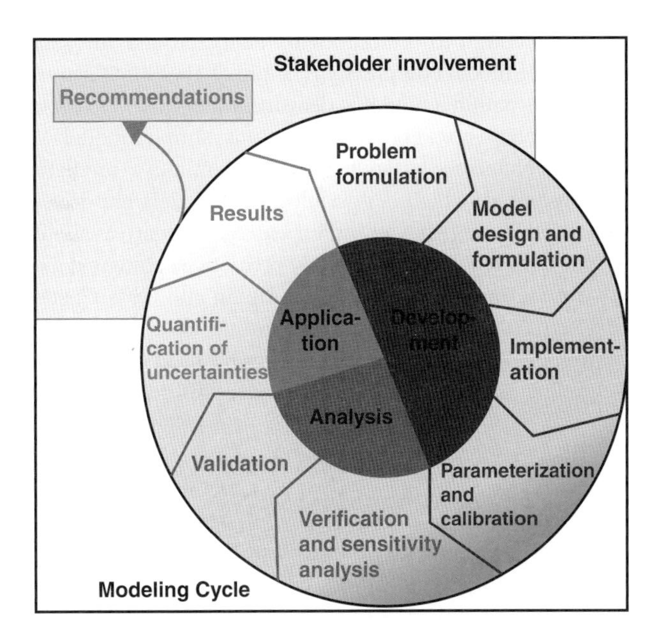

FIGURE 11.3 Tasks of the "Modeling Cycle," that is, of the iterative process of formulating, implementing, testing, and analyzing ecological models (after Schmolke et al. 2010b). Full cycles usually include a large number of subcycles, for example, verification leading to further effort for parameterization or reformulation of the model. The elements of this cycle are used to structure a new standard format for documenting model development, testing, analysis, and application for environmental decision making, TRACE (Schmolke et al. 2010b). (For a color version, see the color plate section.)

rate and the sensitivity of growth rate to changes in mortality or reproduction of certain classes of individuals. Again, matrix models are easy to communicate but, once they are designed to include stochasticity, spatial effects, or density dependence, they have to be run on computers and are, therefore, no longer very different from IBMs. Simple matrix models have a standard format and are relatively easy to parameterize and analyze. They project current average conditions into the future and can, therefore, be used for initial screening, corresponding to lower tier tests in risk assessment, with small or negative population growth rate indicating risk.

IBMs are computer simulation models in which each individual and its life cycle is represented explicitly (see the common shrew model presented above). Population dynamics and growth rates emerge from what individuals do and how they interact with each other and their environment. IBMs are harder to communicate, parameterize, test, and understand than simpler mathematical models, but nevertheless used when one or more of the following factors are assumed to be essential for explaining population dynamics: local interactions, differences among individuals, and adaptive behavior (Grimm and Railsback 2005). IBMs are no longer new but routinely used not only in ecology but also in many other disciplines ranging from behavioral ecology to social sciences, where they are usually referred to as "agent-based" models (Railsback and Grimm 2012). Strategies exist to optimize model complexity (Grimm et al. 2005) and to formulate and communicate IBMs according to a standard format, the ODD ("Overview, Design concepts, Details") protocol (Grimm et al. 2006, 2010).

To use models for pesticide risk assessment, two conflicting criteria for assessing the suitability of models are critical: on the one hand, models need to be complex enough to deliver testable predictions which enable decisions about whether or not the model is a sufficiently good representation of the real world. On the other hand, models need to be simple enough to be thoroughly analyzed and fully understood. Modeling thus requires finding the optimal level of model complexity (Grimm et al. 2005; Grimm and Railsback 2012).

Understanding the main process within a model is decisive, otherwise we would be asking for blind faith in output from the equivalent of a black box. For some questions, simpler models can be sufficient, correctly predicting trends and general mechanisms without making quantitative predictions. For other questions, more accurate predictions are required, which is possible if the models are driven by first principles, such as physiology, stoichiometry, or adaptive behavior, and if enough data are available to directly or indirectly estimate model parameters with sufficient certainty. Highly predictive ecological models have been developed (e.g., Railsback and Harvey 2002; Topping et al. 2009; Stillman and Goss-Custard 2010), but all required more than 5 years before first versions could be used to support decision making. However, once a predictive model exists, it pays off extremely well because it can then be used as a virtual laboratory to answer a wide range of questions regarding population dynamics under different and possibly new environmental conditions.

11.4 MODELING PRACTICE FOR RISK ASSESSMENT

Claims about the high potential of ecological modeling for pesticide risk assessment are not new and have been made for at least 20 years (Barnthouse 1992). In fact, approximately 100 academic publications exist that use population or other ecological models to explore the effects of pesticides at the population level (Schmolke et al. 2010a). Galic et al. (2010) summarize the scientific insights of these studies, which are certainly important and contribute to our understanding of the significance of individual-level effects at the population level. Nevertheless, the use of models is still limited to a few recent exceptions. Why is this so? Schmolke et al. (2010a) found that most models in this field are not fit for being used for pesticide registrations. The main reason is that criteria for being accepted as a scientific publication, such as novelty, focusing on one main aspect, simplicity, or generality, are less relevant for making a model suitable for basing environmental decisions on their output. In most cases, choice of model structure and complexity was not justified, endpoints directly relevant for regulatory risk assessments were not considered, sources of parameter values were unclear, uncertainty of model output was not communicated, and most importantly, little effort was made to demonstrate that the model was a sufficiently good representation of the real population such that insights gained in the model world could be transferred to the real world with sufficient confidence.

This situation is, however, changing in Europe. Two main challenges to make models fit to be used for regulatory risk assessment are (1) the establishment of GMoP, so that both industry and regulators have clear criteria for how to create and assess models, and (2) the lack of researchers who are well-trained both in ecological modeling and risk assessment (Thorbek et al. 2010). Therefore, CREAM (Chemical Risk Effects Assessment Models), a large research and training network funded by the European Commission launched in 2009 (Grimm et al. 2009; http://cream-itn.eu), which includes 13 academic institutions and 10 partners from industry, consulting firms, and regulatory authorities, ran until 2013, and delivered both guidelines for GMoP and more than 20 young researchers trained in modeling and risk assessment. Moreover, models were developed which, for indicator species and risk assessment questions, are good demonstrations for how models can be used for regulatory risk assessments.

Elements of GMoP have long been identified but are still widely ignored. The real challenge is to get these elements accepted and used in practice. Schmolke et al. (2010b) found that for this, regulators or, more generally, decision makers need to be involved, direct benefits for modelers who follow GMoP (which usually requires extra effort) need to be identified, and a consistent terminology needs to be established. Therefore, the basic approach of CREAM in establishing GMoP is to define and use a standardized documentation framework, Transparent and Comprehensive Ecological Modeling (TRACE). Schmolke et al. (2010b) suggest the use of the structure of the iterative modeling cycle (Figure 11.3) as the basis for a general and standardized document structure. As a result, all models that are to be used to support pesticide registration and come with a TRACE documentation as a supplementary document, can be assessed in exactly the same way. Regulators will know that, for example, sensitivity analysis will be described in Section 2.2, or that the conceptual model underlying the model's design can be found in Section 1.2. Modelers, on the other hand, will know that regulators will expect to see, a documentation of sensitivity analysis, for example, so they can use the TRACE format as a checklist. The direct benefit for the modeler is that the TRACE format helps keeping notes in the "modeling notebook," which corresponds to "lab journals" in laboratories, in a format that later can directly be transferred to TRACE documents.

Once a critical number of example TRACE documents exist, by the end of the CREAM project, more specific assessment guidelines can be developed that help standardize the use of ecological models for regulatory risk assessment. This includes the agreement on standard scenarios, species, landscapes, ecoregions, and population-level endpoints. Honey bees and pollinators will play an important role in this context, due to their unique significance for biodiversity and ecosystem services.

11.5 EXISTING MODELS OF POLLINATORS

Quite a few models exist that address various aspects of honey bee behavior and ecology (for an overview, see Section 5.4. in Schmickl and Crailsheim 2007). However, there are surprisingly few sufficiently described models addressing dynamics of non-swarming, managed colonies which include the full life cycle of worker bees from a single hive over several years such that the colony-level effects can be assessed (Table 11.1).

Two of these models are interesting from an academic point of view, but too simple to be tested against observed data (Omholt 1986; Khoury et al. 2011). Nevertheless, theoretical insights can guide the design and

TABLE 11.1
Colony Models That Include the Full Life Cycle of Worker Bees and Run Long Enough, that is, Two or More Years, to Assess Status and Survival of a Model Colony. The Third Column Lists Additional Factors Included in the Model That Can Affect Colony Status and Survival

Reference	Purpose of Model/Question Addressed	Additional Factors
Omholt (1986)	Explain brood-rearing peaks in non-swarming colonies	
DeGrandi-Hoffman et al. (1989)	Simulate honey bee population dynamics to support management	
Martin (2001)	Explain the link between *Varroa* mite infestation and honey bee colony failure, including the effects of virus diseases	*Varroa* and virus infections
Al Ghamdi and Hoopingarner (2004)	Develop a tool for research and management; interaction between *Varroa* and honey bees	*Varroa*
Thompson et al. (2005), (2007)	Explore effect of an insecticide on colony dynamics	Pesticides
Schmickl and Crailsheim (2007)	Explore significance of important feedback loops, pollen supply, and brood cannibalism	Swarming
Becher et al. (2010)	Does temperature during development affect colony survival?	
Khoury et al. (2011)	Impact of increased forager mortality on colony growth and development	

analysis of more complex models. For example, Khoury et al. (2011) implement two feedback mechanisms: between colony size and brood production and between the number of foragers and recruitment to foraging, which have been referred to as "social inhibition" (Leoncini et al. 2004). They found that if forager mortality exceeds a certain threshold, the colony can no longer maintain itself and will decline to extinction. These feedback mechanisms have been observed empirically and the results of Khoury et al. (2011) suggest that their significance should be further tested in more detailed models, containing a colony's age structure, nectar and pollen stores, further feedback mechanisms, and variable environmental drivers.

The model by Thompson et al. (2005) is also simple and considers the abundance of brood, in-hive and forager bees. This model was originally used in combination with a more detailed population model of *Varroa* mites (Wilkinson and Smith 2002), but Thompson et al. left out the *Varroa* part and added assumptions about the effects of a certain type of pesticide (insect growth regulators), based on observations from their own experiments. Such reuse of models for new questions can be problematic, since the model's design may not be appropriate for the new questions. In this case, model resolution is likely to be too coarse to make robust predictions, still, the model serves as a demonstration of how, in principle, individual-level effects of pesticides can be included in colony models of honey bees.

The models presented by Martin (2001) and Al Ghamdi and Hoopingarner (2004) are modifications of BEEPOP (DeGrandi-Hoffman et al. 1989), a simulation model proceeding in time steps of one day and representing cohorts (or age classes) of eggs, brood, and adults of both worker bees and drones (Figure 11.4). BEEPOP distinguishes between in-hive and foraging bees, whereas the other two models do not. Colony dynamics are driven by the queen's egg-laying rate, which is mainly driven by weather, in particular

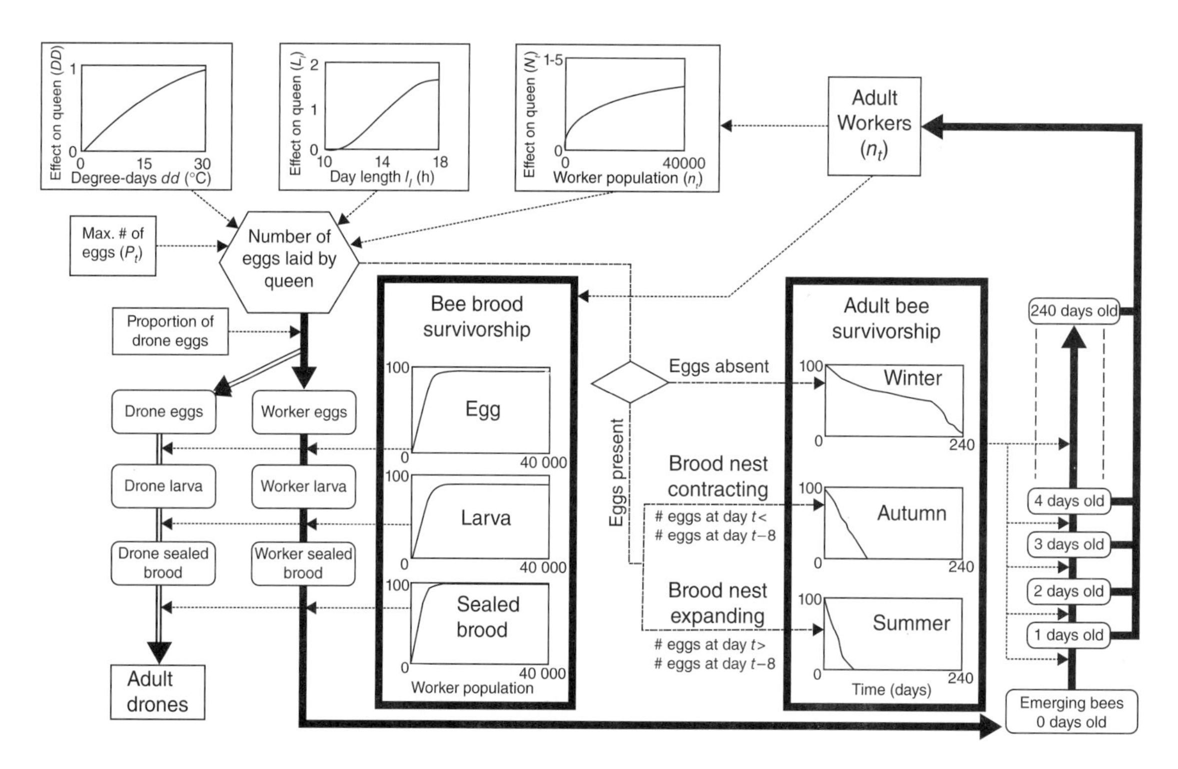

FIGURE 11.4 Conceptual diagram of the colony model of Martin (2001). Solid lines represent the flow of individuals between developmental stages and dotted lines represent influences (from Martin 2001).

temperature and photoperiod. Additionally, these models include feedbacks between egg laying and colony size. Drones are mainly included because mites are more attracted by drone cells and mite reproduction is higher in drone cells, so that the proportion of drone cells has an important impact on the dynamics and effects of *Varroa* infestation.

BEEPOP has been augmented by detailed modules for including effects of pesticides (Bromenshenk et al. 1991). The module BEETOX included a toxicity database for more than 400 chemicals and calculated lethal and sublethal effects for specific exposures; the module BEEKILL allowed the linkage these effects to exposure scenarios and feed the resulting changes in mortality, development, and longevity into the colony model. Unfortunately, details of these modules were not published and the software implementing them, PC BEEPOP, is unlikely to run on modern computers. It also seems that it has never been used for regulatory risk assessment of pesticides, probably because it was very much ahead of its time. Nevertheless, the design of PC BEEPOP is interesting since it allows one to test effects of pesticides on honey bee colonies in a standardized way.

Becher et al. (2010) include the effect of colony size and structure on heating and the resulting temperature in the brood chamber. It had been observed that brood developed under higher temperatures proceeds faster from in-hive tasks to foraging. It turns out, however, that this has little effect on the dynamics and status of the colony. This is a good example of the role of models for relating individual-level effects to colony-level phenomena. Without the model, it would have been impossible to predict this relationship for the temperature effect, simply because colony structure, environmental drivers, and feedback mechanisms are too complex to be even qualitatively assessed just by reasoning. Negative results, as in this case, that is, the working hypothesis is shown to be false, are no less important than positive results.

The most complex colony model is HoPoMo (Schmickl and Crailsheim 2007). In contrast to all other colony models, HoPoMo is not entirely driven by demographic rates, such as egg-laying rate of the queen and age- and task-dependent mortalities. Rather, the current number, stage, age, and task of bees are used to calculate the estimated requirements of the colony for nectar and pollen. Depending on current stocks of these two resources, the proportion of worker bees devoted to different tasks is dynamically reallocated every day. The three different tasks distinguished are nursing, food processing, and foraging. HoPoMo includes a large number of further feedbacks between the current state of the colony, or parts of it, and process rates.

HoPoMo consists of 60 difference equations, which are all well documented and biologically justified. The model has been thoroughly tested, including sensitivity analyses and exploration of certain mechanisms. It reproduces several empirical patterns and correctly predicts at least one feature of real colonies that was not used to calibrate or design the model, but emerged during analysis of the full model: in smaller model colonies, with no more than 20000 brood cells, the number of unsealed brood cells shows oscillations similar to what has been observed in real experimental hives. The model has, however, not yet been used to answer any specific question about how colonies respond to environmental stress, such as exposure to a pesticide.

Two of the colony models in Table 11.1 also consider infestation with *Varroa* mites. Phoretic mites, that is, mites attached to worker bees, enter brood cells about one day before they are sealed and reproduce within these brood cells. Emerging mites try to infest another brood cell or become phoretic, and thereby spread *Varroa* infestation. During the interaction with brood and worker bees, mites transfer viruses, for example, deformed wing virus (DWV), or acute paralysis virus (APV). The model of Martin (2001) integrates honey bee and mite population dynamics and the effects of viruses. It shows, for example, that the less virulent DWV will become more widely spread than APV, and that mite control measures need to be taken before the longer-lived overwintering bees emerge. Further *Varroa* models, which focus on various aspects of *Varroa* population dynamics, but are coupled to much simpler colony models than BEEPOP, include Omholt and Crailsheim (1991), Fries et al. (1994), Martin (1998), Calis et al. (1999), Wilkinson and Smith (2002), and DeGrandi-Hoffman and Curry (2005). For the purpose of pesticide registration, it seems necessary to use models that allow inclusion of *Varroa* infestation because at least in Europe and North America, *Varroa* is a ubiquitous stressor. It remains as an open question, the way in which *Varroa* infestation could or should be taken into account for pesticide registration. Should decisions be made to ensure safety under a worst-case assumption of high infestation where colonies have high risk of collapsing even without exposure, under an assumption of effective *Varroa* management by beekeepers, or should average infestation levels based on national or international surveys be used? These questions cannot be answered scientifically, but robust, well-tested, and predictive colony models which allow inclusion of *Varroa* and possibly other stressors would support decisions by quantitative arguments.

Currently, only the model by Martin (2001) is suitable to consider different, but simultaneous stressors. On the other hand, HoPoMo is a more realistic model and includes feedback mechanisms which seem to be important for the functioning of a colony; in particular, HoPoMo is driven by pollen and nectar stores, demand, and availability in the landscape. If HoPoMo could include a module representing *Varroa* infestation and virus effects, it would currently be the most suitable model for pesticide risk assessment. However, changes in landscape structure, crop plants and their rotation, and agricultural practice also affect honey bee colony performance so that, for registration purposes, a model should also allow such factors to be represented with sufficient detail regarding spatial structure, crop dynamics and rotation, and foraging behavior. Adding such a module to HoPoMo would make an already very complex model even more complex and, therefore, harder to test and understand. Therefore, a colony model that includes *Varroa*, viruses, and foraging in heterogeneous landscapes should preferably be similar in design to the model of Martin (2001) but include the most important feedbacks included in HoPoMo.

A well-tested prototype of such a model, dubbed "BEEHAVE," was developed by M. Becher and coworkers at Rothamsted Research, UK, in 2009–2013. Its purpose is not pesticide registration *per se*, but to explore the possible reasons for honey bee decline and collapse as well as devising strategies for improving honey bee health. For this purpose, the model includes *Varroa*, viruses, and explicit foraging in heterogeneous landscapes. The option to include pesticide effects, or other additional stressors subsequently shown to be important, was considered from the beginning of this modeling project and a design developed to enable this to be relatively straightforward. The model and its computer code and user manual were made available March 2014 (http://beehave-model.net), so that other researchers can test the model independently and use it or the model for various purposes.

As for non-*Apis* pollinators, fewer models exist than for honey bees. The population model of the solitary red mason bee, *Osmia rufa* (L.) (Ulbrich and Seidelmann 2000) shows, however, that if sufficient empirical knowledge of a species ecology and behavior exists, developing a population model is straightforward and can lead to important insights. The purpose of the *Osmia* model was predicting the risk of extinction of this solitary species in different types of habitat, which are characterized by the amount and quality of food they provide. The model is individual-based and focuses on cell construction and maternal investment in brood cells. The life stages distinguished are eggs, larvae, imagines in cocoons, males, pre-nesting females, and nesting females. A key decision of nesting females is the sex determination of their brood. The first brood cells are always daughter cells, but at some point, the mother bee switches to construction of son cells. In the model, it is assumed that this switching depends on the mother's weight, that is, heavier bees produce more daughter cells. Likewise, size of progeny is related to their mother's weight. As a measure of habitat quality, time for cell construction was used as a proxy (Gathmann 1998). In this way, the model can be linked to habitat quality without explicitly representing habitat and foraging. As stressor, parasites were taken into account, with parasitism rates being higher for longer cell construction times. Mean population size and extinction risk were taken as population-level endpoints.

Mitesser et al. (2006) developed a colony model for the halictid bee *Lasioglossum malachurum* to explore the emergence of activity cycles, which are typical for some annual eusocial "sweat bees" (Halictidae). The model is very simple and includes only two state variables, the numbers of workers and of sexuals; the time horizon considered is so short that mortality of sexuals could be ignored. Still, there is no principle reason why it should not be possible to develop an age-structured model, similar to BEEPOP or BEEHAVE that includes the full life cycle.

A very interesting IBM of bumble bees was developed by Hogeweg and Hesper (1983). It includes the full life cycle of individuals and different types of behaviors, and is, like HoPoMo, to a large degree driven by food collection and consumption and time budgets for certain activities. Focus, though, is less on colony dynamics *per se* but on explaining division of labor within the colony and the so-called "dominance interactions," by which this division emerges. This model was about 20 years ahead of its time as IBMs, which go beyond demographic rates and include behavior, have only become more widely used within the last 10 years. It would certainly be worthwhile to reimplement this model and try to adapt it to new questions. Whether or not it would be sufficient to just assume division of labor, or have mechanisms included that allow this division to emerge, remains an open question.

In general, eusocial non-*Apis* pollinators have simpler and smaller colonies. This implies that, although they benefit from cooperative activities, they do not maintain buffer mechanisms and reserves which would be as powerful as in honey bee colonies. They also show greater foraging activity, to compensate for the lack of maintained reserves, potentially increasing the risk of pesticide exposure.

A bottleneck for developing models for non-*Apis* pollinators might be the lack of data about their foraging behavior in real landscapes since exposure to pesticides to a large extent depends on foraging. Detailed foraging models need to be developed and parameterized and tested using corresponding field studies and experiments (Everaars 2012; Everaars and Dormann 2014).

11.6 DISCUSSION

Sophisticated tests and schemes exist to assess the risk that pesticides impose to honey bees. Current regulations and thresholds seem to be conservative but still leave many questions open. The difficulty is that to confirm whether or not the sublethal or lethal effects of pesticides, observed in laboratories or field experiments, translate into a significant risk to the functioning or survival of a colony, controlled, long-term experiments are required to take into account the individual and combined effects of pesticides and other stressors on colonies at the landscape scale. For example, if on a normal day an average of 100 dead bees is found around the hive, and during acute pesticide exposure 300 dead bees are found, is this of any significance to the colony? Likewise, if larvae develop more slowly, or worker bees have a shortened lifespan due to pesticides, how does this affect colony functioning in terms of honey production and pollination? Answering such questions with real experiments might be possible to some degree, but would require enormous resources.

Ecological models could, in principle, compensate for this limitation of empirical approaches. And there are, indeed, fields where models are used to support environmental decision making. For example, recent regulations of wildlife diseases, such as rabies or classical swine fever, are based on predictions of models which are quite similar to the common shrew model presented earlier (Thulke and Grimm 2010). In some federal states of Germany, forest management plans on the timescale of 10–20 years are based on predictions of the individual-based forest model SILVA (Pretzsch et al. 2002). Common features of these and other ecological models used for decision making is that their development took at least 5 years, and their acceptance by the responsible decision makers about 10 years.

Establishing the use of ecological models to assess risk of pollinators, in particular honey bees, can nevertheless be achieved faster. Well-tested and documented models already exist, which can at least be used, preferably in joint workshops, to discuss and learn the use of such models for higher-tier risk assessments. BEEHAVE, the model currently developed in the UK, holds further promise, in particular because it includes the main potential stressors of colonies and foraging in heterogeneous landscapes. Ideally, to make BEEHAVE fit for use with pesticide registrations, it would need to be used in one or more workshops where researchers from all three sectors involved in pesticide risk assessment, industry, regulators, and academia, agree on standard model scenarios, endpoints, and risk assessment schemes. BEEHAVE is described in a standard format (Grimm et al. 2006, 2010), its development and analysis will be available as a TRACE document, and it is implemented in a software platform, NetLogo (Wilensky 1999), that is freely available and easy to learn. BEEHAVE is thus designed to be tested, used, and developed not only by its developers but by the scientific and user community involved in honey bee research and management.

The good news is that honey bee models are less limited by data for parameterization than models of most other species. Experimentally managed colonies are relatively easy to observe in the laboratory and field. Bee behavior has been investigated a lot, and beekeepers accumulated sound empirical knowledge on how colonies respond to environmental events and beekeeping practices. Foraging still is a bottleneck in empirical knowledge, but remote sensing techniques can be used now to follow the flight path of individual foragers (Riley et al. 1996; Osborne et al. 1999). Moreover, in response to the decline or collapse of honey bees in Europe and North America, large international networks such as COLOSS (Neumann and Carreck 2010) compile and analyze huge amounts of data, which can be used to test model predictions.

Ecological models are no silver bullet to solve all problems of pollinator risk assessment, but they are a valuable and needed tool for extrapolating individual-level effects to the colony-level, to overcome important limitations of field studies, and to explore endpoints that quantify adverse effects not only on pollinators *per se* but also on biodiversity and ecosystem services.

REFERENCES

Al Ghamdi A, Hoopingarner R. 2004. Modeling of honey bee and varroa mite population dynamics. *Saudi. J. Biol. Sci.* 11:21–36.

Barnthouse LW. 1992. The role of models in ecological risk assessment: a 1990s perspective. *Environ. Toxicol. Chem.* 11:1751–1760.

Becher MA, Hildenbrandt H, Hemelrijk CK, Moritz RFA. 2010. Brood temperature, task division and colony survival in honeybees: a model. *Ecol. Model.* 221:769–776.

Bromenshenk IJ, Doskocil J, Olbu GJ, DeGrandi-Hoffman G, Roth SA. 1991. PC BEEPOP, an ecotoxicological simulation-model for honey-bee populations. *Environ. Toxicol. Chem.* 10:547–558.

Calis NM, Fries I., Ryrie SC. 1999. Population modelling of *Varroa jacobsoni* Oud. *Apidologie* 30:111–124.

DeGrandi-Hoffman G, Roth SA, Loper GL, Erickson EH. 1989. Beepop—A honeybee population-dynamics simulation-model. *Ecol. Model.* 45:133–150.

DeGrandi-Hoffman G, Curry R. 2005. Simulated population dynamics of Varroa mites in honey bee colonies: Part II—What the VARROAPOP model reveals. *Am. Bee. J.* 145:629–632.

Everaars J. 2012. The response of solitary bees to landscape configuration with focus on body size and nest-site preference. PhD dissertation, Martin-Luther-University.

Everaars J, Dormann CF. 2014. A simulation model for non-*Apis* bees: solitary bees foraging and competing for pollen. Boca Raton (FL): CRC Press. 209–268.

Forbes VE, Hommen U, Thorbek P, Heimbach F, van den Brink P, Wogram J, Thulke H-H, Grimm, V. 2009. Ecological models in support of regulatory risk assessments of pesticides: developing a strategy for the future. *Integr. Environ. Assess. Manag.* 5:167–172.

Fries I, Camazine S, Sneyd J. 1994. Population dynamics of *Varroa jacobsoni*—a model and a review. *Bee World* 75:5–28.

Galic NG, Hommen U, Baveco JM, Van den Brink PJ. 2010. Potential application of population models in the European ecological risk assessment of chemicals: II review of models and whether they can address the protection aims. *Integr. Environ. Assess. Manag.* 6:338–360.

Gathmann A. 1998. Bienen, Wespen und ihre Gegenspieler in der Agrarlandschaft: Artenreichtum und Interaktionen in Nisthilfen, Aktionsradien und Habitatbewertung. Göttingen: Cuvillier Verlag.

Grimm V, Railsback SF. 2005. *Individual-Based Modeling and Ecology*. Princeton University Press, Princeton, NJ.

Grimm V, Railsback SF. 2012. Pattern-oriented modelling: a "multiscope" for predictive systems ecology. *Philos. T. Roy. Soc. B.* 367:298–310.

Grimm V, Revilla E, Berger U, Jeltsch F, Mooij WM, Railsback SF, Thulke H-H, Weiner J, Wiegand T, DeAngelis DL. 2005. Pattern-oriented modeling of agent-based complex systems: lessons from ecology. *Science* 310:987–991.

Grimm V, Berger U, Bastiansen F, Eliassen S, Ginot V, Giske J, Goss-Custard J, Grand T, Heinz SK, Huse G, Huth A, Jepsen JU, Jørgensen C, Mooij WM, Müller B, Pe'er G, Piou C, Railsback SF, Robbins AM, Robbins MM, Rossmanith E, Rüger N, Strand E, Souissi S, Stillman RA, Vabø R, Visser U, DeAngelis DL. 2006. A standard protocol for describing individual-based and agent-based models. *Ecol. Model.* 198:115–126.

Grimm V, Ashauer R, Forbes V, Hommen U, Preuss TG, Schmidt A, van den Brink PJ, Wogram J, Thorbek P. 2009. CREAM: a European project on mechanistic effect models for ecological risk assessment of chemicals. *Environ. Sci. Pollut. Res.* 16:614–617.

Grimm V, Berger U, DeAngelis DL, Polhill G, Giske J, Railsback SF. 2010. The ODD protocol: a review and first update. *Ecol. Model.* 221:2760–2768.

Hogeweg P, Hesper B. 1983. The ontogeny of the interaction structure in bumble bee colonies: a MIRROR model. *Behav. Ecol. Sociobiol.* 12:271–283.

Hommen U, Baveco JM, Galic NG, Van den Brink PJ. 2010. Potential application of ecological models in the European environmental risk assessment of chemicals I: review of protection goals in EU directives and regulations. *Integr. Environ. Assess. Manag.* 6:325–337.

Kevan PG, Greco CF, Belaousoff S. 1997. Log-normality of biodiversity and abundance in diagnosis and measuring of ecosystem health: pesticide stress on pollinators in blueberry heaths. *J. Appl. Ecol.* 34(5):1122–1136.

Khoury DS, Myerscough MR, Barron AB. (2011). A quantitative model of honey bee colony population dynamics. *PLoS ONE.* 6(4):e18491. doi:10.1371/journal.pone.0018491

Leoncini I, Le Conte Y, Costagliola G, Plettner E, Toth AL. 2004. Regulation of behavioral maturation by a primer pheromone produced by adult worker honey bees. *Proc. Nat. Acad. Sci. USA.* 101:17559–17564.

Martin S. 1998. A population model for the ectoparasitic mite *Varroa jacobsoni* in honey bee (*Apis mellifera*) colonies. *Ecol Model.* 109:267–281.

Martin SJ. 2001. The role of Varroa and viral pathogens in the collapse of honeybee colonies: a modelling approach. *J. Appl. Ecol.* 38:1082–1093.

Mitesser O, Weissel N, Strohm E, Poethke H-J. (2006). The evolution of activity breaks in the nest cycle of annual eusocial bees: a model of delayed exponential growth. *BMC. Evol. Biol.* 6:45. doi:10.1186/1471-2148-6-45

Neumann P, Carreck C. (2010). Honey bee colony losses: a global perspective. *J. Apic. Res.* 49:1–6.

Omholt SW. 1986. A model for intracolonial population dynamics of the honeybee in temperate zones. *J. Agric. Res.* 25:9–21.

Omholt SW, Crailsheim K. 1991. The possible prediction of the degree of infestation of honeybee colonies (*Apis mellifera*) by *Varroa jacobsoni* OUD by means of its natural death-rate: a dynamic model approach. *Norw. J. Agric. Sci.* 5:393–400.

Osborne JL, Clark SJ, Morris RJ, Williams IH, Riley JR, Smith AD, Reynolds DR, Edwards AS. 1999. A landscape-scale study of bumble bee foraging range and constancy, using harmonic radar. *J. Appl. Ecol.* 36:519–533.

Pretzsch H, Biber P, Dursky J. 2002. The single tree-based stand simulator SILVA: construction, application and evaluation. *Forest. Ecol. Manag.* 162:3–21.

Railsback SF, Harvey BC. 2002. Analysis of habitat selection rules using an individual-based model. *Ecology.* 83:1817–1830.

Railsback SF, Grimm V. 2012. *Agent-Based and Individual-Based Modeling: A Practical Introduction*. Princeton University Press, Princeton, NJ.

Riley JR, Smith AD, Reynolds DR, Edwards AS, Osborne JL. 1996. Tracking bees with harmonic radar. *Nature.* 379:29–30.

Schmickl T, Crailsheim K. 2007. HoPoMo: a model of honeybee intracolonial population dynamics and resource management. *Ecol. Model.* 204:219–245.

Schmolke V, Thorbek P, Chapman P, Grimm V. 2010a. Ecological modelling and pesticide risk assessment: a review of current modelling practice. *Environ. Toxicol. Chem.* 29:1006–1012.

Schmolke A, Thorbek P, DeAngelis DL, Grimm V. 2010b. Ecological modelling supporting environmental decision making: a strategy for the future. *Trends. Ecol. Evol.* 25:479–486.

Stillman RA, Goss-Custard JD. 2010. Individual-based ecology of coastal birds. *Biol. Rev.* 85:413–434.

Thompson HM, Wilkins S, Battersby AH, Waite RJ, Wilkinson D. 2005. The effects of four insect growth-regulating (IGR) insecticides on honeybee (*Apis mellifera* L.) colony development, queen rearing and drone sperm production. *Ecotoxicology.* 14:757–769.

Thorbek P, Forbes V, Heimbach F, Hommen U, Thulke HH, van den Brink PJ, Wogram J, Grimm V (eds.) 2010. *Ecological Models for Regulatory Risk Assessments of Pesticides: Developing a Strategy for the Future.* CRC Press, Boca Raton, FL (in collaboration with the Society of Environmental Toxicology and Chemistry (SETAC), Pensacola, FL).

Thulke H-H, Grimm V. 2010. Ecological models supporting management of wildlife diseases. In: Thorbek P, Forbes V, Heimbach F, Hommen U, Thulke HH, van den Brink PJ, Wogram J, Grimm V (eds), *Ecological Models for Regulatory Risk Assessments of Pesticides: Developing a Strategy for the Future.* CRC Press, Boca Raton, FL (in collaboration with the Society of Environmental Toxicology and Chemistry (SETAC), Pensacola, FL), pp. 67–76.

Topping CJ, Dalkvist T, Forbes VE, Grimm V, Sibly RM. 2009. The potential for the use of agent-based models in ecotoxicology. In: Devillers J. (ed.) *Ecotoxicology Modeling*, Springer, pp. 205–237.

Ulbrich K, Seidelmann K. 2000. Modeling population dynamics of solitary bees in relation to habitat quality. *Web. Ecol.* 2:57–64.

Wang M, Grimm V. 2007. Home range dynamics and population regulation: an individual-based model of the common shrew. *Ecol. Model.* 205:397–409.

Wang M, Grimm V. 2010. Population models in pesticide risk assessment: lessons for assessing population-level effects, recovery, and alternative exposure scenarios from modelling a small mammal. *Environ. Toxicol. Chem.* 29: 1292–1300.

Wilensky U. (1999). *Center for Connected Learning and Computer-Based Modeling*, Northwestern University, Evanston, IL. http://ccl.northwestern.edu/netlogo/ (accessed March 21, 2014).

Wilkinson D, Smith GC. 2002. A model of the mite parasite, *Varroa destructor*, on honeybees (*Apis mellifera*) to investigate parameters important to mite population growth. *Ecol. Model* 148:263–275.

Williams NM, Crone EE, Roulston TH, Minckley RL, Packer L, Potts SG. 2010. Ecological and life history traits predict bee species responses to environmental disturbances. *Biol. Cons.* 143:2280–2291.

12 Data Analysis Issues

W. Warren-Hicks

CONTENTS

This chapter discusses recommendations from the Workshop participants on existing methods and approaches for statistically assessing exposure and effects to pollinators using both laboratory and field tests. In a few cases, broad suggestions are discussed on how to examine, present, and evaluate data from these tests. Participants identified a need for additional statistical analysis tools for evaluating data from existing study designs and results to aid in the design and conduct of future study protocols. However, neither the discussions of the Workshop nor established guidance documents (e.g., EU's Dir 91/414 and EPA Part 158 Test Guidelines) provide suggestions or case study illustrations detailing appropriate approaches for statistically examining data from both short- and long-term laboratory and field tests. An exploration of analytical methods most appropriate for evaluating data would serve to inform regulatory authorities, agrochemical registrants, and researchers engaged in such studies. This chapter provides a brief overview of the types of statistical issues relevant to evaluating the potential effects of pesticides on pollinators that will be addressed by attendees of the Workshop through a subsequent effort, at a later date (note: details will be provided in a separate document through SETAC publications). The intent is to highlight issues of interest to risk assessors during the data analysis and risk characterization phases specifically with data generated on bees for use in an ecological risk assessment.

12.1 STUDY DURATION

Decisions regarding study duration and dosing time will impact any statistical model applied to the data, including dose–response models, and can have a large impact on the statistical inferences drawn from the data. The duration of some of the proposed laboratory-based chronic studies is 10 days. However, the implications of both longer and shorter durations have not been tested either in terms of their ability to detect subacute or chronic effects or the relevancy of laboratory-based studies to field-based studies of longer duration.

12.2 REPLICATES AND DOSING

Questions around the number of bees per replicate, the number of replicates per dose, is an element in both laboratory and field studies that requires consideration. In laboratory-based studies, key issues include the clear definition of treatment units, estimation and interpretation of between-treatment variance, and temporal variation over the course of the test. In semi-field and field studies, the concept of a replicate and whether

Pesticide Risk Assessment for Pollinators, First Edition. Edited by David Fischer and Thomas Moriarty.

information from multiple hives on the same field can be considered true replication versus pseudoreplication is critical to the calculation of variation in these tests.

Dosing in laboratory-based studies is more standardized than in field studies. Dose levels in a laboratory-based studies are carefully selected to cover the range of possible effects to allow the estimation of a dose–response function. Whereas individual bees may be "dosed" in laboratory-based study, in field studies there can be uncertainty regarding the extent to which bees are actually exposed to test material. Examination of raw data from tests of such a design can result in a visual non-monotonic dose–response relationship. Methods for interpreting this information, and the implications for selection of dose levels, are of interest to the development of subsequently applied statistical models.

12.3 LONG-TERM TESTS

In chronic tests (10/14-day test, semi-field, and field), the test is generally designed to be sensitive to sublethal effects, and consequently treatment levels and duration may be different from lethality tests. The length of the test and its influence on the calculation of statistical endpoints and uncertainty in the model-based endpoints should be examined. However, high variability in measurement endpoints and low replication can confound efforts to detect statistically significant effects. Field studies have the advantage of extending for longer periods than other tests, but the length of these tests should be examined with respect to bee life stage and the extent of an effect that would be necessary to impair the colony as a whole. Consideration may need to be given to cumulative dosing effects in longer-term studies. In addition, how issues of temporal variation, temporal correlation, and trends are assessed for multiple endpoints are areas which should be more standardized to ensure greater consistency and comparability between studies.

12.4 STATISTICAL MODELS

Many methods are available for dealing with dose–response information. Selection of the model structure is important and mathematical approaches for treating study data and resulting curves are issues. Classic probit and logit models are typically chosen, but given biological and experimental variations, the choice of model or experimental design can result in differing LC50 and EC50 estimates. Methods and approaches for dealing with differing results will be addressed in the anticipated analysis.

In brood tests, mortality is expressed as a percentage of the reference population after an adjustment according to the Abbott formula. However, other statistical methods and variance calculations are available, although no sensitivity studies on the test results have been conducted to date to determine the appropriateness of the models used to fit the data. Statistical methods for estimating the probability of survival at a specific age may be appropriate for these data, depending on the experimental design established for the test. In semi-field and field tests, which are typically hypothesis-based as opposed to regression-based study designs, questions include whether there are appropriate time-series models for testing for long-term trends in multiple endpoints, and how non-linear or episodic time-series data are treated. Use of specific statistical models may be more appropriate to evaluate long-term survival and hazard. Examination of survival functions for semi-field tests is an area of future research.

Through the review of several existing datasets, additional areas of analysis may be addressed, including treatment of controls or baseline effects. The anticipated work will examine approaches and interpretation of uncertainty in examining endpoints and output from tests. In addition to examining variability, an evaluation of uncertainty will include examples and case studies for interpreting results in light of the uncertainty estimates.

13 Risk Mitigation and Performance Criteria

E. Johansen, M. Fry, and T. Moriarty

CONTENTS

13.1 THE ROLE OF RISK MANAGEMENT IN POLLINATOR PROTECTION

The risk assessment paradigm discussed at the SETAC Pellston Workshop articulates a process to measure the effects of a compound against the protection goals of a regulatory authority. When sufficient data are available to reasonably predict that the intended use of a plant protection product is inconsistent with protection goals of a regulatory authority, and the use of that product remains beneficial and desirable to stakeholders, then risk managers may seek to either continue to refine the estimate of risk, through higher tier testing or analyses (if this remains an option), or to bring the estimated risks into line with the protection goals through specific mitigation measures affecting the proposed use of that compound. Regulatory agencies rely upon mitigation to balance environmental protection goals with other (stakeholder) demands and incorporate mitigation into their management decisions. Consequently, the role of mitigation is central to the process for pesticide regulation. With the exception of few scenarios[1], most mitigation includes reducing potential exposure. The regulatory agency may mitigate the potential risk by denying use on a particular crop or use site. However, in most cases, mitigation actions are those which modify the manner in which a product is used.

[1] Certain inert ingredients have been shown to [indirectly] increase the potency of a compound; in addition, specific environmental conditions may also modify the behavior, and therefore the potency of a compound.

Pesticide Risk Assessment for Pollinators, First Edition. Edited by David Fischer and Thomas Moriarty.

Stakeholders in the process of risk management include regulatory agencies (national and local), chemical producers, distributors, field advisors, and practitioners (including growers and applicators). At the national level, regulatory authorities are charged with registering pesticide products in a manner consistent with their statutory responsibilities. At the local level, for example, state governments in the United States have their own pesticide registration process, which is equally or more protective than the national level. In other scenarios, in France, for example, specific restrictions can be implemented on the basis of specific cropping or pedo-climatic conditions that may be associated with increased potential risk. At the field level, (additional) mitigation actions can be developed, promoted, and implemented by industry experts, crop specialists, beekeepers, growers, and/or pesticide applicators that extend beyond what is legally required by the regulatory authorities (such as through different management programs).

Mitigation language should be specified in a way that allows for consistent (spatial and temporal) implementation. If mitigation language fails to be clear enough for proper, consistent implementation, then inconsistent protection scenarios may result, and the relationship between the regulatory decision and the protection goals may be lost. Clarity and consistent interpretation are also important because the use of a pesticide product inconsistent with the label directions is in many countries considered a violation of the law that may carry with its prosecutorial action. Insofar that the adjudication of the label violation involves investigation by a third party (usually a local regulatory authority such as in the United States) and arbitration by a civil official, the clarity of the intended use and restrictions associated with a product label is necessary in order to establish misuse. Misuse of a pesticide can also result in severe adverse effects on either human health or the environment.

Regulatory authorities directly or indirectly rely upon feedback information to understand whether assessments and decisions actually support stated protection goals. Feedback information may come in different forms, such as research studies, reports of bee poisoning incidents, or targeted monitoring programs. Feedback information can provide insight into how a product is actually used, unforeseen variables that affect the use of a compound, unforeseen effects of a mitigation action, and/or simply whether the mitigation measures are sufficient to ensure the protection goals. Targeted programs (i.e., investigation designs that time information collection with the actual use of the products) can be expensive but provide high quality data. Investigations such as eco-epidemiological analyses such as those described by Susser (2004) may not be as valuable as targeted monitoring programs, but can provide information on one or several co-variables. Information gained through bee poisoning incident reports may lack some information (such as timing of application, application rate, or analytical analysis) that may be useful in establishing that a particular chemical use resulted in an incident, but may provide information on a specific type of product or use scenario that may be anecdotally linked to an incident. In addition, because incident reports frequently rely upon volunteer reporting, it is difficult to know the degree to which incident reports reflect real world conditions. Therefore, a lack of incident reports may or may not be indicative that the intended (directed) use of a product is safe. Conversely, the lack of incident may not represent the extent of events related to a product, that is, the absence of incident reports cannot be reasonable construed as the absence of incidents. Conversely, the presence of limited incidents may not necessarily indicate whether a risk exists with a product. However, a pattern of incidents related to a specific compound, application method, or crop, for example, may be a clear indication of a risk issue. Nonetheless, information from these feedback sources provides multiple lines of evidence that can be used to inform and modify existing or future assessment or management decisions. Additional discussion may be found in a recent European "OPERA" review (Alix et al. 2011).

It is worth noting that when honey bee workers are killed in the field, the loss of these workers may, to a certain extent, be compensated by the growth of the colony, which may continue to grow and reproduce with little or no impact from the kills. Because most non-*Apis* bees are solitary species, where single females build their nests, lay eggs, and forage for pollen and nectar to feed their offspring, the death of a foraging female or even the incapacity to provision her cells can have adverse effects on queen reproduction and ultimately on the colony (Tasei 2002). Below is a brief discussion of considerations with respect to pesticide risk mitigation for *Apis* and non-*Apis* bees.

13.2 REGULATORY RISK MITIGATION METHODS

The risk assessment should provide a clear description of the risk (i.e., the likelihood and magnitude of an adverse effect) that needs to be mitigated. Knowledge of the chemical physical properties, environmental fate, and ecological effects of a compound are integrated with an understanding of the use of a compound to provide the information necessary to develop potential mitigation options. Specific characteristics of the risks to be mitigated may include the following:

- Whether the risk is related to acute effects on adult bees, chronic effects on adult bees, adverse effects on larval development, or other effects (such as interactive effects of tank-mixes containing insecticides and fungicides).
- Whether the risk is related to honey bees, other species of bees, or both.
- Whether the risk is related to a particular crop or site being treated, to off-target movement of the pesticide to adjacent crops or blooming weeds where bees may be foraging on nectar and/or pollen, or to other concerns (such as contamination of nesting materials used by non-*Apis* bees).
- Whether the risk is related to a particular application mode (systemic or topical) or method (such as spray, or irrigation).
- Whether or how long the pesticide exhibits hazard to bees following application (referred to as extended residual toxicity (RT) in the United States.

Crops Requiring Pollination by Bees: Central to managing risk of pesticides to bees is controlling potential exposure at the time, or under conditions when bees are (likely to be) present in an agricultural setting. One of the most critical issues for risk mitigation is when bees are present at a site for pollination of the crop (Riedl et al. 2006), which may also include bees foraging on understory bloom or in an adjacent or border area. For crops that require pollination by bees, the primary consideration should be to protect bees from pesticide residues that represent a hazard potential. While every attempt should be made to avoid applications of insecticides and fungicides during the pollination period, use of a plant protection product may be needed (or designed for use) when the crop may be most attractive to bees. When developing risk mitigation statements, there are several mitigation options that could be considered:

- *Product Formulation:* Typically there may be several formulations that could be used to treat a crop/pest combination. To the extent possible, formulations should be those that pose the least threat to bees. Formulations that approximate pollen grains (e.g., some microencapsulated products) in terms of particle size can lead to greater exposure as bees may accumulate the product through their normal foraging activity. However, addition of a sticking agent to a foliar application can potentially reduce transfer from the plant to the bee. Granular formulations are typically considered the least hazardous to bees. Seed treatments also provide limited exposure (similar to granular formulations) provided that dust emission (from abrasion during planting) is properly managed. However, dust particles from seed treatments were responsible for a large number of bee poisoning incidents in Germany during 2008 (Pistorius et al. 2009). Soluble and emulsifiable (liquid) formulations are usually safer to bees than wettable powders. Dust and micro-encapsulated formulations may be more hazardous to bees than other formulations, (or routes). For more information on the relative hazard of different formulations, see Johansen and Mayer (1990).
- *Method of Application:* The application method may also be examined to reduce potential environmental exposure. Generally, ground applications result in less off-target drift to both adjacent areas and the understory than aerial applications. Soil incorporated application methods provide limited environmental exposure (via drift); however, since the compound is available to all the growth material, this method may lead to pesticide residues to be expressed in understory bloom. With

respect to aerial application, droplet size can have a marked effect on the extent of drift; in general, larger droplets are less likely to drift compared to finer droplets.

- *Application Parameters:* Limiting the use rate and frequency of application to the minimum required to effectively control the pest or disease organism. Increased application intervals or reduced application rates may lower potential exposure. Application intervals may be related to residue levels in the field that may represent a potential route of exposure (via uptake by the plant or by contact). Products that have demonstrated synergism may be identified or prohibited by a product label.
- *Understory and Adjacent Areas:* The understory of a crop may represent an attractive source of nutrition for the bees separate from, or in addition to, the cultivated crop; and can be a source of either foliar (e.g., from aerial drift) or systemic (when pesticide residues in the soil are taken up by understory flora) exposure to pesticides applied on field. Potential methods of controlling weed bloom include mowing, disking, flailing, or through use of an herbicide. However, it is important to note that eliminating understory forage (as a source of exposure) also forfeits this material as a source of forage or habitat for both pollinators and arthropod fauna. And consequently not considered a sustainable mitigation measure in some European countries.
- *Control Off-Site Movement of the Pesticide:* Buffer zones between application and adjacent areas, particularly if they are attractive to pollinators will reduce potential exposure. Use of low drift spray nozzles, not allow application when wind conditions favor drift onto adjacent crops or weeds that are attractive to bees.
  - Windbreaks may also be employed to reduce drift. Avoid seed dust at sowing (low wind conditions, equip drillers with dust reducing devices).
- *Timing of Application and Environmental Conditions:* Applications may be restricted to times when bee activity is expected to be at a minimum. Honey bees do not forage at night (in temperate regions), and do not begin actively foraging until the temperature reaches at least 55°F (12.8°C). In addition, some flowers close at night, consequently, spray is less likely to land on this portion of the plant, further reducing potential exposure to the bee the following day when foraging begins. This risk mitigation technique is only effective if the pesticide has an intermediate residual hazard to bees of 8 hours or less (evening applications only), has a short residual hazard of less than 4 hours (evening or morning applications), or if flowers are closed during applications.
  - It should be noted though, that other bee species have slightly different activity times, and high temperatures encourage bees to forage earlier in the day or continue to forage later into the evening than usual. Late evening applications are generally less hazardous to bees than early morning applications; environmental conditions such as temperature and dew point may affect the dissipation of a compound (e.g., slow down), thereby extending a compounds residual toxicity. This mitigation option is likely to be of very limited benefit in tropical regions, since the non-foraging period for honey bees in the tropics is very short when compared with temperate regions. For more information on application timing and environmental conditions, see Johansen and Mayer (1990).
- *Tank-mixes:* Tank-mixing may represent an economical option in pest control. However, care should be taken to understand if there are unforeseen effects to non-target organisms from mixing different compounds in a single application. Tank-mixing certain types of compounds may result in interactive effects that can enhance to toxicity of the mixture to bees, such as in the case of pyrethroids and EBI fungicides. (France has recently prohibited tank-mixes of triazole fungicides and pyrethroids.
- *Notification:* Growers may notify beekeepers of anticipated pest control needs. This allows the parties involved to discuss variables and options to reduce potential exposure to bees. While beekeepers may try to protect their stock from an application by covering colonies, doing so for an extended

period of time may be damaging to the colonies, particularly in warm weather. Further, it may be difficult to move managed bees "on demand" since the configuration of the colonies, number of colonies, and the bee activity level effect how quickly stock can be relocated (or protected). (Also, while moving or protecting may be an option for managed bees, it will not protect non-managed bees.)

Crops Not Requiring Pollination by Bees: Pesticide applications to blooming crops, crops with extrafloral nectaries, and pollen shedding crops not requiring pollination that are attractive to bees have also been documented as an important cause of bee poisoning (Riedl et al. 2006). The mitigation options listed above should be considered, but the mitigation statements may need to be modified to address the specific circumstances involved with crops that do not require pollination.

13.3 NON-REGULATORY RISK MITIGATION METHODS

Where limitations exist with regard to the level of risk management that can be reliably and effectively implemented through a national-scale label (regulatory method), implementation of risk management may be possible at the *landscape*, or field level through best management practices (BMPs) employed by the user (non-regulatory). Alternative or additional methods to mitigate risk to pollinating bees may be used in conjunction with measures identified through the product registration and captured on the product label. Beekeepers, growers, and applicators together with integrated pest management (IPM) agents, agricultural extension agents, crop advisors, and pesticide product representatives can exercise field-level knowledge (i.e., practical experience) to achieve maximum protection for both the grower and the beekeeper. Measures that go beyond the product label reflect local knowledge and relationships which foster cooperation that are often the most effective way to manage potential risks.

Among regulatory and non-regulatory methods to mitigate potential risks, communication and cooperation between growers, applicators, and beekeepers is perhaps the most important tool to reduce risk, and ensure that the needs of all of the stakeholders are met. Growers and beekeepers engage in reciprocal, mutually beneficial endeavors and it is to the advantage of each to anticipate and respect the concerns and needs of the other. Growers can learn the pollination requirements of the crops they grow and plan pest control operations with pollination needs in mind. Growers and advisors can proactively manage routine insect pests by developing and monitoring for economic thresholds to initiate appropriate treatment early to reduce pest population and prevent, avoid, or lessen loss without having to rely on higher application rates or intervals that may represent a risk to bees. Such a program is often less hazardous to pollinators and other beneficial insects as well. Applicators can use their knowledge of local weather patterns to time applications in a way that responds to pest pressure and accounts for bee activity, and/or chemical physical properties of the pesticide product. Through communication with growers and applicators, beekeepers should be familiar with pest control problems and programs, to develop mutually beneficial agreements that better ensure the prudent use of insecticides and fungicides. Beekeepers, growers, crop advisors, and applicators should be aware of the toxicity of products being used, and any residual toxicity characteristics. As discussed previously, depending on the size and location of apiaries and weather conditions, some beekeepers can protect honey bee colonies by covering them with wet burlap the night before a crop is treated with an insecticide that has an extended residual hazard. These covers are typically maintained wet and in place for enough time to provide protection from initial hazards. Honey bee colonies should be clearly marked with identification as this facilitates communication.

Apiaries can be situated to isolate them from intensive pesticide application area and to protect them from insecticide and fungicide drift. Establish holding yards for honey bee colonies at least four miles from blooming crops being treated with insecticides that are highly toxic to bees.

Ridge tops are preferable to canyon bottoms, as pesticides can drift down into canyons and flow with morning wind currents.

13.4 SUGGESTED TECHNIQUES TO MITIGATE RISKS TO OTHER SPECIES OF BEES

13.4.1 Nesting and Moving Bees

While shelters for certain species, such as alfalfa leafcutting and bumble bees can be built to be covered, closed, or removed during insecticide applications to reduce the threat of insecticide drift, most non-*Apis* bees, especially soil-nesting species, cannot be relocated as a protection measure. Many non-*Apis* bees will nest in the ground in orchards and even within row crops (Kim et al. 2006). Squash bees (genus *Peponapis*), for example, frequently nest underground at the base of squash and pumpkin plants within the production fields (Shuler et al. 2005), as do *Melissodes* bees in cotton fields (Delaplane and Mayer 2000). Therefore, recommendations made to protect honey bees by closing up or moving hive boxes are of little value for economically important wild bees living in and around crop fields and orchards. Similarly, some alfalfa seed producers in the western United States rely on artificially constructed salt flats to aggregate large numbers of ground-nesting alkali bees (*Nomia melanderi*) for pollination (Cane 2008). The large size of such nesting areas, the long distance these bees can fly (up to 3.2 km (2 miles)), and their potential location away from seed production fields makes it impossible to close off nest entrances to prevent them from foraging in recently sprayed fields.

Blooms of any type, including weedy species that may be available in areas adjacent to the crop site, may serve as nesting sites or as a nutritional source for native pollinators (as it is for managed pollinators as well). To the extent that growers can leave such plants undisturbed and manage pesticide drift, they contribute to the conservation of these native pollinators and the diversity of the farm ecosystem. Approximately 70% of native bees are ground nesters, burrowing into areas of well-drained, bare, or partially vegetated soil. Growers and beekeepers can provide resources for nesting sites for many non-*Apis* species. More information on improving habitat for non-*Apis* pollinators may be found in Vaughan et al. (2007) and Vaughan and Skinner (2008).

13.4.2 Timing of Application

Mitigation of potential exposure through restricting applications to the evening or during periods of cool temperatures was discussed earlier, based upon the premise that honey bees usually do not forage when temperatures are below 13°C (55°F) or between late evening and early morning (Johansen and Mayer 1990), thus giving pesticides with a short residual hazard more time to become inactive or biologically less available. For example, alfalfa leafcutting bees (*Megachile rotundata*) are nearly inactive at 21.1°C (70°F) and completely inactive at 15.6°C (60°F). Both managed alfalfa leafcutting and bumble bees (*Bombus* spp.) can be safeguarded from potential exposures by removing nests prior to pesticide applications. However, this does not reflect the cooler weather tolerance of some temperate species of non-*Apis* bees, such as *Bombus* spp. and *Osmia* spp., both of which are frequently noted for their ability to forage during cool, inclement weather, as well as earlier and later in the day (Thompson and Hunt 1999; Bosch and Kemp 2001). Furthermore, the peak foraging times for bumble bees are very early and late in the day, whereas peak honey bee foraging typically occurs at different periods. Similarly, squash bees (genus *Peponapis*) have been documented to perform a significant amount of pollination in the pre-dawn hours when honey bees are inactive (Sampson et al. 2007). Hence, application of pesticides during the evening, while still preferable, may in fact disproportionately affect certain non-*Apis* species. In some instances, spraying crops that are soon to bloom (e.g., those at budburst) may have a disproportionately higher impact on male solitary bees that emerge before the females and often spend the night in flowers or attached to bud stems.

13.5 PESTICIDE APPLICATION TECHNOLOGIES TO MITIGATE EXPOSURE TO BEES

For compounds that are acutely toxic to bees by contact exposure and a screening-level risk assessment indicates a potential risk to bees via contact exposure, data from a higher-tier test, such as EPA's Tier 2 study to evaluate the toxicity of a pesticide on foliage (e.g., alfalfa) should be used to determine when products should not be applied (e.g., do not apply when bees are actively foraging). To minimize exposure of bees to pesticides, it is important to be aware of weather conditions, particularly wind speed and direction, and avoid applying during those times. Applications at dusk or late evening or early morning prior to dawn when the majority of honey bees are not actively foraging could help minimize contact exposure, depending on the residual time and bioavailability of the pesticide.

13.5.1 Mitigation for Exposure to Seed Treatment Dust

In order to minimize the emission of abraded seed treatment dust during sowing, particularly when seeds dressed with insecticides that are toxic to bees, the following parameters are considered to be particularly relevant.

13.5.2 Seed Coating Quality

Prior to seed treatment, seeds need to be properly cleaned to remove extraneous debris. Thereafter, care should be taken to minimize loose dust in the seed bag. The use of optimized seed treatment recipes is a key parameter to guarantee a high abrasion resistance of the treated seed, while for some treated seeds (e.g., corn), the use of appropriate stickers and film-coatings will further enhance the resistance of treated seeds to abrasion.

13.5.3 Seeding Technology

When seeds are sown using vacuum pneumatic sowing equipment, the use of deflectors, which direct dust downward into the field being planted, has been demonstrated to reduce off-site dust emission. However, even with deflectors, caution should be taken when using this type of sowing equipment in no-till fields, if blooming weeds are present in the field. In this scenario, dust could be deflected directly onto the flowering weeds. Mechanically operated sowing equipment, as well as those using compressed air, are less prone to emit dust into the environment.

13.5.4 Soil Applied Uses

Crops that are not in bloom often harbor blooming weeds or have blooming cover crops. These blooming plants may represent a potential source of pesticide exposure for both honey bees and non-*Apis* bees if the plants are exposed to soil-applied systemic pesticides. Chemigation systems should be maintained in proper working order to ensure that pesticides will not spray, leak, or run off into areas where potential contamination of blooming plants or water sources for bees could occur. Care should also be taken when making granular applications for the same reasons. These potential routes of exposure are probably best addressed through product stewardship, that requires applicator education and post-registration monitoring.

13.5.5 IPM/Crop Rotation

IPM techniques can contribute to the natural reduction of pests by simply employing techniques that reduce the reliance on the broad application of pesticides. When IPM techniques are used, populations of pests can be more easily maintained below detrimental thresholds, thus reducing the need for pesticide treatments, and thus reducing potential exposures to bees.

13.5.6 Landscape Management

Preserved habitats, refuges, food resource, and the like may reduce the dependence of non-target species on commercially cropped areas (Vaughan et al. 2007). Variables such as the nature of the refuge, the proportion or density, location and management of such areas contribute to the effectiveness of the protected area. Initiatives have been undertaken that illustrate the effect of the implementation of flowering strips on pollinating species (e.g., Operation Pollinator developed by Syngenta, http://www.operationpollinator.com) which could provide a useful basis for further recommendations in the future. Further work is needed to actually quantify the benefit in terms of exposure (drift reduction) and impact of the implementation of habitat for non-*Apis* pollinator species. Eventually landscape-level modeling may be used in support of the design of the landscape elements that may be recommended as mitigation measures.

REFERENCES

Alix A, Adams L, Brown M, Campbell P, Capri E, Kafka A, Kasiotis K, Machera K, Maus C, Miles M, Moraru P, Navarro L, Pistorius J, Thompson H, Marchis A. 2011. *Bee Health in Europe: Facts and Figures*. http://www.pollinator.org/PDFs/OPERAReport.pdf (accessed on January 14, 2014).

Bosch J, Kemp W. 2001. *How to Manage the Blue Orchard Bee as an Orchard Pollinator*. Sustainable Agriculture Network, Beltsville, MD.

Cane JH. 2008. A native ground-nesting bee (*Nomia melanderi*) sustainably managed to pollinate alfalfa across an intensively agricultural landscape. *Apidologie* 39:315–323.

Delaplane K, Mayer D. 2000. *Crop Pollination by Bees*. CABI.

Johansen CA, Mayer DF. 1990. *Pollinator Protection—A Bee and Pesticide Handbook*. Wicwas Press.

Kim J, Williams N, Kremen C. 2006. Effects of cultivation and proximity to natural habitat on ground-nesting native bees in California sunflower fields. *J. Kans. Entomol. Soc.* 79:309–320.

Pistorius J, Bischoff G, Heimbach, U, Stähler M. 2009. Bee poisoning incidents in Germany in Spring 2008 caused by abrasion of active substance from treated seeds during sowing of maize. Hazards of pesticides to bees—10th International Symposium of the ICP-Bee Protection Group. Julius-Kühn-Archiv 423:118–126.

Riedl H, Johansen E, Brewer L, Barbour J. 2006. *How to Reduce Bee Poisoning from Pesticides, PNW 591*. Oregon State University.

Sampson BJ, Knight PR, Cane JH, Spiers JM. 2007. Foraging behavior, pollinator effectiveness, and management potential of the new world squash bees *Peponapis pruinosa* and *Xenoglossa strenua* (Apidae: Eucerini). *HortScience* 42:459.

Shuler RE, Roulston TH, Farris GE. 2005. Farming practices influence wild pollinator populations on squash and pumpkin. *J. Econ. Entomol.* 98:790–795.

Susser E. 2004. Eco-epidemiology: thinking outside the black box. *Epidemiology* 15(5):519–520. doi:10.1097/01.ede.0000135911.42282.b4 http://journals.lww.com/epidem/Fulltext/2004/09000/Eco_Epidemiology__Thinking_Outside_the_Black_Box.4.aspx (accessed on January 18, 2014).

Tasei JN. 2002. Impact of agrochemicals on non-*Apis* bees. In: Devillers J, Pham-Delegue MH (eds), *Honey Bees: Estimating the Environmental Impact of Chemicals*. Taylor & Francis, New York, pp. 101–131.

Thompson HM, Hunt LV. 1999. Extrapolating from honeybees to bumblebees in pesticide risk assessment. *Ecotoxicology*. 8:147–166.

Vaughan, M., Shepherd, M., Kremen, C., and Black, S.H. 2007. Farming for Bees: Guidelines for Providing Native Bee Habitat on Farms. The Xerces Society.

Vaughan, M. and Skinner, M. 2008. Using Farm Bill Programs for Pollinator Conservation, Technical Note No. 78. US Department of Agriculture, Natural Resources Conservation Service and Xerces Society for Invertebrate Conservation.

14 Recommendations for Future Research in Pesticide Risk Assessment for Pollinators

CONTENTS

Following the discussions in the preceding chapters, the proposed recommendations are aimed at further improving the risk assessment scheme that could be developed in these proceedings.

14.1 EXPOSURE

14.1.1 Consumption of Guttation Water As a Source of Exposure

Various investigations of residues in guttation droplets collected from seed-treated crop plants revealed the potential for high residue levels to be present in guttation droplets (Girolami et al. 2009; Joachimsmeier et al. 2010; Schenke et al. 2010). Highest residues in guttation water occur immediately after seedling emergence and have been shown to decline with time. Current data suggests that monocotyledons tend to show guttation on a more frequent basis than dicotyledons. Some plants such as sugar beets produce practically negligible guttation. If bee hives are located in the immediate proximity to treated crops (field margin), some individual honey bees have been observed collecting guttation droplets (Girolami et al. 2009). If highly toxic systemic seed treatments or soil applications have been used, some individual forager bees could be potentially exposed to lethal levels of residues in guttation water. However, in currently available colony-level studies, neither

Pesticide Risk Assessment for Pollinators, First Edition. Edited by David Fischer and Thomas Moriarty.

FIGURE 14.1 Guttation water on a strawberry leaf, photo by Mace Vaughan (Xerces Society for Invertebrate Conservation). (For a color version, see the color plate section.)

adverse effects on colonies, nor impact on beekeeping practices have been associated with pesticides in guttation water. Further studies are currently under evaluation, and more research is required to clarify if exposure of systemic pesticides through guttation water needs to be included in the pesticide risk assessment process.

14.1.2 Quantify In-hive Exposure To Larvae, Queens, and Other Hive Members for Use in Screening Assessments

Data on actual exposure of larvae or other hive members could be established by chemical analysis of larval jelly, royal jelly, and bee bread following a field application (such as in a semi-field or field scenario). Spraying a surrogate crop (e.g., *Phacelia* or buckwheat), enclosed in a tunnel containing a hive with minimal pollen and nectar stores would provide an optimal test system to measure in-hive exposure. Larval jelly and bee bread could be sampled from larval cells and analyzed for the appropriate pesticide residues. Data from a series of such tests that capture and represent a range of chemistry classes, and application methods could be averaged to provide a generalized value to represent in-hive "pesticide" exposure (e.g., in larval food) for use in screening-level analyses. Analysis could include both foliarly applied and systemic compounds. For systemic compounds, representative crops could be selected and treated using different delivery routes. Residues in leaves, pollen, and nectar could be sampled over time, and particularly during flowering to determine uptake and decline rates of the pesticide. These data could help refine the default exposure calculation for systemic compounds and would also be helpful in determining the number of samples (e.g., bee bread, larval jelly) that should be analyzed to obtain a robust and repeatable measurement of residue levels, and would also provide information to compare residue levels in pollen to that in other in-hive products, such as bee bread.

14.2 EFFECTS

14.2.1 Role of Inerts and Co-formulants

Although pesticide effects testing typically focuses on the technical grade active ingredient in a relatively pure form (e.g., greater than 95% pure), these compounds are often applied as formulated products that contain other ingredients (e.g., adjuvants or surfactants). In certain cases potential effects from a formulated product may differ from the effects from the active ingredient *per se*. Also, the constituent elements of a formulated product have different chemical and/or physical properties that can cause the formulated product to behave differently in the environment than does the active ingredient, that is, a formulated product may dissipate at a different rate. Consequently, methods for studying these products in an environmentally realistic way can be challenging. As there can be many formulated products associated with an active ingredient, methods are needed for determining which, if any, formulation should be tested in a manner similar to the active ingredient *per se*.

14.2.2 Comparisons Between *Apis* and Non-*Apis* Species

An obvious knowledge gap identified by the participants of the Workshop is data to compare effects between *Apis* and non-*Apis* species. This includes effects in laboratory-based studies and semi-field and full-field studies (exploring both differences in sensitivity and susceptibility). One way to address this uncertainty is to include non-*Apis* bees in semi-field and field studies.

14.2.3 Testing for Sublethal Effects

There is a real need for reliable (field-level) tests for sublethal effects and a means to translate these effects into meaningful measures at the hive level, that is, to establish quantitative linkages between sublethal measurement endpoints on individual bees and more traditional colony-level assessment endpoints. Sublethal effects are most often made at the individual level but even when effects are noted it is difficult to extrapolate these effects to the whole colony. Research is needed to develop reliable test measurements to consistently document sublethal effects on bee behavior. Equally important is a means to translate these effects at the individual level to effects at the colony level. Suggestions for sublethal tests include a standard test for foraging disorientation that might include a "time back to the hive" or a maze at the hive entrance.

14.2.4 Determining the Degree of Adult or Brood Loss That Affects Colony Productivity and Survival

Losses of adult bees in dead bee traps and brood are often noted but the impact of these losses is hard to determine, especially if the losses are transitory. A series of experiments are needed to determine the rate of adult and brood loss necessary to impact colony productivity and pollination and ultimately colony survival. *Apis* colonies have a reserve of worker bees that serve to buffer the effects of temporary losses. However, there remains a fundamental uncertainty regarding the point at which the hive buffer becomes exhausted, and the colony is impaired.

14.2.5 Extrapolating From Semi-Field or Field Scale To Protection Goals

Currently, if any significant effects are observed or measured in semi-field or field studies, then it is predicted that protection goals will unlikely be met. This is due to the inability to confidently extrapolate from effects seen in a semi-field or field study to what may, or may not occur under field conditions. It would be extremely valuable if research could be carried out to link measurement endpoints, derived from a semi-field or field study, with protection goals. This may include not only well-designed testing, but well-designed post-monitoring as well.

There is a need for cost-effective reporting schemes that provide incentives to all parties involved, for example, beekeepers, applicators, and growers, to help increase accurate representation of use and effects of pesticide use in the field. This information would be an important input to the pesticide regulatory framework (i.e., risk assessment and risk management). Furthermore, a common platform for incident reporting between regulatory authorities would facilitate the sharing of incident data and management strategies.

Modeling has been identified as a promising tool for the purpose of risk assessment and risk management. Further research and work on model development for use in pesticide risk assessment for pollinators would help to document and refine modeled biological realism, sensitivity, robustness, parameterization, and calibration. Models could be used to explore potential linkages between measurement endpoints and assessment endpoints or protection goals. Models could also be used in support of extrapolation in time and space of the outcome of a risk assessment based on laboratory studies. Models could also be developed as a support in the design of higher-tier studies and landscape management. Collaboration between modelers and others such as regulators or entomologists would help direct model development and refinement.

The role that landscape management and alternative foraging and habitat resources may play in limiting the impact of pesticides and agronomic practices on pollinators calls for further research in this area. Typically monitoring studies undertaken in agronomic systems proposing diverse options for landscape management would provide feedback and support appropriate recommendations. Such approaches include population ecology, landscape ecology, and exposure characterization. It is noteworthy that the data generated may also feed model development and could thus be generated with the advice of modelers.

14.2.6 Efficacy of Mitigation Techniques

Research is needed to inform whether different risk mitigation techniques are efficacious in reducing the frequency or severity of bee poisoning incidents. For example, research could be carried out that investigates drift reduction technologies or the impact of vegetated buffers in mitigating spray drift or their effectiveness as a refuge and habitat for pollinators.

14.2.7 Investigating Interactive Effects of Pesticides

More research is needed to inform the understanding of interactive effects between crop protection products, particularly between insecticides and fungicides. Evidence of interactions has been observed under laboratory conditions; however, the extent of these interactions in the field remains poorly described. Information on this, including research involving residues occurring in hives is needed to improve our understanding of whether label directions should be revised to restrict or prohibit tank-mixtures of certain pesticides, adjuvants, or surfactants that are applied in conjunction with the pesticide and may be available as an exposure source to bees.

Of critical importance is information on the interaction between in-hive mite control chemicals (acaricides), applied by beekeepers for control of *Varroa* mites, and insecticides or fungicides applied to pollinated crops. Understanding linkages or relationships between these exposure mixtures and honey bee diseases is very important. Research in this area, in addition to that conducted by the US Department of Agriculture would improve the understanding of whether label use directions for in-hive acaricide applications or pesticide applications to flowering crops should be revised.

Since the completion of the SETAC Global Pellston Workshop in January 2011, regulatory agencies, non-governmental agencies, researchers, and others have continued to advance the science of pesticide risk assessment for pollinators; and, some of that work was informed in one way or another by the efforts of the SETAC Global Pellston Workshop. This book captures the thoughts and considerations of those who participated in the Workshop. Further efforts to advance the science, whether in Europe, North America or elsewhere can be found through the appropriate regulatory authorities of countries around the world, or through other stakeholder groups within those countries.

REFERENCES

Girolami VM, Greatti M, Di Bernardo A, Tapparo A, Giorio C, Squartini A, Mazzon L, Mazaro M, Mori N. 2009. Translocation of neonicotinoid insecticides from coated seeds to seedling guttation drops: a novel way of intoxication for bees. *J. Econ. Entomol.* 102(5): 1808–1815.

Joachimsmeier I, Heimbach U, Schenke D, Pistorius J. 2010. Residues of different systemic neonicotinoids in guttation droplets of oil seed rape in a fiedl study. *Julius Kühn-Archiv.* 428: 468–469.

Schenke D, Joachimsmeier I, Pistorius J, Heimbach U. 2010. Pesticides in Guttation Droplets Following Seed Treatment—Preliminary Results from Greenhouse Experiments. Presented at the 20th Annual Meeting of SETAC Europe, Seville, 23–27 May 2010 (abstract book ET05P-TU155, p. 259).

Appendix 1: Elements for a Chronic Adult Oral Toxicity Study

Below are elements of a chronic oral toxicity test proposed by Workshop participants:

- The lifespan of adult honey bees isolated from their colony in laboratory test cages is generally only 2–3 weeks. Control mortality is likely to be unacceptably high before the test ends if you begin with older bees.
- Cages should be well ventilated and sufficiently large to allow the bees to move around freely.
- Minimally, 3 replicates per dose and 10 bees per cage should be used; however, it is important to note that statistical power is based on the number of replicates (treatment units or cages) and not the number of bees within the treatment unit.
- There should be a minimum of five dose rates (treatment levels) to achieve a dose–response curve for the test item and to allow generation of the lethal concentration to 50% of the bees tested, that is, LC50, a no-observed-effect concentration (NOEC), and sufficient doses to verify the LC50 of a toxic reference chemical (e.g., dimethoate is used as a reference toxin in other toxicity tests).
- The test substance should be dissolved in the aqueous sucrose solution (using a maximum of 1% solvent (e.g., acetone) if required.
- If a solvent is required to dissolve the test substance, then a suitable solvent control should be run in addition to a negative control concurrent with the treatments; therefore, both an untreated sucrose (50% w/v) control and, if a solvent has been used to suspend the test item in sucrose, a sucrose-solvent control containing the same maximum concentration of solvent as the test item should be used.
- A protein supplement may be used in the 50% w/v sucrose if this ensures control mortality is acceptable at 10 days.
- As a chronic toxicity test, concentrations or levels should be selected to minimize mortality and facilitate measurement of sublethal effects. A median effect concentration (EC50) based on sublethal effects (e.g., impaired behavior, growth) should be a primary focus of the study.
- Two dosing methods should be considered:
    1. The volume of treated sucrose should be sufficient to allow *ad libitum* feeding for a 24-hour period (continuous dosing).
    2. A small volume of treated sucrose (e.g., 20 μL/bee) should be offered for 2–4 hours each day and then replaced with untreated sucrose (daily dosing). It may be necessary, however, to starve the bees before providing the treated sucrose solution to ensure that the dosed test solution will be completely consumed by the test organisms.
- The amount of treated sucrose offered to the bees and the amount remaining each day should be recorded. The dose consumed should be determined by comparing the weight of the dose remaining in the glass feeders with the weight of a known volume of the test solutions. The composition of the feeders is an important consideration because, depending on the test chemical, material other than glass can interfere with the availability of the test substance.

Pesticide Risk Assessment for Pollinators, First Edition. Edited by David Fischer and Thomas Moriarty.

- During the test period, the bees are kept in the dark (except during observations) in an incubator at 25 ± 2°C and 60–80% relative humidity.
- Mortality and sublethal effects should be assessed at 24-hour intervals after the start of the test for up to 10 days. Sublethal effects should be assessed according to appropriate categories. Control mortality should not be greater than 15%.
- As with any toxicity test protocol, the stability of the test material must be considered when determining the exact methods used in the study. Ideally, nominal concentrations or levels of the test chemical should be verified through analytical measurements.
- The source of the test bees must be recorded, and to the extent possible, disease or parasite loads should be minimized. Any treatments (e.g., antibiotics) other than the chemical of interest must be documented and must be consistent across treatments or controls. To the extent possible, the bees should be from a single colony or derived from colonies with sister queens. As with all studies, bees should be assigned to treatment groups randomly.

Appendix 2: Elements of a Larval Study

CONTENTS

A2.1 INTRODUCTION: IMPORTANCE OF LARVAL TESTING

It may be argued that the loss of a newly emerged bee (day 1) or a capped larvae is of greater proportional effect on the colony than the loss of a mature forager. This is because a mature forager is nearing the end of its lifespan and has already made a substantial contribution to the viability of the hive and "paid back" the investment the colony made in rearing it, whereas a capped larvae or newly emerged adult has not yet paid back this investment by the colony. Further, larvae that die later in development (e.g., at day 7 verses day 1) have consumed more of the colony energy, as input into its development, but have not provided any input back into the colony since they have not become part of the work force of the colony.

Besides the energetic cost to the colony from losing either larvae, pupae or newly emerged bees, there may also be an impact on colony strength. Removing half of larvac, for cxample, may result in there not being enough workers present to build a strong colony, make honey, and survive the winter. Investigating how the impact of removing varying amounts of larvae translates to colony increase/decreases may be an appropriate exercise for modeling (see Chapter 11).

Pesticide Risk Assessment for Pollinators, First Edition. Edited by David Fischer and Thomas Moriarty.

A2.2 PROPOSED ELEMENTS FOR A LARVAL STUDY

- Larvae at the L1 (first instar) stage are fed standardized amounts of a semi-artificial diet. Test items (pesticides or other products of interest) are incorporated into the food at different concentrations within an appropriate range in order to compute the following endpoints for larvae (L1 to L5), pupae (L5 to adult emergence), and adults (emergence to Day 22 post-emergence): LC50, LD50, and NOEC (the NOEC will be the principle target endpoint).
- The reference product is typically dimethoate.

A2.3 LARVAE TERMINATION AND COLLECTION

- For one replicate, larvae are collected from a unique colony. Test colonies have to be healthy and must not show any visible clinical symptoms of pests, pathogens, or toxin stress. Tests should be conducted with summer larvae during a period from the middle of spring to the middle of autumn (the exact time of year varies by location). No *Varroa* treatment with the exception brood removal should be applied within the 8 weeks preceding the beginning of experiments.
- At Day 3, the queen of the chosen colony is confined in its own colony onto a comb. This can be done using an excluder cage into which a comb (dark preferred) containing empty cells is placed or by using a smaller push-in cage (~10 × 10 cm) that can be used to confine a queen on a given section of comb containing empty cells. In both cases, the comb is placed close to other combs containing brood.
- At Day 2, with the verification that there are eggs, the queen is removed from the cage 22–26 hours after she was encaged. To ensure that larvae are available at Day 1 of the study it is recommended to cage the queens of 2–3 colonies in the event a queen is laying few or no eggs. Based on queen vigor, the queen's isolation time can be reduced in order to minimize variability in larval size (age).
- The comb containing the eggs is left caged to prohibit the queen from ovipositing further on the comb on the same position near the brood frames. The eggs develop until the larvae hatch at Day 1.
- At Day 1, the comb containing first instar larvae is transferred from the hive to the laboratory for grafting. As L1 larvae are subject to desiccation, a wetted towel should be placed around the comb.

A2.4 PREPARATION OF REARING MATERIAL

A2.4.1 Rearing Cells

- Larvae (≤1 day old) are reared in polystyrene grafting cups (common among beekeeping equipment supply companies, cells with rounded bottoms are best) having an internal diameter of approximately 9 mm. Before use, the cells are washed and sterilized in 0.4% MBC (methyl benzethonium chloride) water solution or ethanol and rinsed in sterile water then dried in a laminar-flow hood. Each larva is placed in a well of a 48-well tissue culture plate.
- Larvae plates with lids closed, are placed into a larval chamber such as a hermetic chamber (e.g., Plexiglas desiccator, a plastic container, etc.) into which a dish having a potassium sulfate (K_2SO_4) saturated solution is placed to maintain a water-saturated atmosphere (>90% relative humidity). The larval chamber is placed into an incubator at 34.5°C. It is important that this temperature is maintained within a small range since temperature can affect the toxicity of pesticides to immature bees (Medrzycki et al. 2010).

A2.4.2 Larval Food

- The food is composed of three diets for different days of the study with Diet A following the recipe of Vandenberg and Shimanuki (1987) and subsequent diets modified from this basic diet.
- Diet A (Day 1): 50% fresh royal jelly + 50% aqueous solution containing 2% yeast extract, 12% glucose, and 12% fructose. A recipe for 20 g diet contains 10 g royal jelly, 1.2 g glucose, 1.2 g fructose, and 0.2 g yeast extract mixed in 7 mL H_2O.
- Diet B (Day 3): 50% fresh royal jelly + 50% aqueous solution containing 3% yeast extract, 15% glucose, and 15% fructose. A recipe for 20 g diet contains 10 g royal jelly, 1.5 g glucose, 1.5 g fructose, and 0.3 g yeast extract mixed in 7 mL H_2O.
- Diet C (from Days 4 to 6): 50% fresh royal jelly + 50% aqueous solution containing 4% yeast extract, 18% glucose, and 18% fructose. A recipe for 21 g diet contains 10 g royal jelly, 1.8 g glucose, 1.8 g fructose, and 0.4 g yeast extract mixed in 7 mL H_2O.

A2.4.3 General Information Regarding Diet Preparation

Royal jelly can be stored frozen at −20°C in small aliquots to avoid multiple freezing which causes a change in the sugar crystals. It should be thawed by placing it at 4°C overnight, or at room temperature for 1–2 hours. Reverse osmosis water or distilled water should be used, boiled for 10 minutes, and cooled to 45–55°C (cool enough for hands to touch) before using it for mixing. Water, sugars, and yeast should be mixed thoroughly (all solid materials should be broken up with a sterile spatula) in lab ware (preferably glass lab ware such as a beaker) that has been autoclaved. The mixture should be vortexed for 30 seconds. Once the bubbles have settled, the total volume should be adjusted to 10 mL with the prepared water. Finally when the mixture has room temperature, 10 g of royal jelly should be added to the mixture and the mixture vortexed for 30 seconds. The diets prepared for a test should be stored in a refrigerator at ~5–10°C during the test.

A2.4.4 Pupation and Emergence

- At Day 7 (prepupal stage), the plates with open lids are transferred into a pupal chamber (i.e., a hermetic Plexiglas desiccator, a plastic container, etc.). The chamber should be maintained with a saturated atmosphere (~75% relative humidity), this can be achieved by placing a dish with an NaCl saturated solution into the chamber.
- The container is then placed into an incubator at 34.5°C.
- At Day 15, each plate is transferred into an emergence box (~11 × 15 × 12 cm) with a cover that is aerated with wire gauze. The emergence chamber should contain a piece of comb (~3 × 5 cm) that attracts the emerging bees. Emerging bees are fed *ad libitum* with a sucrose syrup solution (50% sucrose/distilled water by volume) that is supplied in a 2 mL eppendorf tube with a hole below. The emergence box is returned to the pupal chamber.

A2.4.5 Grafting and Feeding of Larvae

- The rearing cells in the well plate are prepared by pipetting 20 μL of Diet A into each cell. The comb is placed angular on a clean table and a cold light or LED light is used for illumination to prevent larvae from drying.
- The grafting of the L1 larvae is performed by quick transfer from the comb to each plastic cell cup and placed on the surface of the diet using a grafting instrument of choice (a grafting spoon, paint brush size 00, Chinese grafting tool, etc.).

- If grafting is performed from several combs or a comb is not used for a moment it should be covered by a wetted towel. The grafting should be performed randomly to maintain treatment heterogeneity.
- When a plate is completed with 48 larvae, it is placed into the larval chamber and then into the incubator immediately.
- The larvae are fed once a day (except at Day 2) at the same time of day ($\pm$1 hour) three different diets in different amounts using a stepwise pipette with sterile tips. Prior to administration to the larvae, the diet is warmed to 34.5°C by placing in the incubator 1 hour before feeding. The diet should be pipetted on the inner side wall of the cell to slide slowly down in order to avoid the larvae from drowning. It must be avoided that the diet is placed on the larvae to prevent blocking of the spiracles.

A2.4.6 Experimental Groups

- The experimental unit is a single larva in a cell and a treatment group consists of minimum 24 larvae (half of a 48 tissue culture plate). For each test, the following treatment groups should be used:
  - one control diet without solvent (24 larvae)
  - one control diet with solvent (24 larvae)
  - five test item concentrations (24 larvae each)
  - one reference treatment with dimethoate (24 larvae)

Each test (all eight groups of test larvae) should be replicated across three independent colonies (unrelated queens).

A2.5 PREPARATION OF THE PESTICIDE SOLUTIONS

- The test pesticide is dissolved in water (the preferred solvent) or acetone if the pesticide is not water soluble. If a solvent other than water is used, a second solvent control group must be used consisting of control larvae fed with diet containing the solvent at the same concentration as the treated samples.
- Dilutions of the stock solutions are made with non-chlorinated, sterile drinking water using disposable pipette tips equipped with a filter. The amount of test solution administered must not exceed 10% of the final volume. In all cases, one must include the same final volume of water or solvent in all treatments and controls.

A2.5.1 Treatments

- In acute toxicity tests, larvae are treated at Day 4 with Diet C containing the test item solutions at their respective test concentrations.
- For chronic toxicity tests, larvae are treated daily (except Day 2) with the diets containing the test item solutions at test concentrations. In order to assess the adequate endpoints (NOEC and LC50), run a preliminary experiment with appropriate test material, varying geometrically through 5–10 different concentrations, which can be determined.

A2.5.2 Toxic Reference

- The toxic reference is typically the organophosphate dimethoate:
  - in acute toxicity tests: 3 μg/larva is mixed with Diet C and provided at Day 4,
  - in chronic toxicity tests: it is mixed with the three diets at test concentrations of 20 μg/kg diet.

A2.5.3 Definition of Mortality

- Larva: An immobile larva (not breathing or moving when viewed under a dissecting scope) is recorded as dead. If a larva's mortality is in doubt, examine the larva the following day.
- Pupa: A non-emerged individual at Day 22 is considered as dead during the pupal stage.
- Adult: An immobile adult that does not react to a tactile stimulation is recorded as dead.

A2.5.4 Mortality Assessments

- Larva: Daily (except Day 2) when larvae are fed, all dead larvae are removed for sanitary reasons. Specific mortality checks are made according to the type of test (acute or chronic). In the acute test, where exposure is at Day 4, a first mortality check is made at Day 4 in order to replace the dead larvae before they have started consuming the diet containing the insecticide. Mortality must also be recorded at Days 5, 6, and 7. In the test with chronic exposure mortality is noted at Day 7.
- Pupa: Non-emerged bees are counted at Day 22.
- Adult: Short-term survival: dead adults and living (emerged) adult bees that left their cell and show a normal development are counted at Day 22.
- Long-term survival: living adult bees and dead adults are assessed daily through 10 days post-emergence. Typically, control mortality increases from Days 12 to 14.

A2.5.5 Validity Range of Data

- For the test to be considered valid, bees fed the control diet must adhere to the following:
  - Larvae: ≤10% mortality (number of dead larvae/24)
  - Pupae: ≤20% mortality (number of dead pupae at Day 22/24)
  - Adult: ≤10% mortality (number of dead adults at Day 10 post-emergence/total number of emerged adults)

If the mortality in the control groups is higher than that outlined here, the test is invalidated.
The mortality rate within the dimethoate control should be:

- Acute test: ≥50% mortality at Day 6 for larvae exposed to 3 μg dimethoate/larva at Day 4.
- Chronic test: ≥50% cumulative mortality at Day 7 after exposure to 20 mg dimethoate/kg diet.

The calculated LC50 must be in each case between the concentrations tested; the LC50 must not be extrapolated outside of the tested concentration.

A2.5.6 LD50 and LC50 Calculation

- Mortalities are expressed in percentage of the reference populations after an adjustment according to the Abbott formula (1925):

$$\text{Raw mortalities}: M = \frac{(P - T)}{S} \times 100$$

$$\text{Percent mortalities}: M = \frac{(\%P - \%T)}{100 - \%T} \times 100$$

- *M* is the adjusted mortality expressed in percentage of the initial population, initial number of larvae (24) for a larval mortality, number of living prepupae at Day 7 for pupal mortality, number of emerged (adult) bees at Day 22 for adult mortality
- *P*, mortality due to the treatment
- *T*, control mortality
- *S*, surviving number in control
- %*P*, mortality percentage due to the treatment
- %*T*, control mortality percentage

The results will be analyzed using regression or probit modeling. All raw and adjusted data must appear in the study report. The lethality graphs and their equations must be reported. The results should include LC50 values for 24 and 48 hours expressed in terms of μg per individual (for the acute test), and for an LC50 in μg per liter of solution (ppb) for the chronic test. These calculated variables should include their respective 95% confidence intervals.

A2.5.7 Determination of the NOEC

The NOEC is the highest concentration which does not induce mortality significantly higher than that observed in controls. This analysis is typically performed using a chi-square test (one-tail test, at $\alpha = 0.05$).

REFERENCES

Medrzycki, P, Sgolastra, F, Bortolotti, L, Bogo, G, Tosi, S, Padovani, E, Porrini, C, Sabatini, AG. 2010. Influence of brood rearing temperature on honey bee development and susceptibility to poisoning by pesticides. *J. Apicult. Res.* 49:52–59. http://dx.doi.org/10.3896/IBRA.1.49.1.07

Vandeberg, DJ, Shimanuki, H. 1987. Technique for rearing worker honeybees in the laboratory, *J. Apicult. Res.* 26(2): 90–97.

Appendix 3: Elements of Artificial Flower Test

Artificial flower experiments are performed with a nucleus ("nuc") colony (about 4000 workers and a fertile queen) placed in an outdoor flight cage. Three feeding periods are typically included in the test design. The initial feeding is with an untreated (blank) sucrose solution (500 g/kg) delivered in both the artificial flower feeder and a standard feeder placed in the flight cage; the second feeding is treated sucrose solutions; and the third feeding is again, an untreated (blank) sucrose solution. The foraging activity and the learning performances are evaluated using an artificial flower feeder adapted from the experimental device described by Pham and Masson (1985). The feeder consists of six feeding sites arrayed on a circular tray (50 cm diameter). Each artificial flower feeder is a plastic Petri dish containing glass balls (allowing landing of foragers on the feeding sites) and filled with a sucrose solution that is, or is not treated with the test compound. The sucrose solution in each Petri dish is maintained at a constant level, and on each side of the feeding sites an odorant (e.g., pure linalool) is allowed to diffuse. To limit the influence of visual or spatial cues, the artificial feeder is rotated slowly (e.g., $^{1}/_{3}$ rpm). The device is placed in front of the hive entrance.

The conditioning (pairing odor or sucrose reward) is conducted for 2 hours on the first day. Testing is then carried out on the following days. The testing device is set with three scented devices with food reward alternating with three unscented devices, without any food reward. The testing device is presented for 5 minutes and then replaced by the conditioning device for 15 minutes, with the odor being again associated with a sucrose solution (treated or untreated). For each observation (every 30 seconds over the 5-minute observation period), the number of forager visits on either the scented sites or the unscented artificial flowers is recorded. After each test, the tray is cleaned with ethanol and the Petri dishes are changed to avoid the deposition of marking scent by the forager bees. The volume of sucrose solution up taken by the foragers is measured.

REFERENCE

Pham MH, Masson C. 1985. Analyse par conditionnement associative du mecanisme de la reconnaissance de sources alimentaires par l'beile. *Bu. Soc. Entomol. Fr.* 90:1216–1223.

Pesticide Risk Assessment for Pollinators, First Edition. Edited by David Fischer and Thomas Moriarty.

Appendix 4: Elements of the Visual Learning Test

CONTENTS

Experimental maze tests have been developed to test whether a pesticide compound can disorientate foragers. Orientation performance of bees in a complex maze relies on associative learning between a visual mark and a reward of sugar solution.

The colony is maintained in an outdoor flight cage covered with an insect-proof cloth. The maze consists of a matrix of 4 rows × 5 columns of identical cubic boxes, each side of the box measuring 30 cm; each wall has a 4 cm diameter hole in its center through which bees can move to the adjacent box (Zhang et al. 1996). The boxes are made of white opaque Plexiglas, and a metallic screen (3 × 3 mm mesh) covers the maze.

Bees fly through a sequence of boxes to reach a feeder containing a reward of sugar solution. The path through the maze spans nine boxes, including three decision boxes and six nondecision boxes. A nondecision box has two holes, each in a different wall; one hole where the bee is to enter and another hole through which the bee is expected to leave. A decision box has three holes, each in a different wall. One hole is where the bee enters and the bee is then expected to choose between the other two holes.

During conditioning, bees are collectively trained to associate a mark (designating the correct hole or path) with food. To achieve this, a mark is fixed in front of the correct hole or path as well as the sucrose solution feeder outside the maze for 1 hour. The feeder is then placed in the first box of the path for about 30 minutes, then in the second box of the path for the next 30 minutes, then in the third box for next 30 minutes and so on. The feeder is then moved to the fifth box for about 20 minutes. Finally, the feeder is placed at the end of the path in the reward box (Figure A4.1). Several conditioning periods (3–5) are necessary to train a sufficient number of bees. After the bees have found the food (reward) and have fed, the bees are captured on the sugar syrup feeder and are then placed in rearing cages equipped with a water supply and a sugar syrup feeder (50% w/w). The bees are put back into laboratory and kept at a temperature of 25 ± 2°C in artificial light while they are labeled with colored and numbered tags.

For oral delivery of the test compound, the treatment chemical is added to a sucrose solution (50% w/w). The effect of the treatment solution on performance is then compared with that of an untreated sucrose solution. After a starving period of 1.5–2, each group of tagged foragers receives a volume of the treated sucrose or the control sucrose solution, during daylight and at 25 ± 2°C. The volumes are adjusted for a consumption of syrup estimated to be approximately 10 μL per bee. After complete consumption of the sugar solution, a new starvation period of about 2 hours is initiated. The bees are then provided with an untreated sugar solution *ad libitum* and released to a hive.

After conditioning, the capacity of an individual bee to negotiate a path through the maze is tested. An observer notes the number of correct and incorrect decisions, and then number of turns back. During retrieval

Pesticide Risk Assessment for Pollinators, First Edition. Edited by David Fischer and Thomas Moriarty.

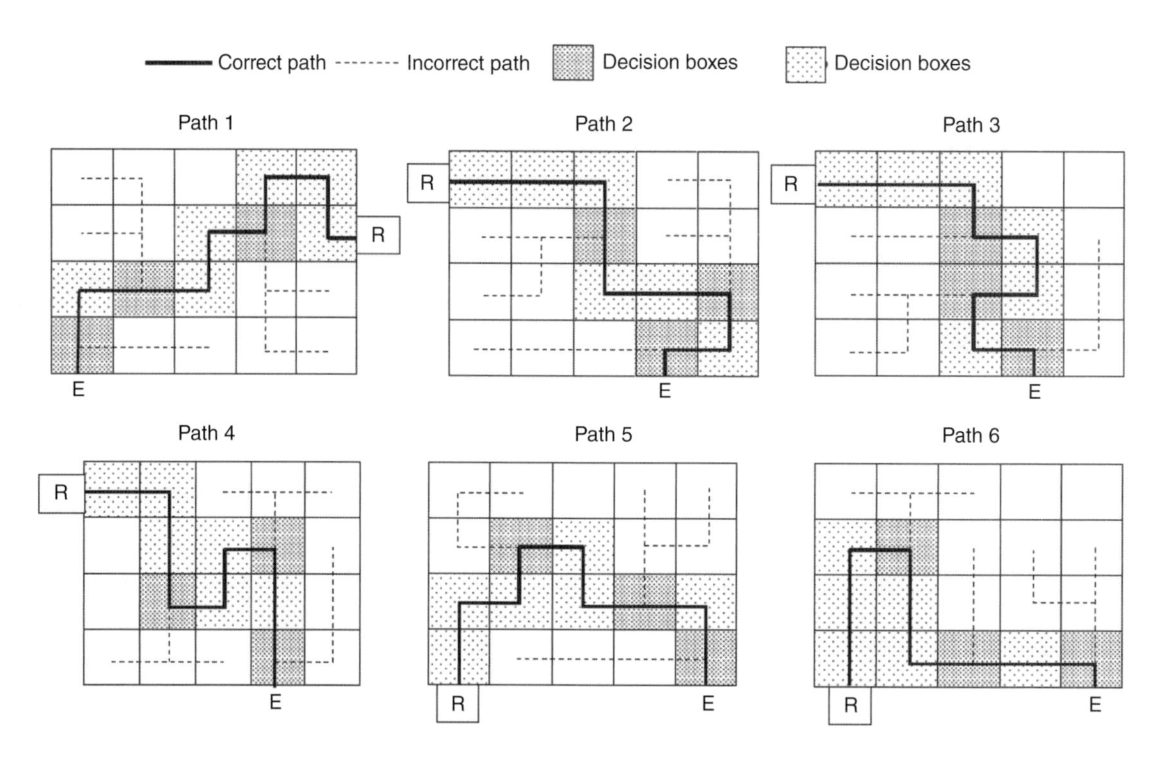

FIGURE A4.1 Maze paths used before, during, and after treatment. Path 1 was used for the conditioning procedure and other paths were used for the retrieval tests. Each path started with the entrance (E), contained three decision boxes, six no-decision boxes, and finished with the reward box (R).

tests, several different paths are used. During a test, only one bee is allowed into the maze at a time and is tested for one of the five path configurations.

Four categories of performances are defined and one of the categories is assigned to each of them:

1. bee moves through the maze and arrives directly at the goal (reward box)
2. bee moves through the maze and arrives at the goal with one or more turns back (bee leaves the box through the hole from which it entered)
3. bee moves through the maze with mistakes (bee making one or more wrong turns at the decision boxes) but arrives at the goal
4. bee does not arrive at the goal within 5 minutes after entering the maze

Performances of control and treated bees are evaluated as the mean of the categories assigned to bees in each group. The time required to reach the goal from the instant of entering the maze is measured for each bee. Flight time is considered only for bees flying through the whole path within 5 minutes.

A4.1 STRENGTHS AND WEAKNESSES

Menzel et al. (1974) have demonstrated that honey bees in flight can associate a visual mark with a reward, and this associative learning is used by bees to negotiate a path in a complex maze (Zhang et al. 1996). After treatment with a sublethal dose of a chemical, the ability of bees to perform the task can be impaired compared with untreated control bees (Decourtye et al. 2009). Work with this type of experimental test has indicated that orientation capacities of foragers in a complex maze can be affected by a pesticide. The maze test relies on the visual learning of foragers in relation to navigation. However, while the maze test has demonstrated effects with pesticides which are neurotoxic, there are insufficient data at this time to determine whether the test will provide useful information for chemicals with other modes of action. Additionally, bee navigation in the field relies upon several guidance mechanisms (e.g., position of sun, magnetism, etc.), unlike in the maze where performance is based on the use of a limited number of pertinent cues. Additional experiments are needed to establish whether effects on maze performance reflect what may actually occur when foragers are exposed to pesticides in the field and are confronted with complex environmental cues.

REFERENCES

Decourtye A, Lefort, S, Devillers J, Gauthier M, Aupinel P, Tisseur M. 2009. Sublethal effects of fipronil on the ability of honeybees (*Apis mellifera* L.) to orientate in a complex maze. In: Oomen PA, Thompson HM (eds), *Hazards of pesticides to bees*. Arno Brynda GmbH, Berlin, pp. 75–83.

Menzel R, Erber, J, Masuhr, T. 1974. Learning and memory in the honey bee. In: Barton Browne L. (ed), *Experimental Analysis of Insect Behavior*. Springer, Berlin, pp. 195–217.

Zhang SW, Bartsch K, Srinivasan MV. 1996. Maze learning by honeybees. *Neurobiol Learn Mem*. 66:267–282.

Appendix 5: Foraging Behavior with Radio Frequency Identification

CONTENTS

Experimental test situations have been designed to explore feeding behavior and social communication (Schricker and Stephen, 1970; Cox and Wilson, 1984; Bortolotti et al., 2003; Yang et al., 2008). These studies generate information on trips between a feeder and a hive, with the variable of pesticide exposure explored. Most test techniques (in this area of exploration) are limited by the number of individuals that can be simultaneously monitored, and by the time devoted to recording individuals. To address these limitations, automated tracking and identification systems have been developed using radio frequency (RF) transponder technology. The use of transponders has the potential to revolutionize the study of insect lifehistory traits, especially in behavioral ecotoxicology.

Different transponder devices have been employed on honey bees, including harmonic radar (e.g., Riley and Smith, 2002) and radio frequency identification (RFID; Streit et al., 2003). Currently, the RFID tags seem to offer unique advantages. Advantages of the RFID technology include the large number of individual insects that can be tracked and the number of detections that can be monitored rapidly and simultaneously (milliseconds) without interference from a variety of matrices (e.g., propolis, glue, plastic, wood) which frequently encumber visual observations. RFID is also less disruptive on bee behavior given the small size of the tags compared to what is needed for harmonic radar tracking.

The tag itself is not equipped with a power source (passive function); rather, it obtains its signal power from the detector (transponder) and causes the tag to emit a unique identification code. The detector (reader) can recognize a virtually unlimited number (18×10^{18} possible identification codes) of individually tagged insects. The RFID technology allows detecting each time a tag-equipped bee is passing in close proximity to the reader (working distance of approximately 3 m). In a study to determine the error rate, Streit et al. (2003) demonstrated that 1 out of 300 tagged bees was not recorded by the RFID readers.

A5.1 EXPERIMENTAL PROCEDURE

The experimental colony is maintained in an outdoor tunnel (8 m $\times$ 20 m, 3.5 m high) covered with an insect-proof cloth and the ground covered with a double layer of plastic. Bees are fed with pollen that is renewed daily. A sucrose solution (50% w/w) is delivered by a feeder positioned 18 m from the hive entrance, in a wooden box (26 $\times$ 26 cm, 30 cm high).

RFID tags (64-bit, 13.56 MHz system, 1.0 $\times$ 1.6 $\times$ 0.5 mm), weighing about 3 mg (3% of bees' weight), represent a relatively low weight given that the honey bee is able to carry up 70 mg of nectar (Ribbands, 1953) and 10 mg of pollen (Hodges, 1952). A tag-equipped bee passing underneath the reader is identified by

Pesticide Risk Assessment for Pollinators, First Edition. Edited by David Fischer and Thomas Moriarty.

the reader that sends the data along with real-time recording to a database. Readers are placed at the entrance of the hive and at the artificial feeder. By passing underneath the reader, both at the hive and at the feeder, the foraging bee is monitored twice, thus determining the direction of travel and the travel time between the two recording points. The reader software records the identification code and the exact time of the detection automatically for 6 days in a database for later analysis of spatial and temporal information. Analyses of the data may provide information on the time spent within the hive; time spent at the feeder; time spent between the feeder and the hive; the number of entries into and exits from the hive; and the number of entries into and exits from the feeder.

RFID devices allow the study of both the behavioral traits and the lifespan of bees, especially under biotic or abiotic stress. However, the large quantity of data obtained with this technique requires an interface for analyzing the data and providing the lifehistory traits of individual bees. Under semi-field conditions, RFID microchips have provided detectable effects due to exposure to an insecticide (Decourtye et al., 2011).

REFERENCES

Bortolotti L, Montanari R, Marcelino J, Medrzycki P, Maini S, Porrini C. 2003. Effects of sub-lethal imidacloprid doses on the homing rate and foraging activity of honey bees. *Bull Insectol.* 56:63–67.

Cox RL, Wilson WT. 1984. Effects of permethrin on the behavior of individually tagged honey bees, *Apis mellifera* L. (Hymenoptera: Apidae). *Environ. Entomol.* 13:375–378.

Decourtye A, Devillers J, Aupinel P, Brun F, Bagnis C, Fourrier J, Gauthier M. 2011. Honeybee tracking with microchips: a new methodology to measure the effects of pesticides. *Ecotoxicology*. 20: 429–437.

Hodges D. 1952. *The Pollen Load of the Honeybee*. London Bee Research Association, London.

Ribbands CR. 1953. *The Behaviour and Social Life of Honeybees*. London Bee Research Association, London.

Riley JR, Smith AD. 2002. Design considerations for an harmonic radar to investigate the flight of insects at low altitude. *Comput. Electron. Agric*. 35:151–169.

Schricker B, Stephen WP. 1970. The effects of sublethal doses of parathion on honeybee behaviour. I. Oral administration and the communication dance. *J. Apicult. Res.* 9:141–153.

Streit S, Bock F, Pirk CWW, Tautz J. 2003. Automatic life-long monitoring of individual insect behaviour now possible. *Zoology*. 106:169–171.

Yang EC, Chuang YC, Chen YL, Chang LH. 2008. Abnormal foraging behavior induced by sublethal dosage of imidacloprid in the honey bee (Hymenoptera: Apidae). *J. Econ. Entomol.* 101(6): 1743–1748.

Appendix 6: Detailed Description of the Proposed Overall Risk Assessment Scheme

CONTENTS

A6.1 SPRAYED TREATMENTS

1. Details of the product and its pattern of use

The most important route of exposure of honey bees to plant protection products for spray applications is by direct exposure to field sprays. In some cases, exposure of bees is not possible and there is no need for a detailed assessment of risks, such as in the case of products used during winter when bees are not foraging, pre-emergent herbicides where plants may not be present to forage on, indoor residential uses and uses in glasshouses where bees are not used for pollination. However, in any scenario where, irrespective of the timing of application, the presence of residues in flowers cannot be excluded the potential for bee exposure should be considered.

The attractiveness of the crop to honey bees may be considered as an entry point for this risk assessment. Useful guidance in this respect may be found in the MRL Working Group publication (EC, 2009) which includes additional criteria to consider, such as the presence of other sources of nectar or pollen in the foraging area. In general, a crop can be considered as unattractive to bees when it is harvested before flowering. Some plants that are intrinsically unattractive to bees may be visited by bees because of extra floral nectaries (e.g., in field beans) or honeydew produced by aphids. As a basis for applying the assessment scheme depicted in Figure A6.1, full details of the product and the intended use must be available. (→ **2**)

2a and 2b. Is exposure of adult and immature stages of bees possible?

Based on the information from the product and the intended application it has to be decided whether exposure of adult bees and immature stages (larvae and pupae, brood) can be excluded. The justification has to take into account all routes of exposure that may be relevant to the intended use, for example, through residues on flowers or in flower matrices (e.g., pollen, nectar), and as for non-*Apis* bees in leaves, soil, etc. (Table A6.1).

The screening step has to be initiated if exposure of adult bees (→ **3a**) or immature stages (→ **3b**) to the active ingredient (a.i.) cannot be excluded. Further risk assessment is not required in cases where exposure can be ruled out for both adults and immature stages of bees (→ **6**).

Pesticide Risk Assessment for Pollinators, First Edition. Edited by David Fischer and Thomas Moriarty.

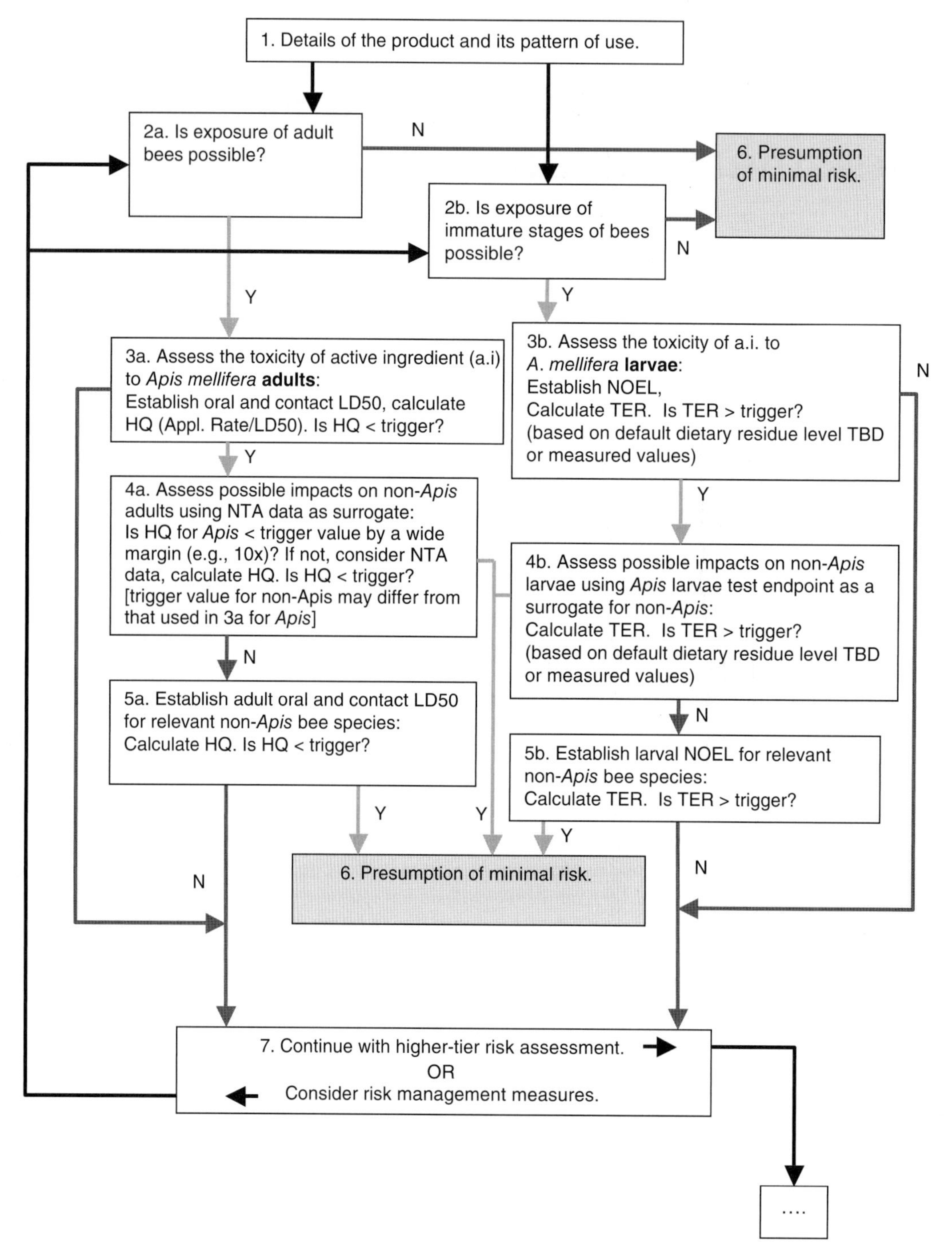

FIGURE A6.1 Insect pollinator screening-level risk assessment process for foliarly applied pesticides. (For a color version, see the color plate section.)

TABLE A6.1
Likelihood of Exposure to *Apis* and Non-*Apis* Bees from Various Routes

	Apis		Non-*Apis*	
Exposure	**Adults**	**Larvae**	**Adults**	**Larvae**
Nectar	+ + +[1]	+	+ to + + +	+
Pollen	+ to + + +	+ +[2]	+ to + + +	+ + to + + +
Water[a]	+ to + +	+[3]	+	+
Nesting material	+[4]	+[4]	+ to + + +[4, 5]	+ to + + +[6, 7, 8]
Exposure to soil	±	–	– to + + +	– to + + +
Foliar residues				
(Contact and direct spray)	+ + +	–	+ + +	– to + + +
Direct spray	+ + +[9]	–	+ + +[9]	–

[a]Collect water for cooling (evaporative cooling; take up into crop, regurgitate it, and flap wings to distribute) and honey production.
[1]Particularly for nurse bees; [2]bee bread; [3]provided by nurse bees; [4]wax; [5]leaves and soil for cement; [6]leafcutting bees; [7]soil used to cap cells; [8]exposure to soil; [9]at flowering.

3a. Assess the toxicity of active ingredient (a.i.) to *Apis mellifera* adults: Establish acute oral and contact LD50, calculate HQ (appl. rate/LD50). Is HQ below the trigger value (e.g., HQ <50)?

Acute oral and contact toxicity of the active ingredient to adult honey bees should be determined in appropriate laboratory tests generating median lethal doses (LD50) for both routes of exposure (Chapter 7). The highest intended field application rate is used to estimate possible exposure in comparison to the most sensitive of these LD50 endpoints. A hazard quotient (HQ) is calculated by dividing the application rate (g a.i./ha) by the most sensitive acute toxicity endpoint (μg/bee). The resulting HQ does not directly specify the relation of exposure level and toxicity since the numerator (application rate in terms of g a.i./ha) and denominator (LD50 in terms of μg/bee) of the HQ are in different units of measurement. Rather, it is used as a preliminary screen to indicate whether a level of exposure may lead to adverse effects (i.e., that a presumption of minimal risk cannot be made) based on empirical incident data. This initial HQ calculation is used as an indicator of risks in the European regulatory process and has been compared to EU incident data. Comparisons of screening-level HQ values with incident data have indicated that adverse effects in the field are not observed when HQ values are greater than 50 (see Mineau et al. 2008). In this flow chart, the screening-level HQ trigger of 50 is given as an example of value that is used in Europe for screening purposes (EC, 2010); however, regulatory authorities must develop their own triggers for moving to more refined assessments. The intent here is to demonstrate that at a screening level, relatively course measures of exposure are used in combination with relatively simple measures of effects to determine whether risk can be presumed low. Where HQ exceeds the trigger value a higher tier risk assessment or consideration of risk mitigation measures is required (→ **7**). Otherwise the risk to adult honey bees (*Apis* bees) may be assessed to be low and consideration of possible effects on non-*Apis* bees is the next step of the screening procedure (→ **4a**).

3b. Assess the toxicity of a.i. to *A. mellifera* larvae: Establish NOEL, calculate TER, is TER >1?

Chronic toxicity of the active ingredient to honey bee larvae should be determined in an appropriate laboratory test generating a NOEC for the brood development including adult emergence weight (Chapter 8). For risk assessment, this toxicity endpoint is compared to the exposure of honey bee larvae via contaminated food items. If chemical or crop specific exposure data are not available, then default exposure estimates may be determined through information from residue analysis data (see Chapter 7 for more details). Toxicity and exposure data (expressed in same measurement units of μg/kg) are related in a TER calculation (TER = NOEC divided by predicted exposure. The resulting TER is compared to an appropriate trigger and any value above that trigger indicates a presumption of minimal risks. In the flow chart, a trigger of 1 is used based on the presumption that maximum residues measured in pollen have not exceeded 100 μg/kg and that using a value of 1000 μg/kg would likely be considered protective. Again, appropriate exposure values and triggers must be determined by the regulatory authority; however, at this level of refinement, potential risks are determined from toxicity data on bee brood and rely on the no-observed-effect concentration.

4a. Assess possible impacts to non-*Apis* adults using NTA data as surrogate: If HQ for *Apis* is between 5 and 50, consider NTA: Calculate HQ, is HQ <2?

When specific data on the toxicity of the compound to adult non-*Apis* bee species are lacking, potential risk may be estimated from the data available on the honey bee and if available in the dossier, the use of data on other non-target arthropods (NTA). A possible tiered approach using these data, to screen for the need of a risk assessment specific to non-*Apis* bees that would use dedicated data is presented thereafter. Initially the HQ calculated under point **3a** using the honey bee LD50 could be used as a trigger of concern for possible effects on non-*Apis* bees. This HQ value would then be compared to a trigger value lower by an order of magnitude to account for interspecies variability of toxicity data. Thus the HQ calculated under point **3a** shall be lower than five for acceptable risks to be concluded for adult honey bees and adult non-*Apis*. The order-of-magnitude increase in the trigger is intended to account for interspecies variability. In the case of $5 < HQ < 50$, data on NTA species would be considered in order to conclude about the level of concern of the product for non-*Apis* bees, taking into consideration the level of risk for NTA species and how representative the test species are of non-*Apis* bees expected to frequent the crop, etc. As an example, in the risk assessment scheme for NTA performed in the EU, the laboratory toxicity endpoint for the most sensitive NTA species is compared to the maximum application rate in an HQ calculation (where the toxicity endpoint is also expressed as a rate (g a.i./ha)) (Candolfi et al. 2001). This HQ is assessed against a trigger value of 2. Where the HQ value for NTA exceeds this trigger value, it is concluded that risk to non-*Apis* cannot be excluded and that risk estimates should be further refined. This refinement could consider the generation of specific adult toxicity data with a non-*Apis* species before a higher tier risk assessment or consideration of risk mitigation measures (→ **5a**). If mitigation measures are considered, then the effect of these measures on potential exposure should be considered using the same process as just described from the point where risk could not be presumed low or minimal.

If this HQ for NTA is below the trigger value, the risk to adult non-*Apis* bees may be considered minimal (→ **6**).

5a. Establish adult oral and contact LD50 for a non-*Apis* bee species: Calculate HQ, is HQ <50?

The screening step **3a** may be repeated using specific toxicity data generated in tests with a non-*Apis* bee species. For further details on laboratory studies on non-*Apis* bees, see Chapter 8. Where the HQ exceeds the trigger value of 50, a higher tier risk assessment or consideration of risk mitigation measures is required (→ **7**). For HQ values below the trigger, risk to larvae of non-*Apis* bees is considered minimal (→ **6**).

5b. Establish larval NOEL for relevant Non-*Apis* bee species: Calculate TER, is TER >10?

The screening step **3b** may be repeated using specific toxicity data generated in tests with a non-*Apis* bee species. For further details on laboratory studies on non-*Apis* bees, see Chapter 8. Where TER is below the trigger value of 10, a higher tier risk assessment or consideration of risk mitigation measures is required (→ **7**). For TER values above 10, the risk to larvae of non-*Apis* bees is considered minimal (→ **6**).

6. Presumption of minimal risk

If exposure can be excluded or the criteria in the screening step are met for both adult bees and larvae, then a minimal risk to honey bees or non-*Apis* bees can be presumed. A minimal risk for honey bees or non-*Apis* bees can also be presumed if treatments in higher tier semi-field and field tests result in no significant difference compared to the untreated control when evaluated against the protection goals. Further risk mitigation measures are not required.

7. Continue with higher tier risk assessment or consider risk mitigation measures and reassess

If in the screening step the criteria for adult bees or larvae are not met, a higher tier risk assessment (depicted in Figures A6.1 and A6.2) should be performed (→ **8**). The screening step may be repeated to consider specific risk mitigation measures that exclude or mitigate exposure (e.g., by reducing the application rate, avoiding the exposure to residues during flowering) (→ **2**). For further considerations of risk mitigation measures, see Chapter 11.

8. Is higher tier risk assessment triggered by failing the screening step with non-*Apis* bees?

Concerns identified in the screening procedures and which are not addressed through mitigation, may then be further refined through semi-field or field tests (→ **9**). If in the screening steps potential risks were identified for non-*Apis* species (adults or larvae) that will be further refined in a higher tier study, then the assessor should consider whether a higher tier study with honey bees would also be representative of the concerns identified for non-*Apis* bees in the screening step (→ **13**).

9. Is higher tier risk assessment triggered by failing the screening step with regard to honey bees?

If in the screening step the criteria for *Apis* (adult bees or larvae) are not met, a semi-field or field test should be performed to further refine potential concerns. (→ **10** or **11**). In transitioning from the use of laboratory-based studies on individual bees to semi-field and field toxicity studies typically conducted at the colony level, test conditions are intended to reflect more realistic exposure conditions. Unlike the lower tier studies, though exposure is incorporated into the results of the semi-field and field studies such that the question being asked is whether there is an adverse effect under the conditions tested. As measurement endpoints are selected in higher tier studies to directly reflect assessment endpoints that are in turn intended to address protection goals, these studies simply answer a yes or no question and do not require risk estimates, that is, no HQ or TER is calculated.

10 and 11. Assess the effects of the a.i. to *A. mellifera* in a semi-field or a field test: Do results indicate minimal risk (no significant difference to control)?

Concerns raised in the screening procedure may be investigated through a semi-field test where possible effects are assessed against the criteria related to the protection goals. This is to say that measurement endpoints should be readily related to assessment endpoints which in turn reflect protection goals. For example, if a protection goal is to ensure pollination services, then having sufficient forage strength in a colony is important. Therefore, adult and larval bee survival is a reasonable assessment endpoint and the number of dead bees in traps or brood termination rates may be reasonable measurement endpoints to reflect that assessment endpoint.

The choice between semi-field and field testing depends on the profile of the product as, for example, the expected duration of exposure, the possibility of occurrence of effects, the nature of the anticipated effects.

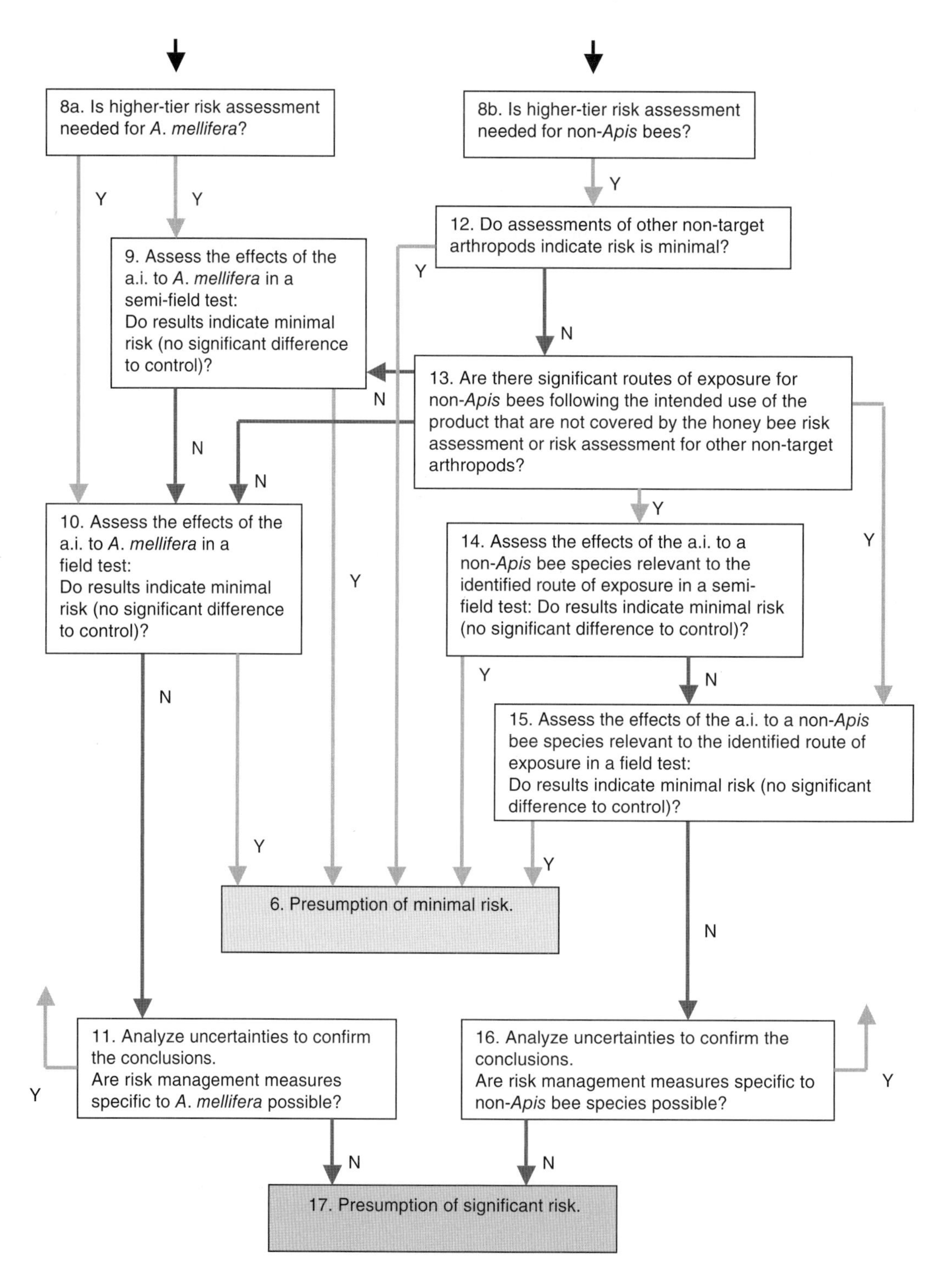

FIGURE A6.2 Higher tier (refined) risk assessment process for foliarly applied pesticides. (For a color version, see the color plate section.)

This choice is a case-by-case decision, but the design of semi-field and field studies should be informed by the results deduced from lower tier testing and other relevant lines of evidence, for example, incident data.

Semi-field testing (cage, tunnel, or tent tests) is a suitable option before full-field testing. The advantage of semi-field testing is that mortality is easier to assess and exposure of bees to the test compound is more readily ensured since bees are confined within a tent and cannot forage elsewhere. In addition, if an accurate quantification of exposure is needed, semi-field studies may provide more reproducible residue levels due to the relative protection from weather conditions.

Semi-field as well as full-field tests aim at evaluating the level of effects that may be expected on bees exposed to the product under realistic conditions, that is, through the crop having been treated at proposed application rates. Because the conditions of exposure of bees are more reflective of actual use conditions, the results of these trials may be directly used in risk assessment (see Chapter 9).

The design of semi-field or field testing may also follow a tiered approach. In the first instance, semi-field tests should be designed in order to maximize the exposure of bees to residues resulting from an application. For sprayed products, the demonstration of acceptable effects in a semi-field or field test performed on a "standard crop" (e.g., wheat) made artificially attractive through a sugar solution and treated at the maximum application rate at flowering may be considered as protective for any crop that may be further treated with the product. Further steps may consider bee-attractive crops treated at flowering (e.g., phacelia), and then the specific crops on which the compound will actually be applied as a highest tier when a treatment at flowering cannot be excluded. Further on, the possibility of an exposure outside the flowering period of the crop through, for example, spray drift onto flowers in vegetated areas or onto flowering weeds within the crop (e.g., understory of orchards), should also be considered in the trials, if triggered by the lower tiers. In the case of soil or seed treatments, it may be more difficult to identify a surrogate (worst case) crop as the exposure results from systemic properties and the attractiveness of the crop to bees. For both sprayed and soil or seed treatments, in the case of systemic activity, if the substance or its residues are persistent and the product may be used on several crops in a rotation, the potential accumulation in soil and subsequent effect on in-plant residues should be considered in the study protocol.

For both semi-field and field trials, it should be demonstrated that the test bees were actually exposed under the environmental conditions (especially weather conditions in case of field trials) of the study. The use of a toxic standard (semi-field trials) or pollen collection and residue analysis, may also help to document exposure. A quantified assessment of the exposure is particularly important for systemic products, as reference substances for systemic products are difficult to define since they too would be dependent on crop properties. There should always be a comparable untreated control in order to provide a reference point against which to compare the test treatments. While positive controls (toxic reference chemicals) are frequently used in laboratory and semi-field studies, they are not typically used in full-field studies. Therefore, it is not possible to demonstrate definitively that the study design is sufficient to detect treatment effects and it is important to document exposure through residue analyses.

For honey bees, suitable methods for semi-field and field trials are discussed in OEPP/EPPO (2010) (see Chapter 9) which have been defined for sprayed treatment and can be adapted to soil or seed treatments (systemic activity). These studies may also be modified for specific assessments with honey bees, for example, repellency and other behavioral effects, effects of aged residues or for specific testing of brood effects. Possible adaptations for non-*Apis* species are discussed in Chapter 9.

The interpretation of semi- and full-field study results is further detailed in Chapter 9. It should rely on a comparison of effects in the test chemical treatments and in the concurrent negative control. If the semi-field test treatment results in no significant difference from untreated controls in lethal and sub-lethal effects (i.e., survival, growth, reproduction, and foraging behavior), a minimal risk is indicated (→ **6**). Otherwise a higher tier evaluation using a field test has to be performed (→ **11**).

12. Are risk mitigation measures specific to *A. mellifera* possible?

If the results of higher tier semi-field and field tests indicate that the protection goals are not met, the assessment scheme may be reiterated considering specific risk mitigation measures mitigating the exposure of honey bees (→ **2**). Note in this respect that semi-field and field tests may be appropriately adapted in order to check for the efficiency of risk mitigation measures to reduce exposure to and subsequent impact from treatment residues on bees.

13. Are there significant routes of exposure for non-*Apis* bees that are not covered by the honey bee risk assessment or risk assessment for other NTA?

In any case when a risk assessment for non-*Apis* bees is triggered and a refined risk assessment is available for honey bees and NTAs, it may be interesting to discuss the extent these risk assessments address as part of the risk issues relative to non-*Apis* species. As an example, concerns with effects on non-*Apis* bees identified at the lower levels may in some cases be addressed by semi-field or field tests with honey bees as for example, where no additional significant routes of exposure for non-*Apis* bees have to be taken into account. Furthermore, higher tier field data generated with NTA species may also address these concerns provided the routes of exposure are comparable to those for non-*Apis* bees (Table 10.3, see Chapter 9). If these data are considered suitable surrogates and if the examination of these data results in no significant risk with regard to the protection goals, then a minimal risk to non-*Apis* bees is indicated (→ **6**). Otherwise semi-field or field tests with non-*Apis* bees should be considered to address the concern (→ **14**).

14 and 15. Assess the effects of the a.i. to a non-*Apis* bee species relevant to the identified route of exposure in a semi-field or a field test: Do results indicate minimal risk (no significant difference to control)?

Potential risks identified in the screening-level assessment may be addressed by appropriately designed semi-field tests where possible effects are assessed against the evaluation criteria related to the protection goals. The derivation of evaluation criteria for specific protection goals is discussed in Chapter 4. For further details on semi-field studies on non-*Apis* bees, see Chapter 9. As previously developed in the case of honey bees, the choice between a semi-field test or a full-field study depends on the outcome of lower tier studies and should also consider choices made for honey bees. If the results of semi-field or field test, in conjunction with information from lower tier studies and other relevant data indicate no significant difference in relevant lethal and sub-lethal effects compared to untreated controls, minimal risk is indicated (→ **6**).

Otherwise, further risk mitigation may be considered or the risk has to be presumed as significant (→ **16**).

16. Are risk mitigation measures specific to non-*Apis* bee species possible?

If the results of higher tier semi-field and field tests on non-*Apis* indicate that the protection goals are not met, the assessment scheme may be reiterated considering specific measures designed to mitigating the exposure of non-*Apis* bees (→ **2**).

Note in this respect that semi-field and field tests may be appropriately adapted in order to check for the efficiency of risk mitigation measures to limit the exposure and potential impact of treatment residues on non-*Apis* bees.

17. Presumption of significant risk

If there are no measures available to sufficiently mitigate the risk to honey bees or non-*Apis* bees indicated by the evaluation of the results of higher tier semi-field and field tests against the protection goals, then a significant risk has to be presumed.

A6.2 SOIL OR SEED TREATMENT WITH SYSTEMIC ACTIVE SUBSTANCES

1. Details of the product and its pattern of use

As a basis for applying the assessment scheme, details of the product and the intended use must be available, especially the crop, the formulation type, type and timing of application, as well as the application rate (g a.i./ha). In addition it has to be determined whether the active ingredient has systemic properties, that is, significant portions of the compound are translocated in the plant resulting in residues of concern in plant matrices like nectar, pollen, and leaves that might lead to exposure of bees (→ **2**). If persistent soil residues may give rise to uptake of the substance by succeeding (rotational) crops, the same considerations with regard to attractiveness of these crops to bees apply as discussed in the description of the risk assessment scheme for spray applications. Restrictions concerning the choice of succeeding crops may be considered as risk mitigation measures.

2a and 2b. Is exposure of adult and immature stages of bees possible?

Based on the information on the product and its intended application it has to be decided whether exposure of adult bees and immature stages (larvae and pupae, brood) can be excluded. The justification has to take into account all routes of exposure that may be relevant to the intended use, for example, through residues on flowers or in flower matrices (e.g., pollen, nectar), and as for non-*Apis* bees in leaves, soil, etc. (Table 10.3).

The screening step should be initiated if exposure of adult bees (→ **3a**) or immature stages (→ **3b**) to the active ingredient cannot be excluded. Further risk assessment is not required in cases where exposure can be ruled out for both adults and immature stages of bees (→ **6**). Special routes of exposure of bees as a result of soil or seed treatment application of active substances with systemic properties may not be covered by the risk assessment scheme for spray application. The exposure of bees to residues of a systemic product may occur through transfer of residues taken up by the roots from the seed coating or soil and distributed to the upper (apical) parts of the plant and in particular in matrices of interest to bees (pollen, nectar, and honeydew) if the crop is visited by bees. The resulting residue of concern may comprise the active substance or systemic soil degradation products or metabolites formed in the plants. Honeydew might not be considered a relevant route because the concentration of a systemic compound translocated in the phloem and reaching honeydew without harming aphids should in principle not be capable of harming bees foraging on the honeydew, unless the compound is highly selective toward non-aphid insects. If there is uncertainty regarding potential residues in honeydew because there is insufficient information on selectivity available in the registration dossier, a dedicated evaluation according to the present risk assessment scheme would be triggered. Information derived from residue studies and plant metabolism studies is in general sufficient to identify if the substance is internally distributed within the plant during its growth, and if it is further degraded into major degradation products. Similarly, possible uptake and distribution in plants of major soil degradation products could be identified in these residue studies as well. The sensitivity (i.e., limits of quantification and detection) of the analytical methods that are used in the residue studies must be checked in order to ensure that they are low enough to detect residue levels that exert toxic effects to bees. If it is uncertain whether the detection methods are sufficiently sensitive, additional investigations have to be considered to demonstrate the absence of residue translocation at potentially toxic levels. Studies that specifically investigate the presence of residues in flowers, nectar, or pollen may be considered as an option for the generation of data to refine the predicted exposure of bees.

Other routes of exposure as a consequence of soil or seed treatment application (e.g., drift of abraded treated seed coating dust into adjacent areas attractive to bees) are not specific to systemic active substances and therefore not addressed in this risk assessment scheme. It should be noted that the emission and dispersion of dusts at sowing is considered as reflecting a poor quality sowing or formulation practices that could be mitigated to reduce potential exposure to a minimum level. Therefore measures aiming at reducing the emission and dispersion of dusts at sowing should be considered.

3a. Assess the toxicity of a.i. to *A. mellifera* adults (oral exposure): Establish Oral LD50, calculate TER, compare TER to an appropriate trigger value

The acute oral toxicity of the active ingredient to adult honey bees should be determined in appropriate laboratory tests generating median lethal doses (LD50) (Chapter 8). The highest intended field application rate is used to estimate possible exposure in comparison to the most sensitive of acute contact and acute oral LD50 endpoints.

For risk assessment, the LD50 is set into relation to the exposure of adult honey bees. For this purpose, a default dietary residue level may be used, as for example the value of 1 mg a.i./kg proposed by the EPPO (2010). Measured residue levels may also be used as a refinement of exposure estimates. As exposure estimates should reflect the maximum expected residue levels for a "worst-case" assessment, the measured residue in plant matrices to be used as a refinement of exposure estimates for TER calculation could, for example, be based on the upper 90th percentile of residue data for the relevant crop for comparison to the most sensitive acute LD50. Toxicity and exposure data expressed in same units are related in a TER calculation (TER = LD50 divided by predicted exposure) where residue concentrations have to be expressed in terms of daily uptake per bee (μg/kg). The calculated TER is assessed against an appropriate trigger value. A trigger value of 10 may, for example, be applied indicating that the predicted exposure is lower than the acute toxicity by at least one order of magnitude and the margin of safety achieved can be regarded as sufficient to cover the uncertainty related to longer exposure periods and possible related increased sensitivity (EPPO 2010).

Where the TER is below the trigger value, a higher tier risk assessment or consideration of risk mitigation measures is required (→ **8**). As a refinement option a prolonged toxicity test in the laboratory may be considered (→ **4a**). Otherwise, the risk to adult honey bees is assessed to be low and an evaluation of possible effects on non-*Apis* bees is the next step of the screening procedure (→ **5a**).

3b. Assess the toxicity of a.i. to *A. mellifera* larvae: Establish NOEL, calculate TER, compare TER to an appropriate trigger value

Chronic toxicity of the active ingredient to honey bee larvae should be determined in an appropriate laboratory test generating an NOEC for the brood development including adult emergence weight (Chapter 8). For risk assessment, this toxicity endpoint is compared to the exposure of honey bee larvae via contaminated food items. If chemical or crop-specific exposure data are not available, then default exposure estimates may be determined through information from residue analysis data (see Chapter 7 for more details). Toxicity and exposure data (expressed in same measurement units of μg/kg) are related in a TER calculation (TER = NOEC divided by predicted exposure. The resulting TER is compared to an appropriate trigger and any value above that trigger indicates a presumption of minimal risks. In the flow chart, a trigger of one is used based on the presumption that maximum residues measured in pollen do not exceed 100 μg/kg and that using a value of 1000 μg/kg would likely be considered protective. Again, appropriate exposure values and triggers must be determined by the regulatory authority; however, at this level of refinement, risks are determined from toxicity data on bee brood and rely on the NOEC.

4a. Assess the oral toxicity of a.i. to *A. mellifera* Adults in a prolonged (10 day) test: Establish oral NOEL, calculate TER, and compare TER to an appropriate trigger value

As a refinement option the NOEL derived from a 10-day toxicity test with oral exposure may be taken into account before embarking on a higher tier risk assessment. The NOEL is related to the potential exposure of adult honey bees via consumption of contaminated food items (default value as for example 1 mg a.i./kg or measured residue data). A TER value is calculated by dividing NOEL by predicted exposure expressed in the same units of measurement. In this case, as the effects are monitored over a 10-day period, the average (or time-weighted average) of residue levels is a more appropriate exposure estimate in a TER calculation. The calculated TER is assessed against an appropriate trigger value. A trigger value of one may be applied as the toxicity endpoint is related to the NOEL. Where the TER is below the trigger value, a higher tier risk assessment or consideration of risk mitigation measures is required (→ **8**). Otherwise the risk to adult honey bees is assessed to be low and consideration of possible effects on non-*Apis* bees is the next step of the screening procedure (→ **5a**).

5a. Assess possible impacts on non-*Apis* adults using NTA Data as surrogate: If TER for *Apis* is Between 10 and 100, consider NTA data

When specific data on the toxicity of the compound to adult non-*Apis* bee species are lacking, risk may be estimated from the data available on the honey bee and if available in the dossier, the use of data on other NTA.

Explore the NTA data package to ascertain whether there is likely to be a significant risk to non-*Apis* bees by considering the characteristics of each species tested, for example, *Aleochara bilineata* may give some evidence concerning soil-dwelling species and Aphidius sp. on nectar-feeding species. Where a risk to non-*Apis* bees (as estimated using NTA) cannot be excluded, more refinement is considered necessary. This refinement could consider the generation of specific adult toxicity data with a non-*Apis* species before a higher tier risk assessment or consideration of risk mitigation measures (→ **6a**). If mitigation measures are considered, then the effect of these measures on potential exposure should be considered using the same process as just described from the point where risk could not be presumed low or minimal. If the risk to NTA is considered to be minimal, the risk to adult non-*Apis* bees may be considered minimal (→ **7**).

6a. Establish adult oral LD50 for a non-*Apis* bee species: Calculate TER, compare TER to an appropriate trigger value

The screening step **3a** may be repeated using specific toxicity data generated in tests with a non-*Apis* bee species. For further details on laboratory studies on non-*Apis* bees, see Chapter 8. For risk assessment, the LD50 endpoint is set into relation to the exposure of adult non-*Apis* bees. For this purpose a worst-case default dietary residue level of 1 mg a.i./kg (EPPO 2010) or measured residue data in relevant food items may be used. Toxicity and exposure data expressed in the same units are expressed as a ratio in a TER calculation (TER = LD50 divided by predicted exposure) where residue concentrations have to be expressed in similar terms, that is, daily uptake per bee. The calculated TER is assessed against an appropriate trigger value. A trigger value of 10 indicating that the predicted exposure is lower than the acute toxicity by at least one order of magnitude may be considered to be appropriate also for non-*Apis* bees. Where TER is lower than the trigger value, a higher tier risk assessment or consideration of risk mitigation measures is required (→ **8**). Otherwise the risk to adults of non-*Apis* bees is considered minimal (→ **7**).

4b. Assess possible impacts on non-*Apis* immature stages: If TER for *Apis* is between 1 and 10, establish larval NOEL for relevant non-*Apis* bee species (→ 5b). Otherwise the risk to immature non-*Apis* bees is considered minimal (→ 7)

Lacking specific data on the toxicity of the compound on immature stages of non-*Apis* bee species, the assessment of possible effects on this group in the screening procedure can utilize data on honey bees. As a trigger of concern for possible effects on non-*Apis* bees the TER calculated under point **3b** using a honey bee larval NOEC is compared to a value higher by an order of magnitude to account for interspecies variability of toxicity data. Where this TER is below a trigger value of 10 a refinement of the screening step may be considered generating specific toxicity data with immature stages of non-*Apis* bee species before a higher tier risk assessment or consideration of risk mitigation measures is required.

5b. Establish larval NOEL for a non-*Apis* bee species: Calculate TER, compare TER to an appropriate trigger value

The screening step **3b** may be repeated using specific toxicity data generated in tests with a non-*Apis* bee species. For further details on laboratory studies on immature stages of non-*Apis* bees, see Chapter 8. Toxicity and exposure data expressed in the same units are expressed as a ration in a TER calculation (TER = NOEC divided by predicted exposure concentration). The calculated TER is assessed against an appropriate trigger value. A trigger value of 10 indicating that the predicted exposure is lower than the acute toxicity by at least one order of magnitude may be considered to be appropriate also for non-*Apis* bees. Where TER is below the trigger value, a higher tier risk assessment or consideration of risk mitigation measures is required (→ **8**). For TER values that are higher than the trigger, the risk to larvae of non-*Apis* bees is considered minimal (→ **7**).

7. Presumption of minimal risk

If exposure can be excluded or the assessment criteria in the screening step are met for both adult bees and larvae a minimal risk to honey bees and non-*Apis* bees can be presumed.

A minimal risk for honey bees and non-*Apis* bees can also be presumed if treatments in appropriate higher tier semi-field and field tests result in no significant difference compared to the untreated control when evaluated against the protection goals. Further risk mitigation measures are not required.

8. Continue with higher tier risk assessment or consider risk mitigation measures and reassess

If in the screening step the assessment criteria for adult bees or larvae are not met, a higher tier risk assessment should be performed (→ **9**). Alternatively the screening step may be repeated considering specific risk measures excluding or mitigating exposure (→ **2**). For further considerations on risk mitigation measures, see Chapter 12.

9. Is higher tier risk assessment triggered by failing the screening step with regard to non-*Apis* bees?

Concerns identified in the screening procedure have to be addressed in semi-field or field tests with honey bees (→ **10**). If in the screening step the criteria for adult bees or larvae are not met with regard to non-*Apis* bees, it must be determined whether a higher tier study with honey bees are sufficient to cover concerns identified for non-*Apis* bees in the screening step (→ **14**).

10. Is higher tier risk assessment triggered by failing the screening step with regard to *A. mellifera*?

If in the screening step the criteria for adult bees or larvae are not met only with respect to honey bees, a semi-field or field test should be performed to address the concern (→ **11** or **12**). (Note that the higher tier part of the risk assessment schemes is identical for both spray and soil or seed treatment application. Due to an additional step in the screening procedure, the numbering of the steps in the higher tier risk assessment scheme for soil or seed treatment application is different (+ 1).)

11. and 12. Assess the effects of the a.i. to *A. mellifera* in a semi-field or a field test: Do results indicate minimal risk (no significant difference to control)?

See 10 and 11 in the risk assessment flowchart for sprayed treatments.

Where in the semi-field test or in the field test treatment results in no significant difference in lethal and sub-lethal effects on survival, growth, reproduction, and foraging behavior compared to untreated control, a minimal risk is indicated (→ **7**). Otherwise a higher tier evaluation a field test has to be performed (→ **12**).

13. Risk mitigation measures specific to *A. mellifera* possible?

Where the results of higher tier semi-field and field tests indicate that the protection goals are not met, the assessment scheme may be reiterated considering specific measures to mitigate the exposure of honey bees (→ **2**). Note in this respect that semi-field and field test may be appropriately adapted in order to check for the efficacy of risk mitigation measures to limit the exposure and subsequent impact on bees.

14. Are there significant routes of exposure for non-*Apis* bees that are not covered by the honey bee risk assessment or risk assessment for other NTA?

In any case when a risk assessment for non-*Apis* bees is triggered and a refined risk assessment is available for honey bees and NTAs, it may be interesting to discuss the extent to which these risk assessments address part of the risk issues relative to non-*Apis* species. As an example, concerns with effects on non-*Apis* bees identified at the lower levels may in some cases be addressed by semi-field or field tests with honey bees, as for example, where no additional significant routes of exposure for non-*Apis* bees have to be taken into account. Furthermore, higher tier field data generated with NTA species may also address these concerns provided the routes of exposure are comparable to those for non-*Apis* bees (Table A6.1). If these data can serve as surrogates and if the examination of these data results in no significant risk with regard to the protection goals, then a minimal risk to non-*Apis* bees is indicated (→ **7**). Otherwise semi-field or field tests with non-*Apis* bees have to be performed to address the concern (→ **15** or **16**).

15. Assess the effects of the a.i. to a non-*Apis* bee species relevant to the identified route of exposure in a semi-field test: Do results indicate minimal risk (no significant difference to control)?

Concerns raised in the screening procedure may be addressed by appropriately designed semi-field or field tests where possible effects are assessed against the criteria intended to reflect the protection goals. The derivation of assessment criteria for specific protection goals is discussed in Chapter 4. For further details on semi-field studies on non-*Apis* bees, see Chapter 9. Where in the semi-field test treatment results in no significant difference in relevant lethal and sub-lethal effects compared to untreated control, a minimal risk is indicated (→ **7**). Otherwise in a higher tier evaluation a field test should be performed (→ **16**).

16. Assess the effects of the a.i. to a non-*Apis* bee species relevant to the identified route of exposure in a semi-field or a field test: Do results indicate minimal risk (no significant difference to control)?

Concerns raised in the screening-level assessment may be addressed by appropriately designed semi-field tests where possible effects are assessed against the evaluation criteria related to reflect the protection goals. The derivation of evaluation criteria for specific protection goals is discussed in Chapter 4. For further details on semi-field studies on non-*Apis* bees, see Chapter 9. As for honey bees, the choice between a semi-field test and a full-field study depends on the outcome of lower tier studies and should also consider decisions for honey bees. If the results of semi-field or field test, in conjunction with information from lower tier studies and other relevant data indicate no significant difference in relevant lethal and sub-lethal effects compared to untreated controls, minimal risk is indicated (→ **7**). Otherwise, further risk mitigation may be considered or the risk has to be presumed as significant (→ **17**).

17. Risk mitigation measures specific to non-*Apis* bee species possible?

Where the results of higher tier semi-field and field tests on *non-Apis* bees indicate that the protection goals are not met, the assessment scheme may be reiterated considering specific measures designed to mitigating the exposure of non-*Apis* bees (→ **2**).

Note in this respect that semi-field and field test may be adapted in order to determine whether risk mitigation measures actually limit the exposure and potential impact on *non-Apis* bees.

18. Presumption of significant risk

If there are no measures available to mitigate the risk to honey bees or non-*Apis* bees indicated by the evaluation of the results of higher tier semi-field and field tests against the protection goals, then a significant risk has to be presumed.

REFERENCES

Candolfi MP, Barrett KL, Campbell PJ, Forster R, Grandy N, Huet MC, Lewis G, Oomen PA, Schmuck R, Vogt H. 2001. *Guidance Document on Regulatory Testing and Risk Assessment Procedures for Plant Protection Products with Non-Target Arthropods, in ESCORT 2 Workshop (European Standard Characteristics of Non-Target Arthropod Regulatory Testing)*, SETAC Publication, Wageningen, 46 pp.

EC. 2009. Draft working document – Guidelines Related to Setting Maximum Residue Limits in Honey – EC Guidance document Part C4.

EC. 2010. Council Directive of 15 July 1991 Concerning the Placing of Plant Protection Products on the Market (Directive 91/414/EEC).

EPPO, 2010. Environmental risk assessment scheme for plant protection products. *EPPO Bull.* 40:1–9.

Mineau P, Harding KM, Whiteside M, Fletcher MR, Garthwaite D, Knopper LD. 2008. Using reports of honey bee mortality in the field to calibrate laboratory derived pesticide risk indices. *Environ. Entomol.* 37(2):546–554.

Glossary of Terms

μg/kg	micrograms per kilogram
μg/L	micrograms per liter
a.i.	active ingredient
BW	body weight
CCD	colony collapse disorder
CFR	Code of Federal Regulations
Colony	a distinct population of bees
CW	concentration in water (μg/L)
EC25	25% effect concentration
EC50	50% (or median) effect concentration
ECOTOX	EPA managed database of ECOTOXicology data
EEC	estimated environmental concentration
EFSA	European Food Safety Authority
EPA	Environmental Protection Agency
EPPO	European and Mediterranean Organization for Plant Protection
EU	European Union
FAO	Food and Agricultural Organization (United Nations)
FIFRA	Federal Insecticide Fungicide and Rodenticide Act
Forager	an adult bee which provides food and water to the colony
g a.i./ha	grams of active ingredient per hectare
GENEEC	GENeric Estimated Environmental Concentration
ha	hectare
HQ	hazard quotient
IAPV	Israeli acute paralysis virus
ICPBR	International Commission for Plant–Bee Relationships
kg	kilograms
kg a.s./ha	kilogram of active substance per hectare
km	kilometers
L	liter
lb a.i./A	pounds of active ingredient per acre
LC50	50% (or median) lethal concentration
LD50	50% (or median) lethal dose
LOC	level of concern
Log	logarithm
LOQ	level of quantitation
m	meters
mg	milligrams
mg/kg	milligrams per kilogram (equivalent to ppm)
mg/L	milligrams per liter (equivalent to ppm)
mi	miles
mL	milliliter

Pesticide Risk Assessment for Pollinators, First Edition. Edited by David Fischer and Thomas Moriarty.

n/a	not applicable
NASS	National Agricultural Statistics Service
NOAEC	no observable adverse effect concentration
Nuc	small colony consisting of 3–5 frames
OECD	Organization for Economic Cooperation and Development
OEPP	Organisation Européenne et Méditerranéenne pour la Protection des Plantes (EPPO)
OPP	Office of Pesticide Programs, US Environmental Protection Agency
PEIP	Pesticide Effects on Insect Pollinators (OECD)
PMRA	Pest Management Regulatory Agency (Canada)
ppb	parts per billion (equivalent to μg/L or μg/kg)
ppm	parts per million (equivalent to mg/L or mg/kg)
PPR	plant protection products and the residues (EPPO)
PRZM	Pesticide Root Zone Model
RQ	risk quotient
SETAC	Society of Environmental Toxicology and Chemistry
TREX	Terrestrial Residue Exposure Model
USDA	United States Department of Agriculture
USEPA	United States Environmental Protection Agency

Index

Pesticide Risk Assessment for Pollinators, First Edition. Edited by David Fischer and Thomas Moriarty.